Analysis on Function Spaces of Musielak-Orlicz Type

Monographs and Research Notes in Mathematics

Lineability
The Search for Linearity in Mathematics
Richard M. Aron, Luis Bernal-Gonzalez, Daniel M. Pellegrino, Juan B. Seoane Sepulveda

Iterative Methods and Preconditioning for Large and Sparse Linear Systems with Applications
Daniele Bertaccini, Fabio Durastante

Monomial Algebras, Second Edition
Rafael Villarreal

Matrix Inequalities and Their Extensions to Lie Groups
Tin-Yau Tam, Xuhua Liu

Elastic Waves
High Frequency Theory
Vassily Babich, Aleksei Kiselev

Difference Equations
Theory, Applications and Advanced Topics, Third Edition
Ronald E. Mickens

Sturm-Liouville Problems
Theory and Numerical Implementation
Ronald. B. Guenther, John. W. Lee

Analysis on Function Spaces of Musielak-Orlicz Type
Osvaldo Méndez, Jan Lang

For more information about this series please visit: https://www.crcpress.com/Chapman--Hall-CRC-Monographs-and-Research-Notes-in-Mathematics/book-series/CRCMONRESNOT

Analysis on Function Spaces of Musielak-Orlicz Type

Osvaldo Méndez
Jan Lang

CRC Press
Taylor & Francis Group
Boca Raton London New York

CRC Press is an imprint of the
Taylor & Francis Group, an **informa** business

CRC Press
Taylor & Francis Group
6000 Broken Sound Parkway NW, Suite 300
Boca Raton, FL 33487-2742

Printed on acid-free paper
Version Date: 20181114

International Standard Book Number-13: 978-1-4987-6260-1 (Hardback)

Visit the Taylor & Francis Web site at
http://www.taylorandfrancis.com

and the CRC Press Web site at
http://www.crcpress.com

To Cristina and Jana.

Contents

Preface ix

1 A path to Musielak-Orlicz spaces 1

1.1 Introduction . . . 1
1.2 Banach function spaces . . . 2
1.2.1 The associate space . . . 5
1.2.2 Absolute continuity of the norm and continuity of the norm . . . 10
1.2.3 Convexity, uniform convexity and smoothness of a norm 21
1.2.4 Duality mappings and extremal elements . . . 25
1.3 Modular spaces . . . 27
1.3.1 Modular convergence and norm convergence . . . 30
1.3.2 Conjugate modulars and duality . . . 33
1.3.3 Modular uniform convexity . . . 38
1.4 The ℓ^{p_n} sequence spaces and their properties . . . 40
1.4.1 Duality . . . 48
1.4.2 Finitely additive measures . . . 55
1.4.3 Geometric properties of ℓ^{p_n} . . . 61
1.4.4 Applications: Fixed point theorems on ℓ^{p_n} spaces . . . 70
1.4.5 Further remarks . . . 82
1.5 Forerunners of the Musielak-Orlicz class: Orlicz spaces, $L^{p(x)}$ spaces . . . 84

2 Musielak-Orlicz spaces 91

2.1 Introduction, definition and examples . . . 91
2.2 Embeddings between Musielak-Orlicz spaces . . . 101
2.2.1 The Δ_2-condition . . . 110
2.2.2 Absolute continuity of the norm . . . 113
2.3 Separability . . . 115
2.4 Duality of Musielak-Orlicz spaces . . . 118
2.4.1 Conjugate Musielak-Orlicz functions . . . 118
2.4.2 Dual of $L^{\varphi}(\Omega)$ . . . 120
2.5 Density of regular functions . . . 124
2.6 Uniform convexity of Musielak-Orlicz spaces . . . 128

2.7 Carathéodory functions and Nemytskii operators on Musielak-Orlicz spaces 132
2.8 Further properties of variable exponent spaces 137
2.8.1 Duality maps on spaces of variable integrability 150
2.9 The Matuszewska-Orlicz index of a Musielak-Orlicz space . . 151
2.10 Historical notes 155

3 Sobolev spaces of Musielak-Orlicz type 161

3.1 Sobolev spaces: definition and basic properties 161
3.1.1 Examples 164
3.2 Separability 165
3.3 Duality of Sobolev spaces of Musielak-Orlicz type 166
3.4 Embeddings, compactness, Poincaré-type inequalities 166

4 Applications 181

4.1 Preparatory results and notation 181
4.2 Compactness of the Sobolev embedding and the modular setting 186
4.3 The variable exponent p-Laplacian 190
4.3.1 Stability of the solutions 198
4.4 Γ-convergence 206
4.5 The eigenvalue problem for the p-Laplacian 210
4.6 Modular eigenvalues 219
4.7 Convergence properties of the eigenvalues and eigenfunctions 228
4.8 Closing remarks 243

Bibliography 245

Author index 253

Subject index 255

Notation index 259

Preface

Many years later, facing the need of a mathematical description of advanced physical problems, analysts were to remember that distant work by W. Orlicz which conduced to Musielak-Orlicz spaces.

The main focus of this book is to present a unified theory of Musielak-Orlicz function spaces while at the same time exploring its ties with other function spaces and its connections with different areas of Analysis. We will place special emphasis on the utility of these spaces in several concrete applications. In particular, we will observe that a wide class of problems arising in nonlinear PDE can be elegantly formulated and efficiently solved if properly translated into the language of Musielak-Orlicz spaces.

To the best of the authors knowledge, the first occurrence of what would later be called Musielak-Orlicz spaces is to be traced back to Orlicz' realization in 1931 [85] of the fact that the exponent p in the definition the L^p spaces introduced by Riesz could be replaced by a function, without losing the essence of the theory. More precisely, in the cited work, Orlicz introduced the class of measurable functions u for which

$$\int_0^1 |u(x)|^{p(x)} dx < \infty \tag{0.1}$$

and suggested that "A variety of results about integrable functions with constant p can be generalized to certain classes of functions that are integrable with a function p." Nineteen years thereafter, Nakano [83] introduced the class of real valued functions u on a domain Ω for which

$$\int_\Omega \varphi(x, |u(x)|) dx < \infty.$$

Here

$$\varphi : \Omega \times [0, \infty) \longrightarrow [0, \infty)$$

is a suitable function, which is called a Musielak-Orlicz function (see Chapter 2 for the precise definition). The class introduced by Nakano is plainly a very natural and elegant generalization of that introduced by Orlicz in [85], and will in fact constitute the main focus of this book.

It is worth mentioning that Orlicz' work [85] was mainly motivated by his interest in [8, Satz a, Satz b] in which Banach addresed the question of summability of the coefficients of lacunary Fourier series, i.e., series of the form

$$\frac{a_0}{2} + \sum_{1}^{\infty} (a_n \cos k_n x + b_n \sin k_n x),$$

where, for all $n \in \mathbb{N}$, it holds that $k_{n+1}/k_n \geq r > 1$. See also [9] for a remarkable comment by Banach on Orlicz work.

It was this endeavor that lead Orlicz to the consideration of spaces of sequences (a_n) for which

$$\sum_{1}^{\infty} |\lambda a_n|^{p_n} < \infty,$$

for some $\lambda > 0$ and a given sequence $(p_n) \subset [1, \infty)$. In fact, the alert reader will realize that these spaces are nothing but the discrete version of the Lebesgue spaces with variable integrability in [85]; this sequence spaces will in the sequel be denoted by ℓ^{p_n}.

After the publication of [85] by Orlicz, fundamental research on Lebesgue spaces with variable exponent remained largely dormant until Nakano's reintroduction of the concept in the general form described above, in 1950. In 1959, Musielak and Orlicz [81] pursued Nakano's idea from a systematic point of view; the product was laid down in final form by J. Musielak himself in 1983. [82].

Parallel to the above historical developments a new class of problems was gradually emerging in the realm of partial differential equations and variational calculus. These problems involved non-standard growth phenomena and already in the late 1920's presaged the practical utility of the theoretical developments pursued by Musielak, Orlicz and Nakano and in particular, of the spaces with variable integrability. Notwithstanding the fact that the history of non-standard growth phenomena is convoluted and not easy to follow chronologically, we will mainly elaborate on progress made after 1990. We refer the interested reader to [24] and the references therein for some articles involving variable exponent spaces published between 1959 and 1990.

In 1997 V. Zhikov [109], motivated by equations arising in electromagnetism was led to the minimization of integrals of the type

$$\int_{\Omega} |\nabla u(x)|^{p(x)} dx;$$

these minimization problems conduce to the corresponding Lagrange-Euler equation:

$$\Delta_{p(\cdot)} u := \operatorname{div}\left(|\nabla u|^{p(\cdot)-2} \nabla u\right) = 0. \tag{0.2}$$

Because of the variability of $p(x)$, (0.2) is an example of differential equation with non-standard growth. The natural habitat for the solutions of such differential equations must take into consideration the dependence of $p(x)$ on the space variable x. It is at this point obvious that the classical L^p theory is not sufficient in this situation and that a condition such as (0.1) for ∇u seems to be desirable as an *a priori* requirement on the solutions, i.e.:

$$\int_\Omega |\nabla u(x)|^{p(x)} dx < \infty.$$

Through these applications, then, there inexorably emerged the need for a deeper understanding of these generalized functional spaces with variable integrability. It was moreover observed by Zhikov that the variability of the exponent $p(x)$ might cause the following undesirable situation to occur:

$$\inf_{u \in W^{1,1}(\Omega)} \int_\Omega |\nabla u(x)|^{p(x)} dx < \inf_{u \in W^{1,\infty}(\Omega)} \int_\Omega |\nabla u(x)|^{p(x)} dx.$$

The preceding discrepancy of the infima is known as Lavrentiev's phenomenon, was observed first by Lavrentiev [69] and can occur in the variable exponent setting unless some conditions are assumed on the exponent $p(x)$. This observation furthered the need of a deeper understanding of the structure of spaces of Musielak-Orlicz type (density of smooth functions, duality, weak convergence etc.). Addresing this need is the driving force of this book.

The book is organized in the following manner: In Chapter 1 we lay the ground work for the concept of Musielak-Orlicz spaces by introducing notation, terminology and by reviewing some aspects of the abstract theory of function spaces. In the next chapter, Musielak-Orlicz spaces are introduced and exhaustively studied. Chapter 3 focuses on the theory of Sobolev spaces in the context of Musielak-Orlicz spaces. Chapter 4 is intended to explore some applications of the theory developed in Chapters 1 to 3.

Chapter 1 is divided into five sections. After a brief note on the notation and terminology to be used in the sequel in Section 1.1, we move to Section 1.2, where the basic abstract theory of function spaces is introduced and some elementary geometric Banach space properties discussed in certain detail.
Section 1.3 focuses on the basic general theory of modular spaces. We emphasize for the most part the aspects of the theory that are necessary to undertake the analysis of spaces of Musielak-Orlicz type. Particular focus is placed on the difference between modular and norm-convergence, conjugation and duality.
Section 1.4 is exclusively devoted to concrete applications of the foregoing theoretical material. In particular, close attention is paid to those aspects of the variable exponent sequence spaces ℓ^{p_n} that are in fact inherent to their modular structure. Specifically, we digress on duality questions of the variable-exponent spaces ℓ^{p_n}, underlining the special case in which $\sup_{\mathbb{N}} p_n = \infty$, that

is in the absence of the Δ_2-condition. Close attention is paid to some geometric properties of the modular whose relevance in the theory of fixed point has only recently been recognized. In this regard it is worth mentioning that the uniform convexity of the ℓ^{p_n} norm is very cumbersome both to verify and to use under the assumption of a variable exponent (p_n). Nevertheless, under mild assumptions on the exponent sequence (p_n), the modular that defines the variable exponent ℓ^{p_n} spaces enjoys some sort of uniform convexity which, while weaker than the uniform convexity property of the norm, is much friendlier to mathematical intuition and it turns out to be sufficient to achieve a generalization of several classical results in fixed point theory, to the variable-exponent setting. It is with this particular type of application in mind that we included Section 1.4.
We conclude our discussion in Chapter 1 in Section 1.5 by introducing Orlicz and $L^{p(x)}$ spaces. The reason for introducing these examples is twofold: they represent concrete instances of the modular structures introduced in Chapter 1 and in addition, could be considered to be forerunners of the Musielak-Orlicz spaces which will be the object of study in Chapter 2.

In Chapter 2 we provide the foundations of the fundamental theory of spaces of Musielak-Orlicz type. Except for some modifications that reflect our personal taste, we mainly follow the classical treatment by J. Musielak [82] and [24]. Sections 2.1 to 2.6 deal with the basic functional-analytic properties of Musielak-Orlicz spaces such as embeddings, separability, absolute continuity of the norm and density of regular functions.
Because of its relevance to the applications that will be discussed in Chapter 4, a discussion of the action of Nemytskii operators on Musielak-Orlicz spaces is included in Section 2.7.
For reasons that will become clear in the next chapter we study in detail some properties of variable exponent spaces and of the Matuszewska-Orlicz index.
We conclude Chapter 2 with a brief historical note about the origins of Musielak-Orlicz spaces.

In Chapter 3 we introduce the Sobolev spaces of Musielak-Orlicz type. In Sections 3.1 to 3.3 we expound their basic properties. The interested reader is referred to [91] for a survey on Sobolev spaces of Orlicz type. The fundamental functional-analytic properties of Musielak-Orlicz-Sobolev-type spaces were studied in [44, 45, 46, 47, 48, 49, 50, 51]. In Section 3.4, we analyze necessary and sufficient conditions on the Musielak-Orlicz function φ, for the validity of the compactness of the Sobolev embedding

$$W_0^{1,\varphi}(\Omega) \hookrightarrow L^{\varphi}(\Omega).$$

Chapter 4 is entirely devoted to applications of the results from the previous chapters. We underline the case of the variable exponent Lebesgue spaces and present some stability results for boundary value problems as well as for eigenvalue problems. Most of these results are based on our recent work [64, 65, 66, 67].

We end the chapter by providing a theoretical treatment for the general eigenvalue problem on spaces of Musielak-Orlicz type.

Chapter 1

A path to Musielak-Orlicz spaces

1.1 Introduction 1
1.2 Banach function spaces 2
1.2.1 The associate space 5
1.2.2 Absolute continuity of the norm and continuity of the norm 10
1.2.3 Convexity, uniform convexity and smoothness of a norm 21
1.2.4 Duality mappings and extremal elements 25
1.3 Modular spaces 27
1.3.1 Modular convergence and norm convergence 30
1.3.2 Conjugate modulars and duality 33
1.3.3 Modular uniform convexity 38
1.4 The ℓ^{p_n} sequence spaces and their properties 40
1.4.1 Duality 48
1.4.2 Finitely additive measures 55
1.4.3 Geometric properties of ℓ^{p_n} 61
1.4.4 Applications: Fixed point theorems on ℓ^{p_n} spaces 70
1.4.5 Further remarks 82
1.5 Forerunners of the Musielak-Orlicz class: Orlicz spaces, $L^{p(x)}$ spaces 84

1.1 Introduction

Here we provide basic information about Banach spaces and Banach function spaces. Proofs of absolutely standard results are often eschewed in favor of references to works in which full details are given.

To keep technicalities at a reasonable level, this monograph will only deal with real vector spaces. The norm on a normed linear space X will usually be written as $\|\cdot\|_X$, or even $\|\cdot\|$ if the context is clear; often $\|\cdot\|_p$ or $\|\cdot\|_{p,I}$ will be used to represent the norm on $L^p(I)$. The dimension of X is denoted by $\dim X$; and B_X (resp. S_X) will stand for the closed unit ball (resp. the unit sphere) in X; $B(x,r)$ will denote the open ball in X with center x and radius r. If A is a subset of a Banach space X with norm $\|\cdot\|_X$, the closure of A in $\|\cdot\|_X$ will be denoted by $Cl^{\|\cdot\|_X}(A)$. By $B(X,Y)$ will be meant the space of all bounded linear maps from a Banach space X to another such space Y, written as $B(X)$ when $Y = X$, and as X^* (the dual of X) when Y is the space of scalars associated with X; the value of $x^* \in X^*$ at $x \in X$ is denoted by $\langle x^*, x\rangle_X$ or $\langle x^*, x\rangle$. Weak convergence in X is represented by a half arrow $\rightharpoonup$, weak*-convergence in X^* by $\overset{*}{\rightharpoonup}$. If X is continuously embedded in Y we write $X \hookrightarrow Y$. The symbol $X \hookrightarrow\hookrightarrow Y$ will be used to indicate that X is compactly embedded into Y. The characteristic function of a set S is written as $\mathbf{1}_S$ and the Lebesgue n-measure of a subset Ω of $\mathbb{R}^n$ will be denoted by $|\Omega|_n$, or $|\Omega|$ if there is no ambiguity. $\mathcal{S}_\infty$ will stand for the vector space of all finite linear combinations of characteristic functions of measurable sets; the vector space of all simple functions $w = \sum_{k=1}^N w_k \mathbf{1}_{E_k}$ with $|E_K| < \infty$ will be denoted with $\mathcal{S}$.

1.2 Banach function spaces

Much of the mathematical work to be presented in this book relies on the basic notion of Banach function space. For the remainder of this section we fix a measure space $(\Omega, \mathcal{A}, \mu)$ with a σ-finite measure μ, i.e., it is assumed that there exists an increasing sequence (Ω_j) such that $\mu(\Omega_j) < \infty$ and

$$\Omega = \bigcup_{j=1}^{\infty} \Omega_j. \tag{1.1}$$

In particular, if $\Omega \subset \mathbb{R}^n$ is a domain, $\mathcal{A}$ will be understood to be the Borel class of subsets of Ω, μ will denote the Lebesgue measure on $\mathcal{A}$ and $\mathcal{M}(\Omega)$ will stand for the set of all Borel measurable, real valued functions on Ω, with the usual identification $u = v$ iff $u(x) = v(x)$ almost everywhere.

Definition 1.2.1. *A linear space $X \subseteq \mathcal{M}(\Omega)$ is called a Banach function space if there is a map $\|\cdot\|_X : \mathcal{M}(\Omega) \to [0, \infty]$ with the properties of a norm and such that*

(i) $u \in X$ if and only if $\|u\|_X < \infty$;

(ii) $\|u\|_X = \||u|\|_X$ for all $u \in \mathcal{M}(\Omega)$;

(iii) if $0 \leq u_k \nearrow u$, then $\|u_k\|_X \uparrow \|u\|_X$;

(iv) if $E \subset \Omega$ and $\mu(E) < \infty$, then $\|\mathbf{1}_E\|_X < \infty$;

(v) if $E \subset \Omega$ and $\mu(E) < \infty$, there is a constant $c(E)$ (i.e., independent of u) such that for all $u \in X$,

$$\int_E |u| \, d\mu \leq c(E) \|u\|_X .$$

In particular, property (iv) implies that any simple function that is a finite linear combination of characteristic functions of sets of finite measure, is in X.
In addition, axiom (iii) implies the following lattice property: If $u \in X$ and $|v| \leq |u|$ *a.e.* in Ω, then necessarily $v \in X$.

The distribution function of an element u of a Banach function space X is the map $\mu_u : [0, \infty) \to [0, \infty]$ defined, for all $\lambda \geq 0$, by

$$\mu_u(\lambda) = \mu\left(\{x \in \Omega : |u(x)| > \lambda\}\right).$$

Two functions $f, g \in X$ are said to be equimeasurable if their distribution functions coincide; X is called rearrangement-invariant if $\|f\|_X = \|g\|_X$ whenever f and g are equimeasurable.
Notice that if u is an element of a Banach function space X, then a fortiori $|u(x)| < \infty$ *a.e.* which follows straight from axiom (v). Indeed, since $|u|$ is assumed to be σ-finite, (1.1) holds and

$$\int_{\Omega_j \cap \{x : |u(x)| = \infty\}} |u(x)| \, d\mu \leq C(\Omega_j \cap \{x : |u(x)| = \infty\}) \|u\|_X < \infty.$$

The claim follows at once. We also underline the following observation:

Lemma 1.2.1. *If a sequence* (u_j) *of measurable functions converges a.e. to* u *and if* $\liminf_{j \to \infty} \|u_j\|_X < \infty$, *then* $u \in X$ *and*

$$\|u\|_X \leq \liminf_{j \to \infty} \|u_j\|_X.$$

Proof. Since u is automatically measurable to show that $u \in X$ it suffices to prove that $\|u\|_X < \infty$. Writing $v_n = \inf_{k \geq n} |u_k(x)|$ then $v_n \nearrow \liminf_{n \to \infty} |u_n| = |u|$, pointwise *a.e.* Hence, by Axiom (iii)

$$\|u\|_X = \lim_{n \to \infty} \|v_n\|_X \leq \lim_{n \to \infty} \| \inf_{k \geq n} |u_k| \|_X \leq \liminf_{n \to \infty} \| |u_n| \|_X < \infty.$$

□

The following lemma, though simple, is of paramount importance in the sequel:

Lemma 1.2.2. *If* $(u_n) \subset X$ *is a (norm)-convergent sequence of functions in a Banach function space* X, *then* (u_n) *is a.e. convergent to its norm limit* $u \in X$.

Proof. Pick $E \subset \Omega$ with $|E| < \infty$; Axiom (v) in Definition 1.2.1 yields:

$$\begin{aligned} \mu\left(\{x \in E : |u_n(x) - u(x)| > \delta\}\right) &\leq \delta^{-1} \int_E |u_n - u| \, d\mu \\ &\leq c(E) \delta^{-1} \|u_n - u\|_X \to 0 \text{ as } n \to \infty. \end{aligned}$$

It follows that (u_n) converges *a.e.* in every subset of finite measure; the result follows from the σ-finiteness. □

Theorem 1.2.3. *[11, Theorem 1.6] Every Banach function space X is a Banach space when endowed with the norm $\|\cdot\|_X$.*

Proof. It suffices to show that every absolutely convergent series $\sum_{n=1}^{\infty} u_n$ is convergent. To this end, let

$$v_m = \sum_{n=1}^{m} |u_n|;$$

it is clear that $v_m \nearrow v = \sum_{n=1}^{\infty} |u_n|$ and that, by assumption,

$$\|v_m\|_X \leq \sum_{n=1}^{\infty} \|u_n\|_X < \infty$$

for each $m \in \mathbb{N}$. Thus, $\liminf_{m\to\infty} \|v_m\|_X < \infty$ and it follows from Lemma 1.2.1 that $v \in X$ and $\|v\|_X \leq \liminf_{n\to\infty} \|v_n\|_X$. On account of Lemma 1.2.2, $\sum_{n=1}^{\infty} |u_n|$ is almost everywhere convergent (to v); therefore, so is $\sum_{n=1}^{\infty} u_n$. We set

$$u = \sum_{n=1}^{\infty} u_n.$$

It will be shown next that $u \in X$ (that is, that $\|u\|_X < \infty$) and that

$$\left\| \sum_{1}^{m} u_n - u \right\|_X \to 0 \text{ as } m \to \infty. \tag{1.2}$$

To this end, we observe that if $w_n = \sum_{k=1}^{n} u_k$ then for fixed $m \in \mathbb{N}$,

$$w_n - w_m \longrightarrow u - w_m \text{ as } n \longrightarrow \infty$$

pointwise *a.e.* and since for $m < n$,

$$\|w_n - w_m\|_X \leq \sum_{k=m}^{n} \|u_k\|_X < \sum_{k=m}^{\infty} \|u_k\|_X$$

it is obvious that

$$\liminf_{n\to\infty} \|w_n - w_m\|_X \leq \sum_{k=m}^{\infty} \|u_k\|_X < \infty$$

and on account of Lemma 1.2.1 it follows that $w_m - u \in X$. In particular, $u \in X$, as was to be shown. Moreover, also by Lemma 1.2.1:

$$\|u - w_m\|_X \leq \liminf_{n\to\infty} \|w_n - w_m\|_X \leq \sum_{k=m}^{\infty} \|u_k\|_X \to 0 \text{ as } j \to \infty. \tag{1.3}$$

It is apparent from inequalities (1.3) that for any natural number m,

$$\|u\|_X \leq \|u - w_m\|_X + \|w_m\|_X < \infty$$

and that (1.2) holds. □

It is routine to check that the classical Lebesgue and Sobolev spaces are Banach function spaces. While every Orlicz space $L^{\phi}(\Omega)$ is rearrangement-invariant (see [1], p. 270 and [5], p. 75), this is not the case, in general, for the Musielak-Orlicz spaces $L^M(\Omega)$.

Theorem 1.2.4. *Let X_1 and X_2 be Banach function spaces over the same measure space $(\Omega, \mathcal{A}, \mu)$. If the set-theoretic inclusion*

$$X_1 \subseteq X_2,$$

then the inclusion is bounded.

Proof. Otherwise one could construct a sequence $(u_n) \subset B_{X_1}$ with $\|u_n\|_{X_2} > n^3$ and on account of Axiom *(ii)* in Definition 1.2.1 it is clear that one can choose $u_n \geq 0$. The series

$$\sum_1^{\infty} n^{-2} u_n$$

is absolutely convergent (and hence convergent) in X_1, say

$$\sum_1^{\infty} n^{-2} u_n = u \in X_1 \subseteq X_2.$$

However, since $n^{-2} u_n \leq u$ Axiom *(iii)* in Definition 1.2.1 yields

$$n \leq n^{-2} \|u_n\|_{X_2} \leq \|u\|_{X_2},$$

which is a contradiction. □

1.2.1 The associate space

We next turn to a fundamental concept in the theory of Banach function spaces, namely, that of the associate space. Let $(X, \|\cdot\|_X)$ be a Banach function space on the measure space $(\Omega, \mathcal{A}, \mu)$. The function

$$\|\cdot\|_{X'} : \mathcal{M}(\Omega) \longrightarrow [0, \infty]$$

$$\|v\|_{X'} := \sup_{u \in S_X} \int_{\Omega} |vu| \, d\mu \tag{1.4}$$

is easily verified to have the properties of a norm, i.e., it satisfies the triangle inequality, is homogeneous and $\|v\|_{X'} = 0 \Leftrightarrow v = 0$.

Theorem 1.2.5. *Given a Banach function space X, the set*

$$X' := \left\{ u \in \mathcal{M}(\Omega) : \sup_{u \in S_X} \int_\Omega |vu|\, d\mu < \infty \right\}$$

is a Banach function space called the **associate space** *of X. Hölder's inequality holds in the form*

$$\int_\Omega |vu|\, d\mu \leq \|v\|_X \, \|u\|_{X'}$$

for all $v \in X$ and $u \in X'$.

Proof. It is a routine matter to check that X' is a linear subspace of $\mathcal{M}(\Omega)$. The fact that the functional

$$u \longrightarrow \|u\|_{X'}$$

has the properties of a norm follows from straightforward calculations. It remains to prove the axioms (i) to (v) in Definition 1.2.1.
Axiom (i) transpires directly from the definition of X'. Property (ii) of Definition 1.2.1 is obvious. For property (iii), let $u \in \mathcal{M}(\Omega)$ and $(u_j) \subset \mathcal{M}(\Omega)$ with $0 \leq u_j \nearrow u$. One has, for arbitrary $v \in S_X$,

$$\int_\Omega u_k |v|\, d\mu \leq \int_\Omega u_k |v|\, d\mu.$$

By virtue of monotone convergence one concludes that for any $v \in S_X$,

$$\int_\Omega u_k |v| d\mu \nearrow \int_\Omega u|v| d\mu. \tag{1.5}$$

If $\|u\|_{X'} = \infty$, then for any $N \in \mathbb{N}$ there exists $v_N \in S_X$ such that

$$\int_\Omega u|v_N|\, d\mu > N$$

and (1.5) implies that

$$\int_\Omega u_k |v_N|\, d\mu > N$$

for large enough k, whence the arbitrariness of N yields that $\|u_k\|_{X'} \nearrow \infty$ as $k \to \infty$.
If on the other hand, $\|u\|_{X'} < \infty$, fix $\delta > 0$ and $v \in S_X$ such that

$$\|u\|_{X'} - \delta < \int_\Omega |u(x)||v(x)|\, dx.$$

It follows from the convergence statement (1.5) that for large enough k one has

$$\|u\|_{X'} - \delta < \int_\Omega u_k|v|d\mu \leq \|u_k\|_{X'},$$

i.e., $\|u_k\|_{X'} \nearrow \|u\|_{X'}$ as claimed.
Property (iv) follows at once, for if $E \subset \Omega$ and $|E| < \infty$ and $v \in S_X$, one has on account of property axiom (v) for X, that

$$\int_\Omega (\mathbf{1}_E v)\, d\mu \leq c(E),$$

which immediately yields the desired result. Property (v) is also simple, for if $E \in \mathcal{A}$ with $\mu(E) < \infty$ and $u \in X'$, then:

$$\int_E |u| d\mu = \|\mathbf{1}_E\|_X \int_\Omega |u| \frac{\mathbf{1}_E}{\|\mathbf{1}_E\|_X} d\mu \leq \|\mathbf{1}_E\|_X \|u\|_{X'}.$$

Hölder's inequality follows directly by the definition of $\|u\|_{X'}$. □

Lemma 1.2.6. *Let $v \in \mathcal{M}(\Omega)$. Then $v \in X'$ if and only if for every $u \in X$, $uv \in L^1(\Omega)$. In particular, it holds the equality:*

$$X' = \left\{ v \in \mathcal{M}(\Omega) : \int_\Omega |vu|\, d\mu < \infty \text{ for all } u \in X \right\}.$$

Proof. Hölder's inequality shows the necessity of the L^1-condition. Let now $v \in \mathcal{M}(\Omega)$. If

$$\|v\|_{X'} = \sup_{u \in S_X} \int_\Omega |uv|\, d\mu = \infty$$

one can extract a sequence $(f_n) \subset S_X$ such that for each $n \in \mathbb{N}$, $\int_\Omega |vu_n| d\mu > n^3$. The series

$$\sum_1^\infty n^{-2}|u_n|$$

is an absolutely convergent series in X, it is therefore convergent to $u \in X$ and by virtue of Lemma 1.2.2 the series converges pointwise *a.e.* to u. Since $|u_n|n^{-2} \leq f$ pointwise *a.e.*,

$$\int_\Omega u|v|d\mu \geq \int_\Omega n^{-2}|u_n||v|d\mu > n,$$

i.e., $uv \notin L^1(\Omega)$. □

Every Banach function space X coincides with its second associate space $X'' := (X')'$, and for any $u \in X$ it holds that $\|u\|_X = \|u\|_{X''}$. The next lemma aims at the proof of this essential fact (see [11, Theorem 2.7] and [77]).

Lemma 1.2.7. *Let $\tilde{\Omega} \subset \Omega$ be domains with $\mu(\tilde{\Omega}) < \infty$ and set*

$$\tilde{L^1} = \left\{ w \in L^1(\Omega) : \operatorname{supp} w \subseteq \tilde{\Omega} \right\}.$$

Then $\tilde{L^1}$ is a Banach space with the norm

$$\|w\| = \int_{\tilde{\Omega}} |w| d\mu$$

and if $X \subseteq \mathcal{M}(\Omega)$ is a Banach function space, then $B_x \cap \tilde{L^1}$ is a closed, convex subset of $\tilde{L^1}$.

Proof. It is routine to verify that $\tilde{L^1}$ is a Banach space and that $B_X \cap \tilde{L^1}$ is convex. For the closedness, observe that if $S_X \cap \tilde{L^1} \ni w_j \to w \in \tilde{L^1}$ in $\tilde{L^1}$, then it can be assumed without loss of generality that $w_j \longrightarrow w$ pointwise *a.e.* in Ω. By assumption

$$\liminf_{j \to \infty} \|w_j\|_X \leq 1.$$

On account of Lemma 1.2.1 one has then $w \in B_X$. □

Theorem 1.2.8. *Any Banach function space is identical with its second associate space X''. In other words, it holds the set theoretical equality $X = X''$ and for any $u \in X$ one has $\|u\|_X = \|u\|_{X''}$.*

Proof. For fixed $u \in X$, $\int_\Omega |uv| d\mu < \infty$ for all $v \in X'$: this follows from Lemma 1.2.6; this yields the set theoretical inclusion $X \subseteq X''$. On the other hand, Hölder's inequality implies

$$\|u\|_{X''} \leq \|u\|_X.$$

For the remaining inclusion we invoke the σ-finiteness assumption (1.1). For each $u \in X''$, $k \in \mathbb{N}$ set

$$u_k = \min\{|u(x)|, k\} \mathbf{1}_{\Omega_k}.$$

Then $0 \leq u_k \nearrow f$ pointwise *a.e.* and each u_k is dominated by a simple function. Since simple functions belong to any Banach function space, it is clear that $u_k \in X \cap X''$. Next, we set about to show that for every $k \in \mathbb{N}$ it holds that

$$\|u_k\|_X \leq \|u_k\|_{X''}. \tag{1.6}$$

Once established, inequality (1.6) in conjunction with Axiom (*iii*) in Definition 1.2.1 will yield

$$\|u_k\|_X \nearrow \|u\|_X \leq \|u\|_{X''}.$$

Clearly, (1.6) holds if $\|u_k\|_X = 0$; it can be supposed in the sequel, therefore, that $\|u_k\|_X \neq 0$. Let $\mathbb{R} \ni \alpha > 1$; Lemma 1.2.7 applies for $\tilde{\Omega} = \Omega_k$ ($\tilde{L^1} = L^1_k$) and

$$\frac{\alpha}{\|u_k\|_X} u_k \in L^1_k \setminus S_X.$$

The Hahn-Banach Theorem, then yields the existence of $\gamma \in L^\infty(\Omega) = \left(\tilde{L^1}\right)^*$ and of a real number β such that, for all $a \in L^1_k \cap S_X$.

$$\int\limits_{\Omega_k} \gamma a \, d\mu < \beta < \int\limits_{\Omega_k} \gamma \frac{\alpha}{\|u_k\|_X} u_k \, d\mu.$$

It follows easily from here that

$$\sup_{a \in B_X \cap L^1_k} \int\limits_{\Omega_k} |\gamma a| \, d\mu \leq \beta \leq \int\limits_{\Omega_k} |\gamma| \frac{\alpha}{\|u_k\|_X} |u_k| \, d\mu.$$

Next, observe that any $w \in B_X$ can be written as the pointwise limit

$$w = \lim_{j \to \infty} w_j(x),$$

where $w_j(x) = \min\{w(x), j\} \mathbf{1}_{\Omega_j}$. The monotone convergence theorem implies the equality

$$\sup_{a \in B_X \cap L^1_k} \int\limits_{\Omega_k} |\gamma a| \, d\mu = \sup_{a \in B_X} \int\limits_{\Omega_k} |\gamma a| \, d\mu.$$

In conclusion:

$$\begin{aligned} \|\gamma\|_{X'} &= \sup_{a \in B_X} \int\limits_{\Omega_j} |\gamma a| \, d\mu \leq \beta \\ &\leq \frac{\alpha}{\|u_k\|_X} \int\limits_{\Omega_j} |\gamma u_k| \, d\mu \end{aligned}$$

and it is concluded from the last inequality that for any $\alpha : \alpha > 1$ it holds

$$\begin{aligned} \|u_k\|_X &\leq \frac{\alpha}{\|\gamma\|_{X'}} \int\limits_{\Omega_j} |\gamma| |u_k| \, d\mu \\ &\leq \frac{\alpha}{\|\gamma\|_{X'}} \|\gamma\|_{X'} \|u_k\|_{X''} \\ &= \alpha \|u_k\|_{X''}; \end{aligned}$$

i.e., (1.6) holds and the theorem is proved. □

In general, the associate space of X is (canonically isomorphic to) a closed subspace of the dual X^*. It is natural to investigate circumstances under which X' coincides with X^*. With this in mind, in the forthcoming section (Definition 1.2.3) we shall introduce the concept of absolute continuity of the norm and study its connection to duality.

Theorem 1.2.9. *The map*

$$\mathcal{I} : X' \longrightarrow X^*$$
$$\langle \mathcal{I}(u), v \rangle = \int_\Omega uv \, d\mu$$

defines an isometry from X' to a closed, norm-fundamental subspace of X^*, i.e., for any $v \in X$ it holds that

$$\|v\|_X = \sup_{\gamma \in B_{\mathcal{I}(X')}} |\langle \gamma, v \rangle|.$$

Proof. It is clear that $\mathcal{I}$ is well defined, linear and injective. By virtue of Hölder's inequality, it is also bounded. In addition, for $u \in X'$ it follows by definition that

$$\|u\|_{X'} = \sup_{\|v\|_X \leq 1} \left| \int_\Omega uv \, d\mu \right| = \|u\|_{X^*}. \tag{1.7}$$

Hence $\mathcal{I}$ is an isometry from X' onto its image $\mathcal{I}(X')$. The latter is closed, for if $\mathcal{I}(v_j) = \gamma_j \to \gamma \in X^*$ in the topology of X^*, then (γ_j) is Cauchy in X^* and it is apparent from (1.7) that (v_j) is Cauchy in X'. Next, we invoke the completeness of X' on account of Theorem 1.2.3 and Theorem 1.2.5. Hence, $v_j \to v \in X'$ as $j \to \infty$ and it follows easily that $\gamma = \mathcal{I}(v)$. Finally, if $v \in X$ Theorem 1.2.8 yields

$$\begin{aligned} \|v\|_X = \|v\|_{X''} &= \sup_{\|u\|_{X'} \leq 1} \left| \int_\Omega vu \, d\mu \right| \\ &= \sup_{\|\gamma\|_{\mathcal{I}(X')} \leq 1} |\langle \gamma, v \rangle| . \end{aligned}$$

The last assertion finished the proof. ☐

1.2.2 Absolute continuity of the norm and continuity of the norm

We begin this section with a note on a measure-theoretic concept. Let $(\Omega, \mathcal{A}, \mu)$ be a measure space as in Section 1.2 we assume in particular that μ is σ-finite (1.1).

Definition 1.2.2. *A sequence (M_n) of measurable sets is said to converge a.e. to the empty set, written $M_n \to \emptyset$ a.e., if the indicator function $\mathbf{1}_{M_n} \longrightarrow 0$ pointwise a.e. in X.*

Lemma 1.2.10. *In the preceding notation, the two following conditions are equivalent:*

(i) $M_n \to \emptyset$ *as* $n \to \infty$,

(ii) $\mu\left(\bigcap_{n=1}^{\infty} \bigcup_{j=n}^{\infty} M_j\right) = 0.$

Proof. The proof is elementary and follows from the following set-theoretic equality:

$$\{x \in \Omega : \mathbf{1}_{M_n}(x) \to 0 \text{ as } n \to \infty\} = \Omega \setminus \bigcap_{n=1}^{\infty} \bigcup_{j=n}^{\infty} M_j.$$

□

Lemma 1.2.11. *Let X be a Banach function space as in Section 1.2, $u \in X$. Then the following conditions are equivalent:*

(i) $\|u\mathbf{1}_{M_n}\|_X \to 0$ *as* $n \to \infty$ *for any sequence* $(M_n) \subseteq \mathcal{A}$ *such that* $M_n \to \emptyset$ *as* $n \to \infty$,

(ii) $\|u\mathbf{1}_{M_n}\|_X \searrow 0$ *as* $n \to \infty$ *for any* **decreasing** *sequence* $(M_n) \subseteq \mathcal{A}$ *such that* $M_n \to \emptyset$ *as* $n \to \infty$.

Proof. The necessity is obvious from Property *(iii)* in Definition 1.2.1. Assuming now *(ii)*, consider a sequence of measurable sets (M_n) with $M_n \to \emptyset$ as $n \to \infty$. Then

$$A_n = \bigcup_{k \geq n} M_k$$

is decreasing and the obvious set-theoretic identity

$$\bigcap_{m=1}^{\infty} \bigcup_{n \geq m} A_n = \bigcap_{m=1}^{\infty} \bigcup_{n \geq m} M_n$$

shows that

$$\mu(\limsup_n M_n) = \mu(\limsup_n A_n) = 0.$$

According to Lemma 1.2.10, then

$$\|u\mathbf{1}_{M_n}\|_X \longrightarrow 0 \text{ as } n \longrightarrow \infty.$$

□

Definition 1.2.3. *Let X be a Banach function space. A function $u \in X$ is said to have absolutely continuous norm if it satisfies either of the equivalent conditions of Lemma 1.2.11. In the sequel, we set*

$$X_a = \{u \in X : u \text{ has absolutely continuous norm}\}.$$

In the event that $X_a = X$, X will be said to have absolutely continuous norm.

Lemma 1.2.12. *If X is a Banach function space, then X_a is a $\|\cdot\|_X$-closed subspace of X. Moreover, if $u \in X_a$, $v \in X$ and $|v| \leq |u|$ a.e., then necessarily $v \in X_a$.*

Proof. The algebraic vector space structure of X_a follows easily by definition. If $u_k \longrightarrow u$ in norm and (E_j) is a sequence of measurable subsets of Ω such that $E_j \to \emptyset$, then

$$\begin{aligned}\|u\mathbf{1}_{E_j}\|_X &\leq \|(u-u_n)\mathbf{1}_{E_j}\|_X + \|f_n\mathbf{1}_{E_j}\|_X \\ &\leq \|(u-u_n)\|_X + \|u_n\mathbf{1}_{E_j}\|_X \to 0 \text{ as } n, j \to \infty.\end{aligned}$$

The second asseveration is a direct consequence of the remarks following Definition 1.2.3. □

Example 1.2.1. $\mathbf{L^p(\Omega)}$: For what is possibly the simplest example, consider the Lebesgue space $L^p(\Omega)$, $1 \leq p < \infty$ on $(\Omega, \mathcal{A}, \mu)$. For a sequence of measurable sets $M_n \searrow \emptyset$ and $f \in L^p(\Omega)$, one has, on account of Lebesgue's dominated convergence:

$$\|f\mathbf{1}_{M_n}\|^p_{L^p(\Omega)} = \int_\Omega |f|^p \mathbf{1}_{M_n}\, d\mu \to 0 \text{ as } n \to \infty.$$

Hence, every $f \in L^p(\Omega)$ has absolutely continuous norm.

Example 1.2.2. $\mathbf{L^\infty(\Omega)}$: If $\Omega \subseteq \mathbb{R}^n$ is a domain and μ is the Lebesgue measure on the Borel-measurable sets, the only function with absolutely continuous norm in the Lebesgue space $L^\infty(\Omega)$, however is the *a.e.* zero function. For, if $f \in L^\infty(\Omega)$,

$$\{x \in \Omega : f(x) > 0\} = \bigcup_{k=0}^{\infty} \left\{x \in \Omega : f(x) > \frac{1}{k}\right\}$$

and consequently, if $\mu(\{x \in \Omega : f(x) > 0\}) > 0$, one has, for some $k_0 \in \mathbb{N}$

$$\mu\left(\left\{x \in \Omega : f(x) > \frac{1}{k_0}\right\}\right) > 0.$$

Let $E_0 = \left\{x \in \Omega : f(x) > \frac{1}{k_0}\right\}$. There exists a compact set $S \subseteq E_0$ with $\mu)(S) > 0$. Consider an increasing sequence of domains (Ω_k), $\Omega_k \subset \Omega$ such that

$$\Omega = \bigcup_{k=1}^{\infty} \Omega_k.$$

Thus, $S_k = S \bigcap (\Omega \setminus \Omega_k) \searrow \emptyset$, however

$$\|f \mathbf{1}_{S_k}\|_{L^\infty(\Omega)} \geq \frac{1}{k_0}.$$

It follows that f does not have absolutely continuous norm, unless $f = 0$ *a.e.*.

Example 1.2.3. *ℓ^∞:* Next, consider the space ℓ^∞ consisting of all bounded sequence of real numbers endowed with the norm

$$\|(a_n)\|_\infty = \sup_{n \in \mathbb{N}} |a_n|. \tag{1.8}$$

Notice that this space can be viewed as a special instance of the general case discussed in this Section 1.2, namely, as a Banach function space based on the measure space $(\mathbb{N}, 2^{\mathbb{N}}, \mu)$, where μ stands for the counting measure and the function norm is given by (1.8). In this context, the class of sequences in ℓ^∞ with absolutely continuous norm is precisely the set of sequences that vanish at ∞. More precisely:

Lemma 1.2.13. *A sequence $(a_n) \in \ell^\infty$ has absolutely continuous norm if and only if*

$$\lim_{n \to \infty} a_n = 0.$$

It is customary to denote the space ℓ_a^∞ with $\mathbf{c}_0$.

Proof. If (a_n) has absolutely continuous norm consider the sequence $M_n = \{x \in \mathbb{N} : x \geq n\}$. Evidently, $M_n \searrow \emptyset$ and it follows automatically from the definition that for any positive ϵ,

$$\|(a_k)_{k \geq n}\|_\infty < \epsilon$$

if n is large enough, which yields $(a_n) \in \mathbf{c}_0$. Conversely, assume that $(a_n) \in \mathbf{c}_0$ and that $M_n \searrow \emptyset$. For any $\delta > 0$, $\sup\limits_{n \geq n_0} |a_n| < \delta$ for some $n_0 \in \mathbb{N}$. By assumption, there exists $n_1 \in \mathbb{N}$ such that

$$n \geq n_1 \Rightarrow \mathbf{1}_{M_n}(j) = 0$$

for $1 \leq j \leq n_0$. Consequently, for $k \geq n_1$

$$\|(a_n)\mathbf{1}_{M_k}\|_\infty = \sup_{j > n_0} |a_j| < \delta.$$

It follows that (a_n) has absolutely continuous norm, as claimed. □

Theorem 1.2.14. *Let u be a function in a Banach function space X. Then*

$$(i) \Leftrightarrow (ii) \Leftrightarrow (iii) \Rightarrow (iv).$$

(i) u has absolutely continuous norm.

(ii) Given any sequence (u_n) of measurable functions on X with

$$u_n \searrow 0 \text{ as } n \to \infty \text{ and } u_n \leq |u| \text{ a.e.},$$

one has $\|u_n\|_X \searrow 0$.

(iii) For any measurable function v and every sequence of measurable functions (v_n) such that

$$v_n \to v \text{ a.e. and } |v_n| \leq |u| \text{ a.e., for all } n, \tag{1.9}$$

one has

$$\|v_n - v\|_X \longrightarrow 0 \text{ as } n \to \infty.$$

(iv) For each $\epsilon > 0$ there exists $\delta > 0$ such that for any measurable set M with $\mu(M) < \delta$ it holds that $\|u\mathbf{1}_M\|_X < \epsilon$.

Proof. $(i) \Rightarrow (iv)$: If (i) holds, (iv) follows by contradiction by taking a sequence (M_n) of measurable sets with $\mu(M_n) < 2^{-n}$ such that $\|u_n\|_X \geq \epsilon$ and observing that

$$\mu\left(\bigcup_k^\infty M_n\right) < 2^{1-k} \longrightarrow 0 \text{ as } n \longrightarrow \infty.$$

Hence $(i) \Rightarrow (iv)$. Assuming next that (ii) holds, take $u \in X$ and consider a sequence (M_n) of measurable sets such that $M_n \searrow \emptyset$. Then, *a.e.* one has

$$u\mathbf{1}_{M_n} \to 0 \text{ as } n \to \infty$$

and $u\mathbf{1}_{M_n} \leq |u|$. The valididty of (ii) yields

$$\|u\mathbf{1}_{M_n}\|_X \to 0 \text{ as } n \to \infty,$$

so that u as absolutely continuous norm. This shows that $(ii) \Rightarrow (i)$.
To see that $(i) \Rightarrow (ii)$, take $u \in X_a$ and (u_n) be a sequence of measurable functions as in (ii).
Since $(\Omega, \mathcal{M}, \mu)$ is assumed to be σ-finite, one can take a sequence of mutually disjoint sets $(\Omega_j) \subset \mathcal{M}$ as in (1.1), i.e., $\mu(\Omega_j) < \infty$ for each $j \in \mathbb{N}$ and $\Omega = \bigcup_1^\infty \Omega_j$. Obviously, one has $(\Omega \setminus \Omega_j) \searrow 0$ as $j \to \infty$; thus for any $\epsilon > 0$ and sufficiently large k one has

$$\|u\mathbf{1}_{\Omega \setminus \Omega_k}\|_X < \frac{\epsilon}{2}.$$

On the other hand, by virtue of axiom (v) of Definition 1.2.1, $|u|$ is integrable on Ω; consequently, an application of Lebesgue's dominated convergence theorem implies that for any $\delta > 0$,

$$\mu(\{x \in \Omega_k : u_k(x) > \delta\}) \leq \frac{1}{\delta} \int_\Omega u_k \mathbf{1}_{\Omega_k}\, d\mu \to 0 \text{ as } k \to \infty.$$

Therefore, for sufficiently large k, property (iv) yields the inequality

$$\|u\mathbf{1}_{\{x\in\Omega_k : u_k(x)>\delta\}}\|_X < \frac{\epsilon}{4}.$$

In all, if $\delta < \epsilon/(4\|\mathbf{1}_{\Omega_n}\|_X)$,

$$\begin{aligned} \|u_k\|_X &\leq \|u_k\mathbf{1}_{\Omega_k}\|_X + \|u_k\mathbf{1}_{\Omega\setminus\Omega_k}\|_X \\ &\leq \delta\|\mathbf{1}_{\Omega_k}\|_X + \|u\mathbf{1}_{\{x\in\Omega_k : u_k(x)>\delta\}}\|_X + \|u_k\mathbf{1}_{\Omega\setminus\Omega_k}\|_X \\ &< \epsilon. \end{aligned}$$

In all, $(i) \Rightarrow (ii)$.
$(i) \Rightarrow (iii)$: Assume u has absolutely continuous norm as let (v_n) and v be measurable functions having property (1.9). Setting

$$w_k(x) = \sup_{j\geq k} |v_j(x) - v(x)|$$

it is readily verified that

$$w_k \searrow 0 \text{ as } k \to \infty$$

and that *a.e.*,

$$0 \leq w_k \leq 2|u|.$$

Notice that by definition, $2|u|$ has absolutely continuous norm; a direct application of (ii) yields then

$$\|v_k - v\|_X \leq \|w_k\|_X \searrow 0 \text{ as } k \to \infty.$$

$(iii) \Rightarrow (i)$: For the sufficiency of condition (iii), select $u \in X$, (M_k), let be a sequence of measurable sets with $M_k \searrow \emptyset$ and set $v_k = u\mathbf{1}_k$. Then $|u| \geq v_k \searrow 0$ *a.e.* and the assumed validity of property (iii) implies $\|v_k\|_X = \|u\mathbf{1}_{M_k}\|_X \searrow 0$ as $k \to \infty$. □

For reasons that will naturally emerge in the sequel, another important subspace of a Banach function space X needs to be introduced:

Definition 1.2.4. *If X is a Banach function space, X_b will stand for the $\|\cdot\|_X$-closure of the set of bounded functions supported in sets of finite measure. In other words*

$$X_b = Cl^{\|\cdot\|_X}\left(\{u \in X : u \text{ bounded and } |supp\ u| < \infty\}\right).$$

The following characterization is almost immediate:

Lemma 1.2.15. *X_b is the $\|\cdot\|_X$-closure of the subspace $\mathcal{S}$ consisting of simple functions supported on sets of finite measure.*

Proof. If v is a bounded function supported on a set of finite measure, let (v_n) be a sequence of simple functions that converges to v uniformly in Ω; such sequence can for example be chosen as

$$v_n = \sum_{j=0}^{2^{2n-1}} 2^{-n} j \mathbf{1}_{v^{-1}(j2^{-n},(j+1)2^{-n})}, \tag{1.10}$$

if $\sup_\Omega v \leq 2^m$. Then, for any $\delta > 0$ set $\epsilon = \frac{\delta}{|\Omega|}$ and for sufficiently large n it holds

$$\|v - v_n\|_X \leq \delta.$$

□

Lemma 1.2.16. *For any Banach function space X, $X_a \subseteq X_b$.*

Proof. If $v \in X_a$ it is easy to verify that

$$v_n(x) = \mathbf{sgn}(v(x)) \min\{|v(x)|, n\mathbf{1}_{\Omega_n}\},$$

where the sequence (Ω_n) is as in (1.1), defines a sequence of bounded functions that are supported in a set of finite measure with $v_n \to v$ pointwise *a.e.* and such that $|v_n| \leq |v|$. On account of Theorem 1.2.14 (*iii*) we conclude: $\|v_n - v\|_X \to 0$ as $n \to \infty$, *i.e.*, $v \in X_b$. □

Definition 1.2.5. *An order ideal of a Banach function space X is a closed, linear subspace W of X such that if $u \in W$ and $|v| \leq |u|$ a.e., then $v \in W$.*

Lemma 1.2.17. *If X is a Banach function space, then both X_a and X_b are order ideals of X.*

Proof. As shown in Lemma 1.2.12, X_a is closed. It remains thus to show the order property required for an order ideal. To this end, observe that for any sequence (E_j) of measurable sets and any $u \in X_a$, if $v \in X$ and $|v| \leq |u|$ *a.e.*, the Banach function space properties of X yield $\|v\mathbf{1}_{E_j}\|_X \leq \|u\mathbf{1}_{E_j}\|_X \to 0$ as $j \to \infty$.

Next, since X_b is norm-closed by definition, it remains to observe that if $u \in X_b$ and $|v| \leq |u|$ *a.e.*, then for any sequence $(u_n) \subseteq \mathcal{S}$ that converges to u in norm one can define pointwise the following sequence of bounded functions, supported on sets of finite measure in Ω:

$$\begin{aligned} v_n(x) &= \left(u_n(x)\mathbf{1}_{u_n \leq v} + v(x)\mathbf{1}_{u_n > v}\right)\mathbf{1}_{v>0} \\ &\quad + \left(v(x)\mathbf{1}_{u_n > v} + u_n(x)\mathbf{1}_{u_n \leq v}\right)\mathbf{1}_{v \leq 0} \quad \text{for each } n \in \mathbb{N}. \end{aligned}$$

It is straightforward to verify that $|v_n - v| \leq |u_n - u|$ *a.e.* and hence that $\|v_n - v\|_X \longrightarrow 0$ as $n \to \infty$. The claim follows now by definition of X_b. □

Theorem 1.2.18. *Let X be a Banach function space. Then the following conditions are equivalent:*

(i) $X_a = X_b$,

(ii) $(X_b)^* = X'$,

(iii) $X^* = X'$,

(iv) $X = X_b = X_a$.

Proof. We will proceed with the proof in the real setting. Lemma 1.2.17 and Theorem 1.2.9 guarantee the continuous inclusion

$$X' \hookrightarrow (X_b)^* .$$

$(i) \Rightarrow (ii)$ For the necessity statement, it remains to verify that under assumption (i) the converse inclusion holds. Let $\gamma \in (X_b)^*$ and recall that the underlying measure μ is assumed to be σ-finite, i.e., there is a mutually disjoint sequence (Ω_j) with $\mu(\Omega_j) < \infty$ such that

$$\Omega = \bigcup_{j=1}^{\infty} \Omega_j.$$

For each j we write $\mathcal{A}_j$ for the σ-algebra of measurable subsets of Ω_j; then γ defines a set function δ_j on $\mathcal{A}_j$ in the nearly obvious manner:

$$\delta_j(E) = \gamma(\mathbf{1}_E).$$

The linearity and continuity of γ guarantee the σ-additivity of δ_j, for if $(E_k) \subset \Omega_j$ is a collection of mutually disjoint μ_j-measurable sets then $\mathbf{1}_{\bigcup_1^\infty E_k} \in X_b = X_a$ by definition of X_b and by assumption (i). Furthermore, for fixed $m \in \mathbb{N}$,

$$\mathbf{1}_{\bigcup_1^\infty E_k} - \mathbf{1}_{\bigcup_1^m E_k} \searrow 0 \text{ a.e. as } m \to \infty$$

and on account of Theorem 1.2.14, (iii) one has

$$\|\mathbf{1}_{\bigcup_1^\infty E_k} - \mathbf{1}_{\bigcup_1^m E_k}\|_X \searrow 0,$$

that is,

$$\begin{aligned}
\delta_j\left(\bigcup_1^\infty E_k\right) &= \gamma\left(\mathbf{1}_{\bigcup_1^\infty E_k}\right) = \gamma\left(\lim_{m\to\infty} \mathbf{1}_{\bigcup_1^M E_k}\right) \\
&= \lim_{m\to\infty} \gamma\left(\mathbf{1}_{\bigcup_1^M E_k}\right) = \sum_1^\infty \gamma(\mathbf{1}_{E_k}) \\
&= \sum_1^\infty \delta_j(E_k).
\end{aligned}$$

Hence, δ_j is a signed measure on $\mathcal{A}_j$. Moreover if one denotes the restriction of μ to Ω_j by μ_j the inequality (valid for for any $E \subset \mathbb{A}_j$)

$$|\delta_j(E)| \le \|\gamma\|_{(X_b)^*} \|\mathbf{1}_E\|_{X_b}$$

shows that $\mu_j(E) = 0 \Rightarrow \delta_j(E) = 0$. An application of the Radon-Nikodym theorem yields, for each $j \in \mathbb{N}$, a function $v_j \in L^1(\Omega_j)$ such that for any $E \in \mathcal{A}_j$:

$$\delta_j(E) = \gamma(\mathbf{1}_E) = \int_{\Omega_j} v_j \mathbf{1}_E \, d\mu.$$

Define v on Ω as $v(x) = v_j(x)$ for $x \in \Omega_j$. Then for any $E \in \mathcal{A}$

$$\gamma(\mathbf{1}_E) = \int_{\Omega} v \mathbf{1}_E \, d\mu.$$

It follows from the preceding asseveration that for any $u \in \mathcal{S}$ one has

$$\gamma(u) = \int_{\Omega} vu \, d\mu. \tag{1.11}$$

In fact, as a consequence of (1.11) one concludes that $v \in X'$. In order to see this, observe that given $u \in X$, a sequence (u_k) of simple functions with $\mathbf{supp} u_k \subseteq \Omega_k$ can be constructed via (1.10), such that

$$0 \le u_k \nearrow |u|$$

pointwise in Ω.
By virtue of (1.11) and Axiom (iii) in Definition 1.2.1,

$$\begin{aligned}\int_{\Omega} |v| u_k \, d\mu &= \int_{\Omega_k} v(\mathbf{sgn} v) u_k \, d\mu \\ &= \gamma((\mathbf{sgn} v) u_k) \le \|\gamma\|_{(X_b)^*} \|\mathbf{sgn} v) u_k\|_X \\ &\le \|\gamma\|_{(X_b)^*} \|u\|_X,\end{aligned}$$

i.e., by monotone convergence,

$$\int_{\Omega} |vu| \, d\mu = \lim_{k \to \infty} \int_{\Omega_k} |v| u_k \, d\mu \le \|\gamma\|_{(X_b)^*} \|u\|_X < \infty. \tag{1.12}$$

Lemma 1.2.6 guarantees now that $v \in X'$. We set about next to prove that (1.11) holds in fact for any $u \in X_b$, then for (u_k) as in the above paragraph

$$\int_{\Omega} vu_k \, d\mu \longrightarrow \int_{\Omega} vu \, d\mu \tag{1.13}$$

on account of (1.12) and dominated convergence. In particular, (seen(1.4), (1.12) and (1.13)):

$$\|v\|_{X'} \leq \|\gamma\|_{(X_b)^*}.$$

We have proved that $(i) \Rightarrow (ii)$.

$(ii) \Rightarrow (iii)$: Claim (iii) follows from (ii) via Theorem 1.2.9 and Definition 1.2.4.

$(iii) \Rightarrow (iv)$: Now, if (iii) holds, let $w \in X$ and consider a sequence $(E_j) \subset \mathcal{A}$ with $E_j \searrow \emptyset$ as $j \to \infty$. It is enough to consider $w \geq 0$ on Ω. For any $\gamma \in B_{X^*}$, by assumption, there exists $u \in X'$ with $\gamma(v) = \int_\Omega v(x)u(x)\,dx$ for any $v \in X_b$ and $\|\gamma\|_{(X_b)^*} = \|u\|_{X'} \leq 1$. Since $w \in X_b$ and $u \in X'$, Lemma 1.2.6 and dominated convergence imply

$$\int_\Omega wu\mathbf{1}_{E_j}\,d\mu \to 0 \text{ as } j \to \infty;$$

it follows that for fixed $\epsilon > 0$ and for sufficiently large $j \in \mathbb{N}$,

$$\left|\int_\Omega wu\mathbf{1}_{E_j}\,d\mu\right| < \epsilon.$$

Each of the sets

$$Y_j = \left\{\gamma \in X^* : \left|\langle\gamma, \mathbf{1}_{E_j}\rangle\right| < \epsilon\right\}$$

is weak* open (in fact, it is a neighborhood of the origin for the weak* topology on X^*) and is in view of the preceding argument, the family (Y_j) constitutes an open cover of the unit ball B_{X^*}. On account of the theorem of Banach-Alaoglu there is a finite subcover, say $(Y_{j_k})_{1\leq k\leq N}$.
For $J > \max\{j_k,\ 1 \leq j \leq N\}$ and any $\gamma \in B_{X^*}$ one thus concludes

$$|\langle\gamma, w\mathbf{1}_{E_J}\rangle| < \epsilon.$$

On the other hand Theorem 1.2.8 guarantees that

$$\begin{aligned}\|w\mathbf{1}_{E_J}\|_X = \|w\mathbf{1}_{E_J}\|_{X''} &= \sup_{\|v\|_{X'}\leq 1}\left|\int_\Omega v(w\mathbf{1}_{E_J})\,d\mu\right| \\ &\leq \epsilon.\end{aligned}$$

In conclusion,

$$\|w\mathbf{1}_{E_k}\|_X \to 0 \text{ as } j \to \infty;$$

(iii), thus, implies (iv), as claimed.
$(iv) \Rightarrow (i)$ This implication is trivial.
The theorem is thus proved. □

As a consequence of theorem 1.2.18 we have the following characterization:

Theorem 1.2.19. *Let X be a Banach function space. Then*
(i) X' and X^* are canonically isometrically isomophic if and only if X has absolutely continuous norm;
(ii) X is reflexive if and only if both X and X' have absolutely continuous norm.

Proof. Assume first that X is reflexive. It was shown in Theorem 1.2.9 that X' is isometric to a closed, norm-fundamental subspace of X^*. In view of Theorem 1.2.18, to conclude that X has absolutely continuous norm, i.e. that $X = X_a$, it suffices to show that X' cannot be a proper subspace of X^*. Assume, on the contrary, that $X' \subsetneq X^*$. Let $0 \neq \gamma \in X^{**}$ be zero on X' (the existence of such γ is guaranteed by the Hahn-Banach Theorem). Since X is reflexive one derives the existence of $z \in X$ such that $\int_\Omega zy\,dx = 0$ for each $y \in X'$. By virtue of Theorem 1.2.9 X' is identified with a norm-fundamental subspace of X^*; in particular, it holds that

$$\|z\|_X = \sup_{y \in X'} \left| \int_\Omega zy\,dx \right| = 0,$$

i.e., $z = 0$ *a.e.*, which forces $\gamma = 0$, a contradiction. Thus, $X^{'} = X^*$ and Theorem 1.2.9 guarantees that X has absolutely continuous norm.
Moreover, the identification $X' = X^*$ in conjunction with the reflexivity of X implies

$$(X')^* = X^{**} \approx X,$$

i.e., $(X')^* \approx (X')'$ (Theorem 1.2.8). Invoking Theorem 1.2.18 applied to the Banach function space X', one readily concludes that X' has absolutely continuous norm, as claimed.
Conversely, if both X and its associate $X^{'}$ have absolutely continuous norm, then by virtue of Theorem 1.2.18 applied to X one has the equality $X^* = X'$; the dual spaces are hence the same, and it follows that

$$X^{**} = (X')^*.$$

Theorem 1.2.18 applies now to $X^{'}$ to yield $(X')^* = X''$ and the latter equals X by Theorem 1.2.9. In all, $X^{**} \approx X$, as claimed. □

In particular, we state the obvious consequence of the preceding Theorem:

Corollary 1.2.20. *Let X be a Banach function space. Then $X^* = X'$ if and only if X has absolutely continuous norm. In particular, if both X and X^* have absolutely continuous norm, X is reflexive.*

Example 1.2.4. The sequence spaces ℓ^p, $1 \leq p < \infty$ have absolutely continuous norm. The preceding Corollary implies then

$$(\ell^p)' = (\ell^p)^* = \ell^{\frac{p}{p-1}},$$

with the convention $\frac{p}{p-1} = \infty$ when $p = 1$. It is easy to check that $(\ell^\infty)' = \ell^1 \subsetneq (\ell^\infty)^*$.

1.2.3 Convexity, uniform convexity and smoothness of a norm

In this section we present some standard definitions and results pertaining to the geometry of Banach spaces.
We commence with the following:

Definition 1.2.6. *A normed space $(X, \|\cdot\|_X)$ is said to be strictly convex if whenever $u \in S_X$ and $v \in S_X$ and $0 < t < 1$, then $\|tu + (1-t)v\|_X < 1$.*

Typical examples of strictly convex spaces include the $L^p(\Omega, d\mu)$ spaces, $1 < p < \infty$, over a measure space $(\Omega, \mathcal{A}, \mu)$. It is well known that $L^\infty(\Omega)$ the vector space of Lebesgue-measurable, complex valued function on a domain Ω furnished with the supremum norm is not strictly convex. We next state the standard definition of uniform convexity in the context of Banach spaces.

Definition 1.2.7. *A normed space $(X, \|\cdot\|)$ is uniformly convex iff for any $\epsilon : 0 < \epsilon \leq 2$ one has*

$$\delta_X(\epsilon) = \inf\left\{1 - \left\|\frac{x+y}{2}\right\| : \|x\| \leq 1\,,\ \|y\| \leq 1\,,\ \|x-y\| \geq \epsilon\right\} > 0.$$

The number $\delta_X(\epsilon)$ is known as the modulus of uniform convexity of X (see, for example, [17] and [52]).

The next result indicates how strict convexity can be used.

Proposition 1.2.21. *Let K be a closed, convex, non-empty subset of a strictly convex Banach space X. Then there is at most one element $x \in K$ such that*

$$\|x\| = \inf\{\|y\| : y \in K\}.$$

If X is reflexive, such an x exists.

Proof. Suppose there exist $x, y \in K$ with $\|x\| = \|y\| = \inf\{\|z\| : z \in K\}$, $x \neq y$. Let $0 < \lambda < 1$: then $\lambda x + (1-\lambda)y \in K$, $\|\lambda x + (1-\lambda)y\| < \|x\|$ and we have a contradiction.

Now let X be reflexive and assume that (x_k) is a sequence in K such that $\lim_{k\to\infty} \|x_k\| = l := \inf\{\|y\| : y \in K\}$. This sequence has a subsequence, again denoted by (x_k), such that $x_k \rightharpoonup x$ for some $x \in X$; and $x \in K$ since K is convex and closed, and hence weakly closed. Since $\|x\| \leq \lim_{k\to\infty} \|x_k\| = l$, the result follows. □

A characterization of strict and uniform convexity can be given by means of differentiability considerations. Let U be an open subset of a Banach space

X and $f : U \to \mathbb{R}$. We say that f is **Gâteaux-differentiable** at $x_0 \in U$ if there exists $x^* \in X^*$ such that

$$\langle x^*, h \rangle = \lim_{t \to 0} \frac{f(x_0 + th) - f(x_0)}{t} \text{ for all } h \in X.$$

The limit above is called the **derivative of f in the direction** h; the functional x^* is often denoted by $\mathrm{grad}\, f(x_0)$ and will be referred to as the **gradient** or **Gâteaux derivative** of f at x_0.
If this limit is uniform with respect to $h \in S_X$, we say that f is **Fréchet-differentiable** at x_0 (and then refer to x^* as the **Fréchet derivative** of f at x_0, written $f'(x_0)$): equivalently,

$$\frac{f(x_0 + h) - f(x_0) - \langle x^*, h \rangle}{\|h\|} \to 0 \text{ as } \|h\| \to 0.$$

Clearly Fréchet-differentiability implies Gâteaux-differentiability; in the reverse direction, it can be shown that if $\mathrm{grad}\, f(x)$ exists throughout some neighborhood of x_0 and is continuous at x_0, then f is Fréchet-differentiable at x_0 and $\mathrm{grad}\, f(x_0) = f'(x_0)$.

The case when $f(x) = \|x\|$ $(x \in X)$ is of particular interest. The norm $\|\cdot\|$ is said to be Fréchet- (resp. Gâteaux-) differentiable if $\|\cdot\|$ is Fréchet- (resp. Gâteaux-) differentiable at every point of $X \backslash \{0\}$: the point 0 is excluded because no norm is differentiable at 0. In particular:

Lemma 1.2.22. *If the norm $\|\cdot\|$ is Gâteaux- differentiable, then for $x_0 \neq 0$,*

$$\big\|\mathrm{grad}\|x_0\|_X\big\|_{X^*} = 1 \text{ and } \langle \mathrm{grad}\|x_0\|, x_0 \rangle = \|x_0\|_X.$$

Given a convex subset C of a Banach space X and a point $x \in C$, a functional $f \in X^*$ is called a **supporting functional of C at x** if $\|f\| = 1$ and $\langle f, x \rangle = \sup \{|\langle f, y \rangle| : y \in C\}$. It emerges that if the norm on X is Gâteaux-differentiable, then $\mathrm{grad}\, \|x\|$ is a supporting functional of B_X at every point x of $B_X \backslash \{0\}$: for a proof of this and of the characterizations given next of spaces with Gâteaux-differentiable norms we refer to [37], Chapter 8.

Theorem 1.2.23. *The following statements about a Banach space X are equivalent:*
(i) The norm on X is Gâteaux-differentiable.
(ii) The dual space X^* is strictly convex.
(iii) Given any $x \in S_X$, there is a unique $f \in S_{X^*}$ such that $\langle f, x \rangle = 1$.

It turns out that uniform convexity of a space X has significant implications for the dual space X^*. To explain this we introduce a new object, the modulus of smoothness of a space.

Definition 1.2.8. *Let X be a Banach space. The* **modulus of smoothness** *of X is the function $\rho_X : (0, \infty) \to [0, \infty)$ defined by*

$$\rho_X(\tau) := \sup \left\{ \frac{\|x + \tau h\| + \|x - \tau h\|}{2} - 1 : x, h \in S_X \right\}.$$

If $\lim_{\tau \to 0} \rho_X(\tau)/\tau = 0$, the space X is said to be **uniformly smooth.**

Note that ρ_X really is a non-negative function since for all $x, h \in X$ we have $2\|x\| = \|x + \tau h + x - \tau h\| \leq \|x + \tau h\| + \|x - \tau h\|$. Evidently the property of uniform smoothness is preserved on passage to a subspace. For the following theorem see, for example, [37, Chapter 8].

Lemma 1.2.24. *Let X be a Banach space. The following statements are equivalent:*
(i) The space X is uniformly smooth.
(ii) The limit

$$\lim_{t \to 0} \frac{\|x + \tau h\| - \|x\|}{t} = \langle \text{grad} \|x\|, h \rangle$$

exists, uniformly for $x, h \in S_X$.
(iii) The norm of X is Fréchet-differentiable on S_X and the map $x \longmapsto \text{grad} \|x\| : S_X \to S_{X^*}$ is uniformly continuous.

If $\|\cdot\|$ satisfies any of the equivalent conditions in the last lemma we shall say that it is **uniformly Fréchet-differentiable.**
Connections between the moduli of convexity and smoothness, together with important implications of these connections, are given in the next theorem.

Theorem 1.2.25. *Let X be a Banach space. Then:*
(i) for all $\tau > 0$,

$$\rho_{X^*}(\tau) = \sup \left\{ \frac{1}{2}\tau\varepsilon - \delta_X(\varepsilon) : 0 < \varepsilon \leq 2 \right\}$$

and

$$\rho_X(\tau) = \sup \left\{ \frac{1}{2}\tau\varepsilon - \delta_{X^*}(\varepsilon) : 0 < \varepsilon \leq 2 \right\};$$

(ii) X is uniformly convex if and only if X^* is uniformly smooth;
(iii) X is uniformly smooth if and only if X^* is uniformly convex.

Proof. *(i)* We establish only the first of the two statements, the proof of the second being similar. Let $\varepsilon \in (0, 2]$ and $\tau > 0$; let $x, y \in S_X$ be such that $\|x - y\| \geq \varepsilon$. By the Hahn-Banach theorem, there exist $f, g \in S_{X^*}$ such that $\langle f, x + y \rangle = \|x + y\|$ and $\langle g, x - y \rangle = \|x - y\|$. Then

$$\begin{aligned} 2\rho_{X^*}(\tau) &\geq \|f + \tau g\| + \|f - \tau g\| - 2 \\ &\geq \langle f + \tau g, x \rangle + \langle f - \tau g, y \rangle - 2 \\ &= \langle f, x + y \rangle + \tau \langle g, x - y \rangle - 2 = \|x + y\| + \tau \|x - y\| - 2. \end{aligned}$$

Thus $2- \|x+y\| \geq \tau\varepsilon - 2\delta_{X^*}(\tau)$, which gives $\delta_X(\varepsilon) + \rho_{X^*}(\tau) \geq \tau\varepsilon/2$, so that $\rho_{X^*}(\tau) \geq \sup\left\{\frac{1}{2}\tau\varepsilon - \delta_X(\varepsilon) : 0 < \varepsilon \leq 2\right\}$.

To establish the reverse inequality, let $\tau > 0$ and $f, g \in S_{X^*}$. Given $\eta > 0$, let $x, y \in S_X$ be such that $\langle f + \tau g, x\rangle \geq \|f + \tau g\| - \eta$ and $\langle f - \tau g, y\rangle \geq \|f - \tau g\| - \eta$. Then

$$\begin{aligned}
\|f+\tau g\| + \|f-\tau g\| - 2 &\leq \langle f, x+y\rangle + \tau\langle g, x-y\rangle - 2 + 2\eta \\
&\leq \|x+y\| - 2 + \tau\|x-y\| + 2\eta \\
&\leq -2\delta_X(\|x-y\|) + \tau\|x-y\| + 2\eta \\
&\leq 2\sup\left\{\frac{1}{2}\tau\varepsilon - \delta_{X^*}(\varepsilon) : 0 < \varepsilon \leq 2\right\} + 2\eta,
\end{aligned}$$

from which the desired inequality follows.

(*ii*) Suppose that X is uniformly convex and let $\varepsilon_0 \in (0, 2]$. Then $\delta_X(\varepsilon) \geq \delta_X(\varepsilon_0) > 0$ if $\varepsilon \in [\varepsilon_0, 2]$. Let $\tau \in (0, \delta_X(\varepsilon_0))$. If $\varepsilon \in [\varepsilon_0, 2]$, then

$$\varepsilon/2 - \delta_X(\varepsilon)/\tau \leq \varepsilon/2 - \delta_X(\varepsilon_0)/\tau \leq \varepsilon/2 - 1 \leq 0.$$

By (*i*),

$$\rho_{X^*}(\tau)/\tau = \sup_{0<\varepsilon\leq\varepsilon_0} (\varepsilon/2 - \delta_X(\varepsilon)/\tau) \leq \sup_{0<\varepsilon\leq\varepsilon_0} \varepsilon/2 = \varepsilon_0/2.$$

Thus $\lim_{\tau\to 0} \rho_{X^*}(\tau)/\tau = 0$: X^* is uniformly smooth.

Conversely, if X is not uniformly convex, then $\delta_X(\varepsilon_0) = 0$ for some $\varepsilon_0 \in (0, 2]$. By (i), if $\tau > 0$ then

$$\rho_{X^*}(\tau) = \sup_{0<\varepsilon\leq\varepsilon_0} (\varepsilon\tau/2 - \delta_X(\varepsilon)) \geq \varepsilon_0\tau/2.$$

Hence $\limsup_{\tau\to 0} \rho_{X^*}(\tau)/\tau \geq \varepsilon_0/2$, which implies that X^* is not uniformly smooth.

(*iii*) The proof is similar to that of (*ii*), again using (*i*). □

Note that from the last theorem it follows that if X is uniformly smooth, then it is reflexive.

For a Hilbert space H it is simple to verify that

$$\rho_H(\tau) = \left(1+\tau^2\right)^{1/2} - 1,$$

so that H is uniformly smooth. This assertion also follows from the fact that H is uniformly convex. It is known (see [70] II, Chapter 1e) that for every Banach space X we have

$$\rho_X(\tau) \geq \left(1-\tau^2\right)^{1/2} - 1.$$

This supports the view that Hilbert spaces are the 'most uniformly smooth' spaces.

1.2.4 Duality mappings and extremal elements

Definition 1.2.9. *A gauge map is a continuous, strictly increasing function*

$$s : [0, \infty) \longrightarrow [0, \infty)$$

with $s(0) = 0$ and such that $\lim_{t\to\infty} s(t) = \infty$.

Definition 1.2.10. *A map $J : X \longrightarrow 2^{X^*}$ from a Banach space X to the power set of its dual is said to be a duality map with gauge function s if for all $x \in X$*

$$J(x) = \{x^* \in X^* : \langle x^*, x\rangle = \|x^*\|\|x\|,\ \|x^*\| = s(\|x\|)\}$$

By the Hahn-Banach theorem, for each $x \in X$ the set $J(x)$ is non-empty and convex. For if $x^*, y^* \in J(x)$ and $\lambda \in (0,1)$, $z^* = \lambda x^* + (1-\lambda)y^*$, then

$$\langle z^*, x\rangle = \lambda \langle x^*, x\rangle + (1-\lambda)\langle y^*, x\rangle = s(\|x\|)\|x\|.$$

Hence $\|z^*\| \geq s(\|x\|)$. Since

$$\|z^*\| \leq \lambda \|x^*\| + (1-\lambda)\|y^*\| = s(\|x\|),$$

we see that $\|z^*\| = s(\|x\|)$, whence $z^* \in J(x)$ and thus the asserted convexity is established.

If X is a Banach space whose dual X^* is strictly convex and J is a duality map on X with gauge function s, then for each $x \in X$, the set $J(x)$ consists of precisely one point. In fact, for each $x \in X$, the points in $J(x)$ lie on the set

$$s(\|x\|)B_{X^*}.$$

If $J(x)$ contained two distinct points, the midpoint of the line segment joining them would be in the convex set $J(x)$, which is impossible as X^* is strictly convex.

In view of this result, a duality map J on X shall be regarded as a map from X to X^* when X^* is strictly convex. In particular, we single out the following consequence:

Lemma 1.2.26. *Let X be a Banach space with strictly convex dual X^* and $J(x)$ be a duality mapping with gauge function $s(t) = t$. Then*

$$J(x) = \text{grad}\,\|x\|_X.$$

The various properties of duality maps that will be used in the sequel are summarized in the next theorem (see [75], p.176), to prepare for which we introduce some standard terminology.

Let X be a real Banach space. A map $T : X \to X^*$ is called **monotone** if

$$\langle Tx - Ty, x - y\rangle \geq 0 \text{ for all } x, y \in X;$$

if

$$\langle Tx - Ty, x - y\rangle > 0 \text{ for all } x, y \in X \text{ with } x \neq y$$

it is said to be **strictly monotone**; it is **coercive** if

$$\langle Tx, x\rangle / \|x\| \to \infty \text{ as } \|x\| \to \infty;$$

and finally it is **demicontinuous** if $x_n \to x$ implies that $Tx_n \overset{*}{\rightharpoonup} Tx$.

Theorem 1.2.27. *Let X be a real, reflexive Banach space with strictly convex dual X^* and let J be a duality map on X with gauge function s. Then J is monotone, coercive and demicontinuous (continuous if X^* is uniformly convex); if, in addition, X is strictly convex, then J is strictly monotone.*

Proof. That J is monotone is clear since for all $x, y \in X$,

$$\begin{aligned}\langle Jx - Jy, x - y\rangle &= s(\|x\|)\|x\| + s(\|y\|)\|y\| - \langle Jy, x\rangle - \langle Jx, y\rangle \\ &\geq (s(\|x\|) - s(\|y\|))(\|x\| - \|y\|) \geq 0.\end{aligned}$$

As $\langle x, Jx\rangle = s(\|x\|)\|x\|$ $(x \in X)$ the coercivity of J is immediate. For demicontinuity, it is enough to deal with the case in which $s(t) = t$ $(t \geq 0)$. Let (x_n) be a sequence in X that converges to x and let $u \in X$: we must show that $\langle u, Jx_n\rangle \to \langle u, Jx\rangle$. Since the sequence (x_n) and the points x, u span a separable subspace of X it is sufficient to deal with the case in which X is separable. As $(\|Jx_n\|)$ is bounded, it follows that, by passage to a subsequence if necessary, $Jx_n \overset{*}{\rightharpoonup} x^*$ for some $x^* \in X^*$ (see, for example, [100, Theorem 4.41A]. Thus

$$\langle Jx_n, x_n\rangle \to \langle x^*, x\rangle, \quad \|x^*\| \leq \liminf_{n\to\infty} \|Jx_n\| = \lim_{n\to\infty} \|x_n\| = \|x\|,$$

while $\langle Jx_n, x_n\rangle = \|x_n\|^2 \to \|x\|^2$. The definition of J now ensures that $x^* = Jx$. Since this argument holds for every weak*-convergent subsequence of (Jx_n), the demicontinuity of J follows. If X^* is uniformly convex, then since $Jx_n \rightharpoonup Jx$ (the reflexivity of X follows from that of X^*) and $\|Jx_n\| = \|x_n\| \to \|x\| = \|Jx\|$, it follows that $Jx_n \to Jx$.

Assume next that X is strictly convex. The strict monotonicity of J will follow from the fact that $\langle Jx - Jy, x - y\rangle = 0$ implies that $x = y$. The exact same argument given above in the proof of the fact that J is monotone also yields: $\|x\| = \|y\|$: suppose that $x \neq y$. Then $x/\|x\| \neq y/\|y\|$; by Proposition 1.2.21, it follows that

$$\|Jx\| = \langle Jx, x/\|x\|\rangle > \langle Jx, y/\|y\|\rangle,$$

whence $\langle Jx, y\rangle < \langle Jx, x\rangle$; similarly, $\langle Jy, x\rangle < \langle Jy, y\rangle$. Hence

$$0 = \langle Jx - Jy, x - y\rangle > \langle Jx, x\rangle + \langle Jy, y\rangle - \langle Jx, x\rangle - \langle Jy, y\rangle = 0.$$

This contradiction completes the proof. ☐

1.3 Modular spaces

In this section we fix a real or complex vector space V. We denote the scalar field with $\mathbb{K}$. We define the concept of modular on V, analyze the basics properties and present concrete examples, aiming to emphasize the usefulness of the theory.

Definition 1.3.1. *An s-convex modular* $(0 < s \leq 1)$ *on a real or complex vector space* V *is a function*

$$\rho : X \longrightarrow [0, \infty]$$

that satisfies the following conditions:

(i) $\rho(x) = 0 \iff x = 0$

(ii) $\rho(\alpha x) = |\alpha|\rho(x)$ *for any* $x \in V$, $|\alpha| = 1$

(iii) $\rho(\alpha x + (1-\alpha)y) \leq \alpha^s \rho(x) + (1-\alpha)^s \rho(y)$ *for all* $x, y \in V$ *and* $\alpha \in (0, 1]$.

In particular, if $s = 1$ *the modular is said to be convex. A convex modular* ρ *on a vector space* V *is left-(right-) continuous if for any* $x \in V$ *the map*

$$\alpha \longrightarrow \rho(\alpha x)$$

is left-(right-) continuous on $[0, \infty)$ *(on* $(0, \infty)$ *for left continuity); if* ρ *is both left- and right-continuous we refer to it as a continuous modular.*

Example 1.3.1. If $(\Omega, \mathcal{A}, \mu)$ is a measure space as in Section 1.2 and $p \in [1, \infty)$, then

$$\rho(u) = \int_{\Omega} |u|^p \, d\mu \tag{1.14}$$

defines a convex modular on $L^p(\Omega, d\mu)$.

Example 1.3.2. Any norm on a real or complex vector space X is a convex modular on X.

It follows from the above axioms that for each fixed $x \in$ the map

$$\alpha \longrightarrow \rho(\alpha x)$$

is non-decreasing on $[0, \infty)$, for if $0 \leq \alpha < \beta$ and $x \in V$, then convexity yields:

$$\rho(\alpha x) = \rho(\alpha\beta^{-1}\beta x) \leq \alpha\beta^{-1}\rho(\beta x) \leq \rho(\beta x).$$

Consequently:

$$\rho(\alpha x) = \rho(|\alpha|x) \leq |\alpha|\rho(x) \text{ if } |\alpha| \leq 1 \tag{1.15}$$

$$\rho(\alpha x) = \rho(|\alpha|x) \geq |\alpha|\rho(x) \text{ if } |\alpha| \geq 1. \tag{1.16}$$

To the effect of characterizing the modular space associated to ρ we prove the following lemma:

Lemma 1.3.1. *Let ρ be a modular on a vector space V and $x \in V$. Then conditions (i) and (ii) below are equivalent:*

(i) $\rho(\lambda x) < \infty$ *for some* $\lambda > 0$.

(ii) $\lim_{\lambda \to 0^+} \rho(\lambda x) = 0$.

Proof. If (i) holds for $\lambda > 0$, and $0 < \lambda_j \to 0$ as $j \to \infty$, then there exists $J \in \mathbb{N}$ such that $j \geq J \Rightarrow \lambda_j < \lambda$. Since $\rho(0) = 0$, convexity yields, for $j \geq J$:

$$\rho(\lambda_j x) \leq \lambda_j \lambda^{-1} \rho(\lambda x) \to 0 \text{ as } j \to \infty.$$

The arbitrariness of the sequence (λ_j) yields $(i) \Rightarrow (ii)$.
Conversely if (ii) holds there must be $\theta > 0$ such that $\rho(\lambda) < 1$ for all $\lambda < \theta$, whence (i) holds. □

The preceding lemma justifies the equality of the two sets in the next definition:

Definition 1.3.2. *For a modular ρ on a vector space V we set:*

$$\begin{aligned} V_\rho \quad &:= \{x \in V : \rho(\lambda x) < \infty \text{ for some } \lambda > 0\} \\ &= \left\{ x \in V : \lim_{\lambda \longrightarrow 0^+} \rho(\lambda x) = 0 \right\}. \end{aligned}$$

Proposition 1.3.2. *Let ρ be a convex modular on a linear space X and define*

$$\|x\|_\rho := \inf \{\lambda > 0 : \rho(x/\lambda) \leq 1\} \ \ (x \in V_\rho).$$

Then:

(i) $\left(V_\rho, \|\cdot\|_\rho\right)$ *is a normed linear space.*

(ii) If $\rho(x) \leq 1$, *then* $\|x\|_\rho \leq 1$.

(iii) If ρ is left-continuous, then $\|x\|_\rho \leq 1$ if and only if $\rho(x) \leq 1$. If ρ is continuous, then $\|x\|_\rho < 1$ if and only if $\rho(x) < 1$; and $\|x\|_\rho = 1$ if and only if $\rho(x) = 1$.

Remark 1.3.3. The norm $\| \cdot \|_\rho$ is called the Luxemburg norm and was introduced by Luxemburg [77] in 1955.

Proof. (i) Let $x, y \in V_\rho$ and $\lambda \in \mathbb{K}$. From (1.15) and (1.16) it follows that $\lambda x \in V_\rho$. Since ρ is convex,

$$\rho\left(\lambda\left(x+y\right)\right) \leq \frac{1}{2}\left\{\rho(2\lambda x) + \rho(2\lambda y)\right\} \to 0 \text{ as } \lambda \to 0,$$

and so $x + y \in V_\rho$. Thus V_ρ is a linear space.

To show that $\|\cdot\|_\rho$ is a norm, note first that plainly $\|x\|_\rho < \infty$ for all $x \in V_\rho$, and $\|0\|_\rho = 0$. Let $x, y \in V_\rho$ and $\lambda \in \mathbb{K}$. Then

$$\|\lambda x\|_\rho = \inf\{\beta > 0 : \rho(\lambda x/\beta) \le 1\} = |\lambda| \inf\{\beta > 0 : \rho(x/\beta) \le 1\} = |\lambda|\,\|x\|_\rho .$$

Now let $\lambda > \|x\|_\rho$ and $\mu > \|y\|_\rho$, so that $\rho(x/\lambda) \le 1$ and $\rho(y/\mu) \le 1$. By the convexity of ρ,

$$\rho\left(\frac{x+y}{\lambda+\mu}\right) \le \frac{\lambda}{\lambda+\mu}\rho\left(\frac{x}{\lambda}\right) + \frac{\mu}{\lambda+\mu}\rho\left(\frac{y}{\mu}\right) \le 1.$$

Thus $\|x+y\|_\rho \le \lambda+\mu$, and the triangle inequality follows. Finally, if $\|x\|_\rho = 0$, then $\rho(\alpha x) \le 1$ for all $\alpha > 0$, and hence, for all $\lambda > 0$ and $\beta \in (0,1]$,

$$\rho(\lambda x) \le \beta\rho(\lambda x/\beta) \le \beta.$$

This shows that $\rho(\lambda x) = 0$ for all $\lambda > 0$, and so $x = 0$.
(ii) If $\rho(x) \le 1$, then from the definition of $\|x\|_\rho$ it is immediate that $\|x\|_\rho \le 1$.
(iii) Suppose that ρ is left-continuous and that $\|x\|_\rho \le 1$. Then for all $\mu \in (0,1)$ we have $\rho(\mu x) \le 1$. Then $\rho(x) = \lim_{\mu \to 1-} \rho(\mu x) \le 1$.

Now assume that ρ is continuous. If $\|x\|_\rho < 1$, then for some $\lambda < 1$ we have $\rho(x/\lambda) < 1$, so that in view of the convexity of ρ,

$$\rho(x) \le \lambda\rho(x/\lambda) \le \lambda < 1.$$

Conversely, if $\rho(x) < 1$, then by the continuity of ρ, $\rho(\gamma x) < 1$ for some $\gamma > 1$. Thus $\|\gamma x\|_\rho \le 1$ and $\|x\|_\rho \le 1/\gamma < 1$. The rest is obvious. □

Corollary 1.3.4. *Let ρ be a left-continuous modular on a modular space V. Then:*

(*i*) *If $\|x\|_\rho \le 1$, then $\rho(x) \le \|x\|_\rho$.*

(*ii*) *If $\|x\|_\rho > 1$, then $\rho(x) \ge \|x\|_\rho$.*

(*iii*) *For any $x \in V$, $\|x\|_\rho \le \rho(x) + 1$.*

Proof. For (i), observe that if $0 < \|x\|_\rho \le 1$ then the convexity of ρ yields

$$\rho(x) = \rho\left(\frac{\|x\|_\rho}{\|x\|_\rho} x\right) \le \|x\|_\rho \rho\left(\frac{x}{\|x\|_\rho}\right)$$
$$\le \|x\|_\rho;$$

on the other hand if $1 < \lambda < \|x\|_\rho$ one has

$$\rho\left(\frac{x}{\lambda}\right) > 1$$

and by convexity one gets $1 < \lambda^{-1}\rho(x)$, which forces $\rho(x) \ge \|x\|_\rho$, as claimed.
(iii) follows immediately from (i) and (ii). □

1.3.1 Modular convergence and norm convergence

Though the modular concepts are generally speaking, mathematically weaker than those associated to the norm, they are simpler to handle from the computational point of view and occasionally, sufficiently sharp to handle certain mathematical situations that arise in the treatment of Musielak-Orlicz spaces. Many instances of such a nature will come to light in subsequent chapters; in the meantime we lay the ground work with the pertinent definitions and provide some simple examples to clarify their meaning. Let V be a real or complex vector space and

$$\rho : V \longrightarrow [0, \infty]$$

be a convex modular on V.

Definition 1.3.3. *A sequence $(x_k) \subset V$ is said to be modular-convergent to $x \in V$ ($x_n \overset{\rho}{\to} x$) iff there exists $\lambda > 0$ for which*

$$\lim_{k \longrightarrow \infty} \rho(\lambda(x_k - x)) = 0.$$

The following terminology is standard and will be used extensively in the sequel:

Definition 1.3.4. *(see [54])*

(i) A sequence $\{x_n\} \subset V_\rho$ is said to be ρ-convergent to $x \in V_\rho$ and write $x_n \overset{\rho}{\to} x$, if and only if $\rho(x_n - x) \to 0$. Note that the ρ-limit is unique if it exists.

(ii) A sequence $\{x_n\} \in V_\rho$ is called ρ-Cauchy if $\rho(x_n - x_m) \to 0$ as $n, m \to \infty$.

(iii) A set $C \subset V_\rho$ is called ρ-closed if for any ρ-converging-sequence $\{x_n\} \subset C$ to x one has $x \in C$.

(iv) A set $C \subset V_\rho$ is defined to be ρ-bounded if $\delta_\rho(C) = \sup\{\rho(x - y) : x, y \in C\} < \infty$.

Theorem 1.3.5. *Let ρ be a convex modular on a vector space V. Then*

(i) Modular limits are unique. More precisely, if ρ is a convex modular and $x_n \overset{\rho}{\to} x$ and $y_n \overset{\rho}{\to} y$ then $x = y$.

(ii) If $x_n \overset{\rho}{\to} x$ and $y_n \overset{\rho}{\to} y$, then $x_n + y_n \overset{\rho}{\to} x + y$.

(iii) If $c \in \mathbb{R}$ and $x_n \overset{\rho}{\to} x$, then $cx_n \overset{\rho}{\to} cx$.

Proof. The proofs are straightforward. We present the details of the proof of (i). By assumption, $\rho(\lambda_1(x-x_n)) \to 0$ and $\rho(\lambda_2(y-x_n)) \to 0$ as $n \to \infty$ for some $\lambda_1 > 0$, $\lambda_2 > 0$. It is clear that there is no loss of generality by assuming $\lambda_2 > \lambda_1$. Obviously, as $n \to \infty$, convexity yields:

$$\rho(\lambda_1(x_n - y)) = \rho\left(\frac{\lambda_1}{\lambda_2}\lambda_2(x_n - y)\right) \longrightarrow 0.$$

A further application of convexity yields

$$\rho\left(\frac{1}{2}\lambda_1(x-y)\right) \le \frac{1}{2}\rho(\lambda_1(x-x_n)) + \frac{1}{2}\rho(\lambda_1(x_n-y)).$$

Since the right-hand side tends to zero as $n \to \infty$ one must a fortiori have $x = y$. □

Theorem 1.3.6. *If ρ is a convex, continuous modular on the vector space X, then norm convergence on the space X_ρ implies modular convergence.*

Proof. By convexity and continuity, for any sequence $(u_n) \subset V_\rho$ converging to u in norm, one has:

$$\begin{aligned}\rho(|u_n - u|) &= \rho\left(\frac{\|u-u_n\|_\rho |u-u_n|}{\|u-u_n\|_\rho}\right)\\ &\le \|u-u_n\|_\rho \rho\left(\frac{|u-u_n|}{\|u-u_n\|_\rho}\right) = \|u-u_n\|_\rho,\end{aligned}$$

from which the theorem follows automatically. □

Theorem 1.3.7. *Let ρ be a convex, continuous modular on the vector space X. Then, on the space X_ρ, modular convergence is equivalent to (Luxemburg) norm convergence if and only if for any sequence $(x_k) \subset V$,*

$$\lim_{k\to\infty} \rho(x_k) = 0 \Rightarrow \lim_{k\to\infty} \rho(2x_k) = 0.$$

Proof. If the sequence (x_k) converges modularly to 0, then, by assumption, for any $m \in \mathbb{N}$ one has $\rho(2^m x_k) \le 1$ for $k \ge k_0$ for some k_0 depending on m. By definition, this statement implies

$$\|x_k\|_\rho \le 2^{-m}$$

for $k \ge k_0$, which in conjunction with the preceding theorem proves the claim. □

Example 1.3.3. Let $X = C([0,1])$ stand for the vector space of all real-valued continuous functions on $[0,1]$ furnished with the standard real vector space structure. Consider the modular

$$\begin{aligned}&\rho : X \to [0,\infty)\\ &\rho(u) = \int_0^1 (e^{|u(x)|} - 1)dx\end{aligned}$$

and for $n \in \mathbb{N}$ set $u_n = \frac{1}{2} n \mathbf{1}_{(2^{-n}, 2^{-n+1})}$. Clearly $(u_n) \subset X_\rho$. Then

$$\rho(u_n) = \int_{2^{-n}}^{2^{-n+1}} (e^{2^{-1}n} - 1)dx = 2^{-n}\left(e^{2^{-1}n} - 1\right)$$

whereas

$$\rho(2u_n) = \int_{2^{-n}}^{2^{-n+1}} (e^n - 1)dx = 2^{-n}(e^n - 1).$$

It is immediate from the above equalities that $u_n \xrightarrow{\rho} 0$ as $n \to \infty$, while (u_n) does not converge in norm.

Theorem 1.3.8. *Any sequence (x_k) for which $\rho(\lambda x_k) \to 0$ for all $\lambda : \lambda > 0$, is norm-convergent.*

Proof. For such a sequence (x_n) and any $\lambda > 0$ there exists $N > 0$ such that $n \geq N$ implies $\rho(\lambda x_n) = \rho\left(\frac{x_n}{\lambda^{-1}}\right) < 1$, which by definition yields $\|x_n\| \leq \lambda^{-1}$ for $n \geq N$. □

Definition 1.3.5. *The modular ρ on the vector space V is said to satisfy the Δ_2 condition if there exists a constant $K \geq 1$ such that for any $x \in V$*

$$\rho(2x) \leq K\rho(x).$$

For example, if $1 < p < \infty$, the modular ρ_p introduced in (1.14) satisfies the Δ_2 condition; on the other hand, the modular on $C([0,1])$ introduced in Example 1.3.3 does not satisfy the Δ_2 condition.
In view of Theorems 1.3.6 and 1.3.7 we have the following result:

Corollary 1.3.9. *If a modular $\rho : V \longrightarrow [0, \infty]$ on a vector space V satisfies the Δ_2 condition then, on V_ρ, norm convergence is equivalent to modular convergence.*

Theorem 1.3.10. *A left-continuous modular on a vector space V is lower semicontinuous on V_ρ, i.e. for any sequence $(x_n) \in V_\rho$ such that $x_n \to x$ in norm as $n \to \infty$, one has*

$$\rho(x) \leq \liminf_{n\to\infty} \rho(x_n).$$

Proof. Let $0 < \delta < 1$, then on account of convexity:

$$\rho((1-\delta)x) \leq \delta\rho\left(\frac{1-\delta}{\delta}(x - x_n)\right) + (1-\delta)\rho(x_n).$$

Taking the lim inf on both sides and as a direct consequence of Theorem 1.3.6 it is immediate that

$$\rho((1-\delta)x) \leq (1-\delta)\liminf_n \rho(x_n).$$

The claim follows now directly from the left-continuity assumption on ρ. □

1.3.2 Conjugate modulars and duality

In this section we retain the preceding terminolgy. In particular, V stands for a vector space over $\mathbb{K} = \mathbb{R}$ or $\mathbb{K} = \mathbb{C}$ and

$$\rho : V \longrightarrow [0, \infty]$$

denotes a convex modular on V. A generic element in the dual V^* of V will be denoted by x^* and the action of x^* on $x \in V$ by $x^*(x) = \langle x^*, x \rangle$.
If

$$\|x\|_\rho := \inf \{\lambda > 0 : \rho(x/\lambda) \leq 1\} \ (x \in V_\rho)$$

stands for the Luxemburg norm, it as been proved in Theorem 1.3.2 that

$$(V_\rho, \| \cdot \|_\rho)$$

is a normed space over $\mathbb{K}$. As usual, let

$$(V_\rho, \| \cdot \|_\rho)^* = (V_\rho^*, \| \cdot \|_\rho^*)$$

stand for the dual space, that is to say, for any $x^* \in V_\rho^*$ the norm is given by the usual definition

$$\|x^*\|_\rho^* = \sup\{|\langle x^*, x \rangle| : \|x\|_\rho \leq 1\}.$$

It will be shortly observed that ρ engenders a modular ρ^* on the dual V_ρ^*. We will address the natural question as to the relationship between the ρ^* Luxemburg norm $\| \cdot \|_{\rho^*}$ and the dual norm $\| \cdot \|_\rho^*$ on the vector space V_ρ^*. We open the discussion with the following theorem:

Theorem 1.3.11. *Let ρ be a convex modular on a vector space V with corresponding Luxemburg norm $\| \cdot \|_\rho$. Let x^* be a linear functional on V. Then the following conditions are equivalent:*

(i) $x^* \in V_\rho^*$ (i.e., x^* is bounded on V_ρ).

(ii) There is $c \geq 0$ such that

$$|\langle x^*, x \rangle| \leq c(\rho(x) + 1).$$

Proof. Assume (i) holds and fix $x \in V_\rho$. It follows from convexity that

$$\rho\left(\frac{x}{1 + \rho(x)}\right) \leq 1,$$

which on account the definition of the Luxemburg norm implies the bound

$$\|x\|_\rho \leq (\rho(x) + 1). \tag{1.17}$$

Thus, for any $x^* \in V_\rho^*$ it holds that

$$|\langle x^*, x \rangle| \leq \|x^*\|_\rho^* \|x\|_\rho \leq \|x^*\|_\rho^* (\rho(x) + 1).$$

In all, (ii) holds with $c = \|x^*\|_\rho^*$.
Assume next $x^* \in V^*$ satisfies (ii) for some $c > 0$. Fix $\delta > 0$. For $x \in V_\rho$ with $\|x\|_\rho = 1$ the definition of the Luxemburg norm yields the existence of $\lambda : 0 < \lambda < 1 + \delta$ such that

$$\rho(\lambda^{-1}x) \leq 1.$$

Based on the assumption it is concluded that

$$|\langle x^*, \lambda^{-1}x\rangle| \leq c\left(\rho(\lambda^{-1}x) + 1\right) \leq 2c.$$

In all,

$$|\langle x^*, \lambda^{-1}x\rangle| \leq 2(1+\delta)c$$

whence x^* is actually bounded on V_ρ, as claimed. This concludes the proof. □

Theorem 1.3.12. *If*

$$\rho : V \longrightarrow [0, \infty]$$

is a convex modular on V, then the function

$$\begin{aligned} \rho^* &: V_\rho^* \longrightarrow [0, \infty) \\ \rho^*(x^*) &= \sup_{x \in V_\rho} \{|\langle x^*, x\rangle| - \rho(x)\} \end{aligned} \tag{1.18}$$

is a convex, left-continuous semimodular on V_ρ^*, that is,

(i) For any $x^* \in V_\rho^*$, one has $\rho^*(x^*) = \rho^*(-x^*)$.

(ii) $\rho^*(sx^* + (1-s)y^*) \leq s\rho^*(x^*) + (1-s)\rho^*(y^*)$ for any $x^* \in V_\rho^*$, $y^* \in V_\rho^*$ and $0 \leq s \leq 1$.

(iii) For any $x^* \in V_\rho^*$, $\lim_{\lambda \to 1^-} \rho^*(\lambda x^*) \leq \rho^*(x^*)$.

(iv) If

$$\rho^*(\lambda x^*) = 0 \text{ for all } \lambda > 0, \tag{1.19}$$

then $\rho^*(x^*) = 0$.

Proof. It is immediately seen that by definition, $\rho^* \geq 0$ on V_ρ^*, $\rho^*(0) = 0$ and $\rho^*(x^*) = \rho^*(-x^*)$ for any $x^* \in V_\rho^*$. It follows by definition that

$$\langle x^*, x\rangle \leq \rho^*(x^*) + \rho(x).$$

If condition (1.19) holds for $x^* \in V_\rho^*$, then for $x \in V_\rho$ with $\rho(\alpha x) < \infty$, $\alpha > 0$, one has

$$\begin{aligned} \lambda\alpha|\langle x^*, x\rangle| = \lambda|\langle x^*, \alpha x\rangle| &\leq \rho^*(\lambda x^*) + \rho(\alpha x) \\ &\leq \rho(\alpha x), \end{aligned}$$

which readily yields $\langle x^*, x\rangle = 0$ by letting $\lambda \to \infty$. Since x is arbitrary, one concludes that $x^* = 0$ and hence that ρ^* is a semimodular. Next, if $0 \leq s \leq 1$ it holds that

$$|\langle sx^* + (1-s)y^*, x\rangle| - \rho(x) \leq s|\langle x^*, x\rangle| - s\rho(x) + (1-s)|\langle y^*, x\rangle| - (1-s)\rho(x).$$

Convexity follows immediately from the last inequality. For left-continuity, it is easily observed that for any x^* as above and any $x \in V_\rho$, as $\lambda \to 1$, $0 < \lambda < 1$, one has

$$|\langle \lambda x^*, x\rangle| - \rho(x) \nearrow |\langle x^*, x\rangle| - \rho(x),$$

the latter easily yields that for arbitrary $\epsilon > 0$ there exists $\lambda_0 : 0 < \lambda_0 < 1$ such that for $0 < \lambda_0 < \lambda < 1$ it holds:

$$|\langle x^*, x\rangle| - \rho(x) < |\langle \lambda x^*, x\rangle| - \rho(x) + \epsilon$$

that is,

$$\rho^*(x^*) \leq \sup_{0<\lambda<1} \rho^*(\lambda x^*) = \lim_{\lambda\to 1^-} \rho^*(\lambda x^*).$$

On the other hand, for $0 < \lambda < 1$ convexity immediately yields, for any $y \in V_\rho$:

$$\langle \lambda x^*, y\rangle - \rho(y) \leq \langle x^*, \lambda y\rangle - \rho(\lambda y);$$

the latter clearly shows that

$$\lim_{\lambda\to 1^-} \rho^*(\lambda x^*) \leq \rho^*(x^*).$$

□

Corollary 1.3.13. *In the above notation, the functional*

$$\|\cdot\|_{\rho^*} : V_\rho^* \longrightarrow [0,\infty)$$
$$\|x^*\|_{\rho^*} = \inf\left\{\lambda > 0 : \rho^*\left(\frac{x^*}{\lambda}\right) \leq 1\right\}$$

is a norm on V_ρ^*.

Theorem 1.3.14. *If ρ is a convex, left-continuous modular on a vector space V, then $\rho^{**} = \rho$ on V_ρ (where as usual, V_ρ is identified with a subspace of $(V_\rho)^{**}$.)*

Proof. The fact that ρ^{**} is a convex, left-continuous semimodular on V_ρ follows along the same lines as in Theorem 1.3.12. By definition, for any $x \in V_\rho \hookrightarrow V_\rho^{**}$,

$$\begin{aligned}\rho^{**}(x) &= \sup_{x^*\in V_\rho^*,\, \rho^*(x^*)<\infty} \{|\langle x^*, x\rangle| - \rho^*(x^*)\}\\ &\leq \rho^*(x^*) + \rho(x) - \rho^*(x^*)\\ &= \rho(x).\end{aligned}$$

The second inequality above follows straight from (1.18).
On the other hand, it is well known that the set

$$X = \bigcup_{r \in \mathbb{R}} \{(x, r) \in V_\rho \times \mathbb{R} \ , \ r \geq \rho(x)\} ,$$

being the epigraph of a convex, lower semicontinuous function, is convex and closed in $V_\rho \times \mathbb{R}$. If there were $z \in V_\rho$ with $\rho^{**}(z) < \rho(z)$, then

$$(z, \rho^{**}(z)) \notin X;$$

and the Hahn-Banach's theorem would yield $w^* \in V_\rho^*$ and real numbers a, b such that, for all $x \in V_\rho$ it would hold that

$$\langle x^*, x\rangle - a\rho(x) < b < \langle x^*, z\rangle - a\rho^{**}(z).$$

Notice that by assumption, $a > 0$. It is easy to see that the last inequality implies

$$\rho^* \left(\frac{x^*}{a}\right) \leq \frac{b}{a} < \rho^* \left(\frac{x^*}{a}\right),$$

which is a contradiction. Thus, $\rho^{**}(x) = \rho(x)$ for any $x \in V_\rho$, as claimed. □

Theorem 1.3.15. *If ρ is a convex, left-continuous modular on V, the following inequalities hold for any $x^* \in V_\rho^*$:*

$$\|x^*\|_{\rho^*} \leq \|x^*\|_\rho^* \leq 2\|x^*\|_{\rho^*} \tag{1.20}$$

Proof. Take $x^* \in (V_\rho)^*$. From Definition 1.19, Proposition 1.3.2-(iii) and the definition of the norm on $(V_\rho)^*$ it is obvious that

$$\begin{aligned}
\left\| \frac{x^*}{\|x^*\|_{\rho^*}} \right\|_\rho^* &= \sup_{x \in V_\rho, \rho(x) \leq 1} \left(\left| \left\langle \frac{x^*}{\|x^*\|_{\rho^*}}, x \right\rangle \right| \right) \\
&\leq \sup_{x \in V_\rho, \rho(x) \leq 1} \left(\rho(x) + \rho^* \left(\frac{x^*}{\|x^*\|_{\rho^*}} \right) \right) \\
&= \leq 1 + \rho^* \left(\frac{x^*}{\|x^*\|_{\rho^*}} \right) \leq 2,
\end{aligned}$$

which is the right-hand side of (1.20).
On the other hand, if $x \in V_\rho$ and $\rho(x) > 1$ then by Proposition 1.3.2-(iii) one has $\|x\|_\rho > 1$ and hence,

$$1 = \rho \left(\frac{x}{\|x\|_\rho} \right) \leq \frac{1}{\|x\|_\rho} \rho(x).$$

This yields

$$\|x\|_\rho - \rho(x) \leq 0 \text{ if } \|x\|_\rho > 1.$$

Consequently, if $\|x^*\|_\rho^* \leq 1$, then

$$\sup_{x \in V_\rho\,,\, \rho(x) > 1} \{|\langle x^*, x\rangle| - \rho(x)\} \leq 0$$

and it is readily concluded that

$$\rho^*(x^*) = \sup_{x \in V_\rho\,,\, \rho(x) \leq 1} \{|\langle x^*, x\rangle| - \rho(x)\}.$$

In view of the above observation, if $0 \neq x^* \in (V_\rho)^*$,

$$\begin{aligned} \rho^*\left(\frac{x^*}{\|x^*\|_\rho^*}\right) &= \sup_{x \in V_\rho\,,\, \rho(x) \leq 1} \left\{\frac{1}{\|x^*\|_\rho^*}|\langle x^*, x\rangle| - \rho(x)\right\} \\ &\leq \sup_{x \in V_\rho\,,\, \rho(x) \leq 1} \{\|x\|_\rho - \rho(x)\} \end{aligned}$$

and since $\|x\|_\rho \leq 1$ if $\rho(x) \leq 1$ it follows that

$$\rho^*\left(\frac{x^*}{\|x^*\|_\rho^*}\right) \leq 1.$$

By definition of $\|\cdot\|_{\rho^*}$ it is plain that

$$\|x^*\|_{\rho^*} \leq \|x^*\|_\rho^*.$$

The last inequality completes the proof of (1.20). □

Definition 1.3.6. *A modular ρ on a vector space V is said to be sequentially weakly lower semicontinuous on V_ρ if for any sequence $(x_n) \subset V_\rho$ that converges weakly to $x \in V_\rho$ ($x_n \rightharpoonup x$) one has*

$$\rho(x) \leq \liminf_n \rho(x_n).$$

Theorem 1.3.16. *A modular ρ on a vector space V is sequentially weakly lower semicontinuous on V_ρ.*

Proof. The proof is straightforward. Consider $x \in V_\rho$ and $(x_n) \subseteq V_\rho$ with

$$x_n \rightharpoonup x \text{ as } n \to \infty.$$

Notice that for any linear functional $x^* \in V_\rho^*$

$$|\langle x^*, x_n\rangle| \leq \sup_{y^* \in V_\rho^*} |\langle y^*, x_n\rangle|.$$

Hence

$$\lim_{n\to\infty} |\langle x^*, x_n\rangle| \leq \liminf_n \sup_{y^* \in V_\rho^*} |\langle y^*, x_n\rangle|.$$

We conclude

$$\begin{aligned}\rho(x) = \rho^{**}(x) &= \sup_{x^* \in V_\rho^*} \{|\langle x^*, x\rangle| - \rho^*(x^*)\} \\ &\le \sup_{x^* \in V_\rho^*} \left\{\lim_{n\to\infty} |\langle x^*, x_n\rangle| - \rho^*(x^*)\right\} \\ &\le \liminf_n \sup_{y^* \in V_\rho^*} \{|\langle y^*, x_n\rangle| - \rho^*(y^*)\} \\ &= \liminf_n \rho^{**}(x_n) = \liminf_n \rho(x_n).\end{aligned}$$

□

1.3.3 Modular uniform convexity

Bearing some recently discovered applications in mind, this section is devoted to the consideration of some geometric aspects of modular spaces. More precisely we delve into the idea of uniform convexity. It will not have escaped the attention of the reader that due to the lack of homogeneity, it is expectable that even the formulation of the concept of uniform convexity of a modular poses challenges and must necessarily entail technical difficulties beyond those encountered when dealing with the uniform convexity of a norm. Far from constituting a mere abstract mathematical construction, modular uniform convexity has deep consequences and tangible applications, some examples of which will be presented in the sequel (Section 1.4.4).
The relevance of the notion of modular uniform convexity is reflected in the fact that it was fully investigated in Orlicz function spaces [24, 82]. The concept of modular uniform convexity was introduced and studied by Nakano [83]. The definition of two different types of modular uniform convexity can be found in [7, 55]. The reader will find it easy to realize that the notions introduced hereafter are natural generalizations of those studied in Section 1.2.3.
It is worthwhile to note that if the modular ρ is associated to a Musielak-Orlicz function φ (Chapter 2) the interplay between the uniform convexity of the Luxemburg norm $\|\cdot\|_\rho$ and the function is well understood. The reader is referred to Theorem 2.6.1 in Section 2.6 for the details. In view of the usual concept (Definition 1.2.7) it is natural to state the following definitions:

Definition 1.3.7. *A convex modular ρ on a vector space V is said to be uniformly convex if for every $\epsilon > 0$ there exists $\delta = \delta(\epsilon) > 0$ such that for every $u \in V$ and $v \in V$ with $\rho(u) = 1$ and $\rho(v) = 1$ and $\rho(u-v) > \epsilon$ it holds that $\rho\left(\frac{u+v}{2}\right) < 1 - \delta(\epsilon)$.*

Definition 1.3.8. *Let ρ be a convex modular on a real vector space V, $r > 0$, $\epsilon > 0$. Set*

$$D(r,\epsilon) = \left\{(u,v) \in V_\rho \times V_\rho : \rho(u) \le r\,,\ \rho(v) \le r\,,\ \rho\left(\frac{u-v}{2}\right) \ge \epsilon r\right\} \quad (1.21)$$

and

$$\delta(r,\epsilon) = \inf\left\{1 - \frac{1}{r}\rho\left(\frac{u+v}{2}\right) : (u,v) \in D(r,\epsilon)\right\}. \tag{1.22}$$

If $D(r,\epsilon) = \emptyset$ *we define* $\delta(r,\epsilon) = 1$.

Definition 1.3.9. *The modular* ρ *is said to be type 2-uniformly convex* $UUC2$ *(see [55]) if for each* $s \geq 0$, $\epsilon > 0$, *there exists* $\eta(s,\epsilon) > 0$ *such that for any* $r > s > 0$

$$\delta(r,\epsilon) > \eta(s,\epsilon).$$

Definition 1.3.10. *A convex modular* ρ *on a real vector space* V *is said to be strictly convex if for any* $u \in V$, $v \in V$ *with* $\rho(u) = \rho(v)$

$$\rho\left(\frac{u+v}{2}\right) = \frac{1}{2}\left(\rho(u) + \rho(v)\right) \iff u = v.$$

Lemma 1.3.17. *If a convex modular* ρ *on a vector space* V *is* $UUC2$, *then it is strictly convex.*

Proof. Assume ρ satisfies the above hypothesis, that $u \in V$, $v \in V$, $\rho(u) = \rho(v) = r$ and that

$$\rho\left(\frac{u+v}{2}\right) = \frac{1}{2}\left(\rho(u) + \rho(v)\right).$$

Assuming $u \neq v$ one can find $\epsilon : 0 < \epsilon < 1$ such that,

$$\rho\left(\frac{u-v}{2}\right) \geq \epsilon r.$$

By virtue of the property $UUC2$ one has, for every $r' > r$ and $\eta(r,\epsilon)$ as in Definition 1.3.9

$$\rho\left(\frac{u+v}{2}\right) < xr'(1 - \eta(r,\epsilon)).$$

Letting $r' \to r$ it follows

$$r = \frac{1}{2}\left(\rho(u) + \rho(v)\right) = \rho\left(\frac{u+v}{2}\right) \leq r(1 - \eta(\epsilon, r)).$$

Hence, $u = v$. □

We next move to yet another definition: Our aim is to consider the modular **uniform convexity type**-property defined as follows:

Definition 1.3.11. *Let* ρ *be a convex modular on a vector space* V *over* $\mathbb{R}$ *or* $\mathbb{C}$.

(i) *The modular ρ is said to be strictly convex (in short (SC)) if for every $x \in V, y \in Vx$ such that $\rho(x) = \rho(y)$ and*

$$\rho\left(\frac{x+y}{2}\right) = \frac{\rho(x)+\rho(y)}{2},$$

we have $x = y$.

(ii) *[83] The modular ρ is said to be uniformly convex in every direction (in short (UCED)) if and only if for any $z_1, z_2 \in V_\rho$ such that $z_1 \neq z_2$ and $R > 0$, there exists $\delta = \delta(z_1, z_2, R) > 0$ such that*

$$\begin{cases} \rho(x - z_1) & \leq & R \\ \rho(x - z_2) & \leq & R \end{cases} \Longrightarrow \rho\left(x - \frac{z_1+z_2}{2}\right) \leq R(1-\delta),$$

for any $x \in V$. V is said to be (UUCED) if $\delta(z_1, z_2, R) \geq \delta(z_1, z_2, \bar{R})$, whenever $R \leq \bar{R}$.

The idea of uniform convexity in every direction was introduced by Garkavi [41] while studying the concept of Chebyshev centers in Banach spaces. It is a matter of routine to check that if a modular ρ is $(UCED)$ then ρ is also strictly convex.

1.4 The ℓ^{p_n} sequence spaces and their properties

In the following discussion we address a concrete example of the mathematical structures introduced in the preceding Sections 1.2 and 1.3 and provide some applications of them. Most of what follows will be further generalized in Section 1.4 in the context of Musielak-Orlicz spaces.
Among the first modular spaces to enter the mathematical stage were the variable exponent sequences spaces (see [85]).
We fix the measure space $(\mathbb{N}, 2^{\mathbb{N}}, \mu)$, where μ stands for the counting measure on $\mathbb{N}$. The collection of all sequences of real numbers coincides with the vector space $\mathcal{M}(\mathbb{N})$ of $2^{\mathbb{N}}$-measurable real-valued functions on $\mathbb{N}$.

Proposition 1.4.1. *Consider a sequence*

$$(p_n) \subset (1, \infty). \tag{1.23}$$

Then the functionals

$$\rho_{1,p_n}((a_n)) = \sum_{n=1}^{\infty} \frac{|a_n|^{p_n}}{p_n} \tag{1.24}$$

and

$$\rho_{2,p_n}((a_n)) = \sum_{n=1}^{\infty} |a_n|^{p_n} \tag{1.25}$$

are convex, left-continuous modulars on $\mathcal{M}(\mathbb{N})$.
The functional

$$\rho_{\infty}((a_n)) = \sup_{n\in\mathbb{N}} |a_n| \tag{1.26}$$

is a convex, continuous modular on $\mathcal{M}(\mathbb{N})$.

Proof. The proof is straightforward. The fact that the functionals (1.24) and (1.25) are convex modulars can be proved directly from the definition; left-continuity follows from the monotone convergence theorem.
The modular properties for (1.26) follow by direct calculation. In this case, continuity is obvious from the identity, valid for all $\lambda > 0$:

$$\sup_{n\in\mathbb{N}} \lambda |a_n| = \lambda \sup_{n\in\mathbb{N}} |a_n|.$$

□

To easier relate these ideas to the theory in Section 1.3 write

$$V = \mathcal{M}(2^{\mathbb{N}})$$

and introduce the following notation in accordance to Definition 1.3.2: for fixed $(p_n) \subset (1, \infty)$ and $j = 1, 2$,

$$V_{\rho_{j,p_n}} = \left\{(a_n) \in V_{\rho_{j,p_n}}(\lambda(a_n)) < \infty \text{ for some } \lambda > 0\right\}. \tag{1.27}$$

On account of Proposition 1.3.2

$$\|(a_n)\|_{\rho_{j,p_n}} := \inf\{\lambda > 0 : \rho_{j,p_n}((a_n)/\lambda) \leq 1\} \tag{1.28}$$

defines a norm on $V_{\rho_{j,p_n}}$, for $j = 1, 2$.

Lemma 1.4.2. *For any sequence of real numbers* (a_n) *and any* $\lambda \in \mathbb{R}$, *it holds that* $V_{\rho_{1,p_n}} = V_{\rho_{2,p_n}}$. *In fact, for any* $\lambda > 0$,

$$\rho_{1,p_n}(\lambda(a_n)) < \infty \Leftrightarrow \rho_{2,p_n}(\lambda(a_n)) < \infty \tag{1.29}$$

and the modulars ρ_{1,p_n} *and* ρ_{2,p_n} *yield equivalent Luxemburg norms on the space (1.27).*

Proof. It is straightforward to check that the function

$$x \longrightarrow x^{\frac{1}{x}} \tag{1.30}$$

maps the interval $(1, \infty)$ onto the interval $[1, e^{\frac{1}{e}}]$. Accordingly, for a given sequence $(p_n) \subset [1, \infty)$, if $(a_n) \in \mathcal{M}(\mathbb{N})$ and $\lambda > 0$, it follows from

$$\sum_{1}^{\infty} p_n^{-1} \left(\frac{|a_n|}{\lambda}\right)^{p_n} = \sum_{1}^{\infty} \left(\frac{|a_n|}{p_n^{p_n^{-1}} \lambda}\right)^{p_n} \leq 1$$

that

$$\sum_1^\infty \left(\frac{|a_n|}{e^{e^{-1}}\lambda}\right)^{p_n} \le \sum_1^\infty \left(\frac{|a_n|}{p_n^{p_n^{-1}}\lambda}\right)^{p_n} \le 1.$$

In other words, by definition of norm, one has:

$$\|(a_n)\|_{\rho_{2,p_n}} \le e^{\frac{1}{e}}\lambda$$

and invoking the definition of Luxemburg norm again, the above inequality implies $\|(a_n)\|_{\rho_{2,p_n}} \le e^{\frac{1}{e}}\|(a_n)\|_{\rho_{1,p_n}}$.
Furthermore, for any $\lambda > 0$ such that $\sum_1^\infty \left(\frac{|a_n|}{\lambda}\right)^{p_n} \le 1$ the bounds for the function (1.30) yield:

$$\sum_1^\infty p_n^{-1}\left(\frac{|a_n|}{\lambda}\right)^{p_n} \le 1$$

whence it is clear that $\|(a_n)\|_{\rho_{1,p_n}} \le \|(a_n)\|_{\rho_{2,p_n}}$ and the proof of the equivalence $\|\cdot\|_{\rho_{1,p_n}} \approx \|\cdot\|_{\rho_{2,p_n}}$ is thus completed. □

Definition 1.4.1. *Let $(p_n) \subset (1,\infty)$ be a sequence. In the sequel we will write ℓ^{p_n} for the vector space (1.27) endowed with either of the equivalent norms $\|\cdot\|_{\rho_{1,p_n}}$ or $\|\cdot\|_{\rho_{2,p_n}}$. When the distinction between the two equivalent norms is unnecessary (as it will be in most of the sequel) we will simply write $\|\cdot\|_{\ell^{p_n}}$.*

We next observe that in the above setting, for $j = 1, 2$, the functionals

$$\|\cdot\|_{j,p_n} : V \longrightarrow [0,\infty]$$

defined by

$$\|(a_n)\|_{\rho_{j,p_n}} = \begin{cases} \|(a_n)\|_{\rho_{j,p_n}} & \text{if } (a_n) \in V_{\rho_{p_n}} \\ \infty & \text{otherwise} \end{cases}$$

have the properties of a norm. It is straightforward to verify Axioms $(i) - (v)$ of Definition 1.2.1, whence ℓ^{p_n} is a Banach function space for either functional. In summary:

Theorem 1.4.3. *ℓ^{p_n} is a Banach function space for the Luxemburg norms (1.28).*

Similarly, it can be immediately verified that V_{ρ_∞} coincides with the set of all bounded sequences (a_n) and that the Luxemburg norm is given by

$$\|(a_n)\|_{\rho_\infty} = \sup_{n\in\mathbb{N}} |a_n| = \|(a_n)\|_\infty$$

and that it is a Banach function space with the Luxemburg norm.
In particular if the sequence (p_n) is constant (i.e., $p_n = p$ for all n) the space ℓ^{p_n} defined in (1.27) coincides with the well-known sequence space ℓ^p.

Theorem 1.4.4. *If (r_n) is a sequence subject to condition (1.23) and $r_n \leq s_n$ for every $n \in \mathbb{N}$ then the embedding*

$$\ell^{r_n} \hookrightarrow \ell^{s_n}$$

is continuous.

Proof. If $(a_n) \in \ell^{r_n}$, then for some $\lambda > 0$ it holds that $\rho_{2,r_n}\left(\frac{a_n}{\lambda}\right) \leq 1$, from which it can be plainly observed, since $r_n \leq s_n$ for all n, that for any $n \in \mathbb{N}$:

$$\left|\frac{a_n}{\lambda}\right|^{s_n} \leq \left|\frac{a_n}{\lambda}\right|^{r_n}; \tag{1.31}$$

this inequality leads to

$$\rho_{2,s_n}\left(\frac{a_n}{\lambda}\right) \leq 1.$$

It is clear from the latter that $(a_n) \in \ell^{s_n}$ and that $\|(a_n)\|_{\ell^{s_n}} \leq \lambda$. The arbitrariness of λ yields

$$\|(a_n)\|_{\ell^{s_n}} \leq \|(a_n)\|_{\ell^{r_n}},$$

as claimed. Alternatively, Theorem 1.4.4 follows directly from Theorem 1.2.4, Theorem 1.4.3 and the set-theoretic inclusion $\ell^{r_n} \subseteq \ell^{s_n}$ implied by inequality (1.31). □

Lemma 1.4.5. $\ell^{p_n} \hookrightarrow \ell^{\infty}$ *boundedly.*

Proof. For if $(a_n) \in \ell^{p_n}$ then for some $\lambda > 0$

$$\sum |\lambda a_n|^{p_n} < \infty.$$

The set-theoretic inclusion follows at once, since $\lambda|a_n| < 1$ for n large enough.

Moreover if $\|a_n\|_{p_n} = 1$ then $\sum |a_n|^{p_n} \leq 1$ and from here it is easily obtained that $\|a_n\|_{\ell^\infty} \leq 1 = \|a_n\|_{\ell^{p_n}}$.

Obviously then the bound $\|a_n\|_{\ell^\infty} \leq \|a_n\|_{\ell^{p_n}}$ holds and the boundedness of the embedding follows. □

The next theorem provides a natural characterization the Δ_2 condition of either of the modulars ρ_{j,p_n}, $j = 1, 2$, in terms of the behavior of the variable exponent (p_n).

Theorem 1.4.6. *For a sequence $(p_n) \subset [1, \infty)$, the following conditions are equivalent:*

(i) $p_+ = \sup_{n \in \mathbb{N}} p_n < \infty$

(ii) The modulars ρ_{1,p_n} and ρ_{2,p_n} introduced in Lemma 1.4.2 satisfy the Δ_2-condition

(iii)

$$V_{\rho_{1,p_n}} = \left\{ (a_n) \in X : \sum_{k=1}^{\infty} \frac{|\lambda a_k|^{p_k}}{p_k} < \infty \text{ for some } \lambda > 0 \right\}$$
$$= \left\{ (a_n) \in X : \sum_{k=1}^{\infty} \frac{|a_k|^{p_k}}{p_k} < \infty \right\}$$

(iv)

$$V_{\rho_{2,p_n}} = \left\{ (a_n) \in X : \sum_{k=1}^{\infty} |\lambda a_k|^{p_k} < \infty \text{ for some } \lambda > 0 \right\}$$
$$= \left\{ (a_n) \in X : \sum_{k=1}^{\infty} |a_k|^{p_k} < \infty \right\}$$

Proof. If $p_+ < \infty$ and (a_n) is a sequence of real numbers, then

$$\rho_{1,p_n}(2(a_n)) = \sum_{n=1}^{\infty} \frac{|2a_n|^{p_n}}{p_n} \leq 2^{p_+} \sum_{n=1}^{\infty} \frac{|a_n|^{p_n}}{p_n} = 2^{p_+} \rho_1((a_n))$$

and similarly

$$\rho_{2,p_n}(2(a_n)) = \sum_{n=1}^{\infty} |2a_n|^{p_n} \leq 2^{p_+} \sum_{n=1}^{\infty} |a_n|^{p_n} = 2^{p_+} \rho_2((a_n)).$$

Hence, $(i) \Rightarrow (ii)$. If ρ_{1,p_n} satisfies the Δ_2 condition, fix $(a_n) \in V_{1,p_n}$ such that $\rho_{1,p_n}(\lambda(a_n)) < \infty$ for $\lambda > 0$. Either $\lambda \geq 1$, in which case

$$\rho_{1,p_n}((a_n)) \leq \rho_{1,p_n}(\lambda(a_n)) < \infty$$

or there exists $t \in \mathbb{N}$ such that $2^t\lambda > 1$. In the latter case, the Δ_2 condition yields

$$\rho_{1,p_n}((a_n)) \leq \rho_{1,p_n}(2^t(\lambda a_n)) \leq C^t \rho_{1,p_n}(\lambda(a_n)) < \infty.$$

This proves the non-trivial inclusion in (iii).
If (iii) holds, then (iv) is a direct consequence of (1.29).
Finally, if (iv) is true, the assumption $p_+ = \infty$ leads to the existence of a subsequence $(p_{n_j})_j$ of $(p_n)_n$ with $p_{n_j} > j$; then the sequence (a_n) defined as

$$a_n = \begin{cases} 1 \text{ if } n = n_j \\ 0 \text{ otherwise} \end{cases}$$

is in $V_{\rho_{2,p_{n_j}}}$, since $\sum_{n=1}^{\infty} |2^{-1}a_n|^{p_n} < \infty$; yet

$$\rho_{2,p_{n_j}}((a_n)) = \infty.$$

This contradiction finishes the proof of the theorem. □

If the exponent sequence (p_n) increases fast enough, then the variable exponent space ℓ^{p_n} is none other than ℓ^∞. In fact, the following lemma characterizes ℓ^∞ as a variable exponent space:

Lemma 1.4.7. *Let (p_n) be a sequence such that $p_n \to \infty$ as $n \to \infty$. If there exists $\alpha : 0 < \alpha < 1$ such that*

$$\sum_{n=1}^{\infty} \alpha^{p_n} := M < \infty,$$

then ℓ^{p_n} and ℓ^∞ coincide and their norms are equivalent. In other words, $\ell^{p_n} = \ell^\infty$ with equivalent norms if and only if $\mathbf{1}_{\mathbb{N}} \in \ell^{p_n}$.

Proof. Let $a_n \in \ell^{p_n}$. Take $\lambda > \|a_n\|_{\ell^{p_n}}$. Then

$$\sum_{n=1}^{\infty} \left(\frac{|a_n|}{\lambda}\right)^{p_n} < 1 \;\Rightarrow\; \sup \frac{|a_n|}{\lambda} \leq 1 \;\Rightarrow\; \|a_n\|_{l^\infty} \leq \lambda,$$

which automatically yields the inequality:

$$\|a_n\|_{\ell^\infty} \leq \|a_n\|_{\ell^{p_n}}.$$

Conversely, if $(a_n) \in l^\infty$ and $\lambda = M_0\|a_n\|_{l^\infty}/\alpha$, where $M_0 = \max(1, M)$ it follows that

$$\sum_{n=1}^{\infty} \left(\frac{|a_n|}{\lambda}\right)^{p_n} = \sum_{n=1}^{\infty} \left(\frac{|a_n|\alpha}{M_0\|a_n\|_{l^\infty}}\right)^{p_n}$$
$$\leq \sum_{n=1}^{\infty} \left(\frac{\alpha}{M_0}\right)^{p_n} \leq \frac{1}{M_0} \sum_{n=1}^{\infty} \alpha^{p_n} = 1.$$

Evidently, then,

$$\|a_n\|_{l^{p_n}} \leq \frac{M_0}{\alpha} \|a_n\|_{l^\infty}.$$

□

Example 1.4.1. Consider the sequence

$$p_n = \log_3 \log_3 n.$$

An elementary calculation shows that for any $s \in \mathbb{R}$ the series

$$\sum_{3}^{\infty} (\log_3 n)^s \tag{1.32}$$

diverges. For any $k \in \mathbb{N}$ one has $k^{p_n} = (\log_3 n)^{\log_3 k}$ and it follows easily from here and from (1.32) that there is no positive constant λ such that

$$\sum_{1}^{\infty} \lambda^{p_n} < \infty;$$

the space ℓ^{p_n} is thus strictly contained in ℓ^∞. We remark in passing the elementary fact that Lemma 1.4.7 shows the equivalence between the statements $\ell^{p_n} \approx \ell^\infty$ and $\mathbf{1}_\mathbb{N} \in \ell^{p_n}$.

Though in the preceding example $\mathbf{1}_\mathbb{N} \notin \ell^{p_n}$, the family

$$\mathcal{F} = \{E \subset \mathbb{N} : \mathbf{1}_E \in \ell^{p_n}\}$$

is non empty: for example

$$E = \left\{3^{3^n}, \ n \in \mathbb{N}\right\} \in \mathcal{F}.$$

Recall from Section 1.2.2 that given a Banach space X, X_a stands for the subspace of X consisting of the functions of absolutely continuous norm. Lemma 1.2.13 provides a clear-cut description of the subspace ℓ^∞_a. It is an elementary fact, on the other hand, that for $1 \leq q < \infty$ the classical sequence spaces ℓ^q have absolutely continuous norm, i.e., $\ell^q_a = \ell^q$. This is not the case for the variable exponent sequence spaces. In fact, the existence of functions lacking absolutely continuous norm is equivalent to the absence of the Δ_2-condition. More precisely:

Lemma 1.4.8. *$\ell^{p_n}_a = \ell^{p_n}$ if and only if $p_+ = \sup_n p_n < \infty$ (that is if and only if ρ_{p_n} satisfies the Δ_2 condition, on account of Theorem 1.4.6). In particular, if $p_+ = \sup_n p_n = \infty$, then $\ell^{p_n}_a \subsetneq \ell^{p_n}$.*

Proof. If $p_+ = \sup_n p_n < \infty$ then $(a_n) \in \ell^{p_n}$ if and only if $\rho_{p_n}((a_n)) < \infty$; hence, given $\epsilon : 0 < \epsilon < 1$, there exists $N > 0$ for which $n \geq N$ implies

$$\sum_n^\infty |a_n|^{p_n} < \epsilon^{p_+}.$$

Thus

$$\sum_n^\infty \left(\frac{|a_n|}{\epsilon}\right)^{p_n} \leq \frac{1}{\epsilon^{p_+}} \sum_n^\infty |a_n|^{p_n} \leq 1,$$

that is $\|(a_j)\mathbf{1}_{j \geq n}\|_{p_n} < \epsilon$. Assume now that $\sup p_n = \infty$. If

$$\ell^{p_n} \approx \ell^\infty,$$

which is the case if and only if there exists $0 < q < 1$ such that $\sum q^{p_n} < \infty$, the conclusion is obvious on account of Lemma 1.2.13 ($\ell^\infty_a = \mathbf{c}_0 \subsetneq \ell^\infty$). Otherwise, Lemma 1.4.7 asserts that $\sum q^{p_n} = \infty$ for each $0 < q < 1$. Without loss of generality one can suppose that $p_n \nearrow \infty$. Then there exist sequences $0 < a_n \searrow 0$ and $0 < n_k \nearrow \infty$ such that

$$1/k \leq \sum_{i=n_k}^{n_{k+1}-1} a_i^{p_i} \leq 2/k, \qquad \text{for each } k, \tag{1.33}$$

and $p_{n_k} \geq k$. The existence of such $\{a_n\}$ and $\{n_k\}$ is a direct consequence from the conditions $p_n \nearrow \infty$ and $\sum q^{p_n} = \infty$ for each $0 < q < 1$. To see this we observe that there must exist J_1 such that

$$\frac{1}{2} < \sum_{j=1}^{J_1} \left(\frac{1}{2}\right)^{p_j}$$

and that the right-hand side above is at most

$$\sum_{k=0}^{\infty} \left(\frac{1}{2}\right)^{p_1+k} = \left(\frac{1}{2}\right)^{p_1-1} \leq 1,$$

since $1 < p_1$. Set $n_1 = 1$, $n_2 = J_1 + 1$, $a_i = \frac{1}{2}$ for $1 \leq i \leq J_1$. Assume the terms n_j and a_i satisfying (1.33) have been chosen for $j = 1, 2, ...k$. Then n_{k+1} can be selected as follows: Since

$$\sum_{j=n_{k+1}}^{\infty} \left(\frac{1}{k+1}\right)^{p_j} = \infty,$$

there is $J_{k+1} \in \mathbb{N}$ such that

$$\begin{aligned} \frac{1}{k+1} &< \sum_{j=n_k}^{J_{k+1}} \left(\frac{1}{k+1}\right)^{p_j} \\ &\leq \sum_{j=0}^{\infty} \left(\frac{1}{k+1}\right)^{p_{n_k}+j} \\ &\leq \left(\frac{1}{k+1}\right)^{p_{n_k}} \frac{k+1}{k} \leq \frac{2}{k+1} \quad \text{for } k > 1, \end{aligned}$$

and we set $n_{k+1} = J_{k+1} + 1$ and $a_i = \frac{1}{k+1}$ for $n_k \leq i < n_{k+1}$. Next, take any $0 < d < 1$. By virtue of (1.33) we have:

$$\begin{aligned} \sum_{k=1}^{\infty} \sum_{i=n_k}^{n_{k+1}-1} (a_i d)^{p_i} &\leq \sum_{k=1}^{\infty} d^{p_{n_k}} \sum_{i=n_k}^{n_{k+1}-1} a_i^{p_i} \\ &\leq \sum_{k=1}^{\infty} d^k \sum_{i=n_k}^{n_{k+1}-1} a_i^{p_i} \leq e \sum_{k=1}^{\infty} d^k \frac{2}{k} < \infty. \end{aligned}$$

Consequently, $\sum_{i=k}^{\infty} a_i^{p_i} = \infty$ for each $k \geq 1$ and $\sum (a_i d)^{p_i} < \infty$ from which follows that $(a_i) \in \ell^{p_n} \setminus \ell_a^{p_n}$. □

1.4.1 Duality

We next set about to discuss the duality properties of the variable exponent sequence spaces ℓ^{p_n} introduced in the previous section.
For $(p_n) \subset (1,\infty)$, the conjugate variable exponent of (p_n) is the sequence q_n defined by

$$q_n = \frac{p_n}{p_n - 1}.$$

Lemma 1.4.9. *For $1 < q < \infty$, $\frac{1}{p} + \frac{1}{q} = 1$ and a fixed $s > 0$, one has:*

$$\frac{s^q}{q} = \sup\left\{ts - \frac{1}{p}|t|^p \, , \, t \in \mathbb{R}\right\}. \tag{1.34}$$

Proof. The concavity of the function $x \to \ln x$ implies the inequality

$$ts \leq \frac{1}{p}|t|^p + \frac{1}{q}|s|^q, \tag{1.35}$$

valid for all $s \in \mathbb{R}, t \in \mathbb{R}$.
Here, equality holds if and only if $t = s^{q-1}$. A straightforward argument then yields the desired result. □

Lemma 1.4.10. *For $n \in \mathbb{N}$ and $s_1, ..., s_n \in \mathbb{R}$*

$$\sum_{k=1}^{n} \frac{|s_k|^{q_k}}{q_k} = \sup\left\{\sum_{k=1}^{n} t_k s_k - \sum_{k=1}^{n} \frac{|t_k|^{p_k}}{p_k} \quad , \; t_1, t_2, ..., t_n \in \mathbb{R}\right\}.$$

Proof. It follows from Lemma 1.4.9 and a simple argument. □

Following the terminology in Section 1.2 we denote the associate space to a Banach function space X with X'.

Proposition 1.4.11. *If $(p_n) \subset (1,\infty)$ set $q_n = \frac{p_n}{p_n-1}$. Then ℓ^{q_n} is continuously embedded in the associate space $(\ell^{p_n})'$ and hence in $(\ell^{p_n})^*$, as a consequence of Theorem 1.2.9, i.e.,*

$$\ell^{q_n} \hookrightarrow (\ell^{p_n})' \hookrightarrow (\ell^{p_n})^*.$$

Proof. Consider $(x_n) \in \ell^{q_n}$. By definition there exists $\lambda_2 > 0$ with

$$\sum_{k=1}^{\infty} \frac{|\lambda_2 x_k|^{q_k}}{q_k} < \infty.$$

Select sequences $\mathbf{y} = (y_n) \in \ell^{p_n}$ and (λ_j) such that $1 \leq \lambda_j < \|\mathbf{y}\|_{\rho_{1,p_n}} + j^{-1}$ satisfying the inequality

$$\rho_{1,p_n}((y_n)\lambda_j^{-1}) = \sum_{k=1}^{\infty} \left(\frac{|y_k|}{\lambda_j}\right)^{p_k} p_k^{-1} \leq 1,$$

which in concert with Lemma 1.4.9 yields

$$
\begin{aligned}
\left|\sum_{k=1}^{\infty} x_k y_k\right| &\leq \frac{\lambda_j \lambda_2}{\lambda_j \lambda_2} \sum_{k=1}^{\infty} |x_k||y_k| \\
&\leq \frac{\lambda_j}{\lambda_2}\left(\sum_{k=1}^{\infty} \frac{(\frac{|y_k|}{\lambda_j})^{p_k}}{p_k} + \sum_{k=1}^{\infty} \frac{|\lambda_2 x_k|^{q_k}}{q_k}\right) \\
&\leq \frac{\|(y_n)\|_{\rho_{1,p_n}} + j^{-1}}{\lambda_2}\left(1 + \sum_{k=1}^{\infty} \frac{|\lambda_2 x_k|^{q_k}}{q_k}\right).
\end{aligned}
$$

It is thus clear that for any $(y_n) \in \ell^{p_n}$

$$\sum_{k=1}^{\infty} |x_k y_k| < \infty.$$

Lemma 1.2.6 yields that $(x_n) \in (\ell^{p_n})'$ and Theorem 1.2.9 asserts then that the linear functional

$$x^* : \ell^{p_n} \longrightarrow \mathbb{R}$$

$$x^*((y_n)) = \sum_{k=1}^{\infty} x_k y_k \tag{1.36}$$

is bounded, i.e., $x^* \in (\ell^{p_n})^*$. □

Lemma 1.4.12. *(Orlicz, 1931, [85]) Let (p_n) be a sequence of exponents subject to (1.23) and $(x_j) \subset \mathbb{R}$ be a sequence such that*

$$\sum_{n=1}^{\infty} x_n y_n < \infty \tag{1.37}$$

for all sequences $(y_n) \in \ell^{p_n}$. Then the (obviously linear) functional on ℓ^{p_n} defined by

$$x^*((y_n)_n) = \sum_{n=1}^{\infty} x_n y_n \tag{1.38}$$

is bounded i.e., $x^* \in (\ell^{p_n})^*$. Moreover, $(x_n) \in \ell^{\frac{p_n}{p_n - 1}}$.

Proof. We remark the elementary fact that assumption (1.37) is equivalent to the following: for any $(y_n) \in \ell^{p_n}$, $\sum_{n=1}^{\infty} |x_n y_n| < \infty$. Indeed, this follows from the equality:

$$\sum_{n=1}^{\infty} |x_n y_n| = \sum_{n=1}^{\infty} x_n \mathbf{sgn}(x_n)|y_n|.$$

On account of Lemma 1.2.6 it is readily concluded that under the given hypothesis, $(x_n) \in (\ell^{p_n})'$. As a consequence of Theorem 1.2.9 it follows that the functional defined in (1.38) is bounded, as claimed. □

The following corollary is a direct consequence of Lemma 1.2.6, Proposition 1.4.11 and Lemma 1.4.12.

Corollary 1.4.13. *Let $(p_n) \subset (1, \infty)$. Then the associate space of ℓ^{p_n} is $\ell^{\frac{p_n}{p_n-1}}$, i.e. $(\ell^{p_n})' = \ell^{\frac{p_n}{p_n-1}}$.*

We now focus on the question of duality, starting with the following elementary observation:

To simplify the notation, the next lemma holds for any of the norms $\|\cdot\|_{\rho_{j,p_n}}$, $j = 1, 2$, thus represented by $\|\cdot\|_{\ell^{p_n}}$.

Lemma 1.4.14. *Let $\mathbf{e}_j \in \ell^{p_n}$ be the sequence whose only nonzero entry is the jth entry, which is equal to 1. Then the linear span of $\{\mathbf{e}_j\, ,\, j \in \mathbb{N}\}$ is dense in $\ell^{p_n}_a$.*

Proof. Obvious, for if $(a_n) \in \ell^{p_n}_a$, then:

$$\left\| (a_n) - \sum_{1}^{k} a_j \mathbf{e}_j \right\|_{\ell^{p_n}} = \|(a_n^k)\|_{\ell^{p_n}} \longrightarrow 0 \text{ as } k \longrightarrow \infty.$$

□

Corollary 1.4.15. *Let $\mathbf{e}_j$ be the sequence defined by $(\mathbf{e}_j)_n = 0$ if $j \neq n$ and $(\mathbf{e}_j)_j = 1$. If the sequence (p_n) satisfies $p_+ = \sup_n p_n < \infty$, then the linear span of the set $\{\mathbf{e}_j\, ,\, j \in \mathbb{N}\}$ is dense in ℓ^{p_n}.*

Proof. The proof follows immediately from Lemma 1.4.8. □

Lemma 1.4.16. *Let $x^* \in (\ell^{p_n})^*$. Then, there exists a sequence $(x_k) \in \ell^{q_n}$ such that:*

(i) For some $\alpha \geq 0$:

$$\sum_{1}^{\infty} \frac{|x_k|^{q_k}}{q_k} \leq \alpha \|x^*\|_{(\ell^{p_n})^*}.$$

(ii) For any $(a_n) \in \ell^{p_n}$ the following bound holds:

$$\sum_{k=1}^{\infty} |x_k a_k| < \|x^*\|_{(\ell^{p_n})^*} \|(a_n)\|_{\ell^{p_n}}.$$

(iii) For any $(a_n) \in \ell^{p_n}_a$,

$$x^*((a_n)) = \sum_{k=1}^{\infty} x_k a_k.$$

(iv) If $x^*_{(x_n)}$ is the functional on ℓ^{p_n} defined by

$$\ell^{p_n} \ni (y_n) \to \sum_{k=1}^{\infty} y_n x_n,$$

then $x^* - x^*_{x_n}$ vanishes on $\ell^{p_n}_a$.

Proof. Let $x_k = x^*(\mathbf{e}_k)$. Without loss of generality, to fix ideas one can assume that

$$\sup_{\|x\|_{\rho_{1,p_n}} \leq 1} |\langle x^*, x\rangle| \leq 1.$$

Fix $N \in \mathbb{N}$ and let $t_1, t_2, ..., t_N$ be a collection of real numbers. Then the sequence

$$w = (t_1, t_2, ..., t_N, 0, 0, 0, ...)$$

is in ℓ^{p_n} and plainly

$$x^*(w) = \sum_1^N x_k t_k,\ \rho_{1,p_n}(w) = \sum_1^N \frac{|t_k|^{p_k}}{p_k}.$$

On account of Theorem 1.3.11 (applied to the modular ρ_{1,p_n}), it follows that for any $N \in \mathbb{N}$,

$$\sum_1^N x_k t_k - \sum_1^N \frac{|t_k|^{p_k}}{p_k} \leq 1$$

and therefore 1 is an upper bound for the set on the right-hand side of Lemma 1.4.10. Consequently, by virtue of Lemma 1.4.10 one has the bound

$$\sum_1^{\infty} \frac{|x_k|^{q_k}}{q_k} \leq 1.$$

For any $x^* \in (\ell^{p_n})^*$ the preceding reasoning applies to $x^*/\|x^*\|_{(\ell^{p_n})^*}$; one then has

$$\sum_1^{\infty} \frac{\left(\frac{|x_k|}{\|x^*\|_{(\ell^{p_n})^*}}\right)^{q_k}}{q_k} \leq 1,$$

and claim (i) follows immediately by definition of the Luxemburg norm. Statement (ii) follows from (i) and Proposition 1.4.11. Select now $a \in \ell^{p_n}_a$. Then

$$\left| x^*\left((a_n)\right) - x^*\left(\sum_{n=1}^{N} a_n e_n\right)\right| \leq \|x^*\|_{(\ell^{p_n})^*} \left\| \sum_{n=N+1}^{\infty} a_n e_n \right\|_{\ell^{p_n}} \to 0 \text{ as } N \to \infty.$$

It follows that the series $\sum_1^{\infty} a_k x_k$ converges and its sum is precisely $x^*((a_n))$. □

Theorem 1.4.17. *If* $1 < p_n \leq \sup_{n\in\mathbb{N}} = p_+ < \infty$, *then the topological dual of* ℓ^{p_n} *is isomorphic to* ℓ^{q_n}, *i.e.*

$$(\ell^{p_n})^* \simeq \ell^{q_n}. \tag{1.39}$$

Proof. It is clear from the preceding paragraph that it remains to show that under the given assumptions, any $x^* \in (\ell^{p_n})^*$ is of the form (1.36). The assumption $p_+ < \infty$ implies that $(y_k) \in \ell^{p_n}$ if and only if $\rho_{p_n}((y_n)) < \infty$. For each $j \in \mathbb{N}$, let $\mathbf{e^j}$ be the sequence defined by $\mathbf{e^j_n} = (e^j_n)$ with $e^j_n = 0$ if $n \neq j$ and $e^j_j = 1$. For a fixed $x^* \in (\ell^{p_n})^*$ with $\|x^*\|_{(\ell^{p_n})^*} \leq 1$ set $x^*(e_j) = x_j$, $j \in \mathbb{N}$. It transpires from (1.34), that for fixed $n \in \mathbb{N}$,

$$\begin{aligned}\sum_{k=1}^{n} \frac{|x_k|^{q_k}}{q_k} &= \sup\left\{\sum_{k=1}^{n} y_k x_k - \sum_{k=1}^{n} \frac{|y_k|^{p_k}}{p_k} \ , \ y_k \in \mathbb{R} \ k = 1, 2, ...n\right\} \\ &\leq \sup\left\{x^*(y) - \rho_p(y) \ , \ y \in \ell^{p_n}\right\} \\ &\leq 1.\end{aligned}$$

The last inequality follows from the fact that $\|x^*\|_{(\ell^{p_n})^*} \leq 1$ and from (1.17). In all, $(x_k) \in \ell^{q_n}$ and $\|(x_k)\|_{\ell^{q_n}} \leq 1$. Proposition 1.4.14 implies at once that for any $(y_n) \in \ell^{p_n}$,

$$x^*((y_n)) = \sum_{k=1}^{\infty} x_k y_k.$$

Claim (1.39) follows immediately from Theorem 1.3.15. □

A phenomenon not encountered in the constant exponent case is worth exploring at this point, namely, the fact that different sequences (p_n) and (q_n) could give rise to the same variable exponent sequence space ℓ^{p_n}. The next series of theorems aims at exploring this question.

Theorem 1.4.18. *Let* (p_n) *be a sequence of real numbers with* $1 \leq \alpha \leq \inf_n p_n$ *and let* (q_n) *stand for the sequence of Hölder conjugates of* $(\frac{p_n}{\alpha})$, *i.e.*, $\frac{\alpha}{p_n} + \frac{1}{q_n} = 1$. *Then* ℓ^{p_n} *is isomorphic to* ℓ^α *iff and only if there exists* $\gamma \in \mathbb{R}$: $0 < \gamma < 1$ *for which*

$$\sum_{n=1}^{\infty} \gamma^{q_n} < \infty. \tag{1.40}$$

Proof. For any $n \in \mathbb{N}$, any $0 < \gamma < 1$ and α, p_n and 1_n as in the preceding paragraph, in accordance with Young's inequality it holds that:

$$(|a_n|\gamma)^\alpha \leq \frac{\alpha}{p_n}|a_n|^{p_n} + \frac{1}{q_n}\gamma^{q_n},$$

assuming, thus that $\sum_1^\infty \gamma^{q_n} < \infty$ it is concluded that

$$\sum_{n=1}^{\infty} |a_n|^\alpha \leq \frac{1}{\gamma^\alpha}\left(\sum_{n=1}^{\infty} |a_n|^{p_n} + \sum_{n=1}^{\infty} |\gamma|^{q_n}\right). \tag{1.41}$$

Inequality (1.41) settles the validity of the set-theoretic inclusion

$$\ell^{\alpha} \subseteq \ell^{p_n}$$

under assumption (1.40). The sufficiency of condition (1.40) in the statement of Theorem 1.4.18 follows from Theorem 1.2.4, Theorem 1.4.3 and Theorem 1.4.4.

For the converse, we suppose that

$$\ell^{p_n} \subseteq \ell^{\alpha}. \tag{1.42}$$

We first claim that (1.42) implies that $\ell^{\frac{p_n}{\alpha}} \subseteq \ell^1$. Indeed, if $0 < \lambda < 1$ and

$$\sum_{n=1}^{\infty} |\lambda a_n|^{\frac{p_n}{a}} < \infty$$

it is immediately clear that

$$\sum_{n=1}^{\infty} |\lambda |a_n|^{\frac{1}{a}}|^{p_n} \leq \sum_{n=1}^{\infty} |\lambda^{\frac{1}{a}} |a_n|^{\frac{1}{a}}|^{p_n} < \infty;$$

likewise, for a sequence (a_n) such that

$$\sum_{n=1}^{\infty} |a_n|^{\frac{p_n}{a}} < \infty$$

it follows that $(|a_n|^{\frac{1}{\alpha}}) \in \ell^{p_n}$. The preceding two statements prove that if $(a_n) \in \ell^{\frac{p_n}{\alpha}}$, then $(|a_n|^{\frac{1}{\alpha}}) \in \ell^{p_n}$ and under assumption (1.42) one can readily conclude that $(|a_n|^{\frac{1}{\alpha}}) \in \ell^{\alpha}$. Equivalently, for some $\epsilon > 0$ it must hold that

$$\sum_{n=1}^{\infty} |\epsilon |a_n|^{\frac{1}{a}}|^{a} < \infty.$$

Consequently,

$$\sum_{n=1}^{\infty} |a_n| < \infty$$

for any sequence $(a_n) \in \ell^{\frac{p_n}{\alpha}}$, as claimed.
In particular $\mathbf{1}_{\mathbb{N}}$, *i.e.*, the sequence (b_n) identically equal to 1, satisfies the conditions of Lemma 1.4.12 (with exponents $\frac{p_n}{p_n - a}$) which asserts then that $\mathbf{1}_{\mathbb{N}} \in \ell^{\frac{p_n}{p_n - \alpha}}$. By definition then, there exists a real number γ, $0 < \gamma < 1$ such that

$$\sum_{n=1}^{\infty} \gamma^{\frac{p_n}{p_n - \alpha}} < \infty.$$

□

Along similar lines one has the following :

Theorem 1.4.19. *Let* (p_n) *be a sequence with* $1 \leq p_n \leq b < \infty$*. Then* ℓ^{p_n} *is isomorphic to* ℓ^b *if and only if there exists* $0 < \gamma < 1$ *such that*

$$\sum_{n=1}^{\infty} \gamma^{\frac{b}{b-p_n}} < \infty.$$

Proof. Suppose such a real number γ exists and let $(a_n) \in \ell^b$. By definition, there exists $\lambda > 0$ such that $\sum_{n=1}^{\infty} |\lambda a_n|^b < \infty$. Inequality (1.35) with $t = \gamma^{p_n}$, $s = (\lambda |a_n|)^{p_n}$, $p = \frac{b}{b-p_n}$ and $q = b/p_n$ applied for each $n \in \mathbb{N}$ yields:

$$\begin{aligned}(\gamma\lambda|a_n|)^{p_n} &\leq \frac{b-p_n}{b}\gamma^{\frac{b}{b-p_n}} + \frac{p_n}{b}(\lambda|a_n|)^b \\ &\leq \gamma^{\frac{b}{b-p_n}} + (\lambda|a_n|)^b.\end{aligned}$$

In all

$$\sum_{n=1}^{\infty} (\gamma\lambda|a_n|)^{p_n} \leq \sum_{n=1}^{\infty} \gamma^{\frac{b}{b-p_n}} + \sum_{n=1}^{\infty} (\lambda|a_n|)^b < \infty.$$

Theorem 1.2.4, Theorem 1.4.3 and Theorem 1.4.4 yield $\ell^{p_n} \approx \ell^b$, as claimed. On the other hand, if ℓ^b is continuously embedded in ℓ^{p_n}, take any sequence $(b_n) \in \ell^{\frac{b}{p_n}}$; given that $\frac{b}{p_n} < b$, it must necessarily hold, (Theorem 1.4.6) that

$$\sum_{n=1}^{\infty} |b_n|^{\frac{b}{p_n}} < \infty.$$

It is plain then that $(|b_n|^{\frac{1}{p_n}}) \in \ell^b \hookrightarrow \ell^{p_n}$. Consequently, invoking again Theorem 1.4.6 one concludes that $(b_n) \in \ell^1$.
By definition of the sequence (b_n), it is immediate that $(a_n) \in \ell^{p_n}$.
Conversely, assume that for given (p_n) and b as in the statement of the theorem, ℓ^{p_n} is isomorphic to ℓ^b, *i.e*, $\ell^b \subseteq \ell^{p_n}$. Consider a sequence $(x_n) \in \ell^{\frac{b}{p_n}}$. By definition, either $\sum_{n=1}^{\infty} |x_n|^{\frac{b}{p_n}} < \infty$ or there exists $\lambda_0 : 0 < \lambda_0 < 1$ for which $\sum_{n=1}^{\infty} |\lambda_0 x_n|^{\frac{b}{p_n}} < \infty$. It is immediately derived from the first case that $(|x_n|^{\frac{1}{p_n}}) \in \ell^b$; since $\lambda_0 < \lambda_0^{\frac{1}{p_n}}$, in the second case one has

$$\begin{aligned}\sum_{n=1}^{\infty} \left(\lambda_0 |x_n|^{\frac{1}{p_n}}\right)^b &\leq \sum_{n=1}^{\infty} \left(|\lambda_0 x_n|^{\frac{1}{p_n}}\right)^b \\ &< \infty,\end{aligned}$$

i.e., if $(x_n) \in \ell^{\frac{b}{p_n}}$, then necessarily $(|x_n|^{\frac{1}{p_n}}) \in \ell^b \subseteq \ell^{p_n}$. In all, we have the logical implication $\ell^b \subseteq \ell^{p_n} \Rightarrow \ell^{\frac{b}{p_n}} \subseteq \ell^1$.

Since $1 \leq \frac{b}{p_n}$ Theorem 1.4.18 applies to $\ell^{\frac{b}{p_n}}$, implying the existence of a real number $\gamma : 0 < \gamma < 1$ such that

$$\sum_{j=1}^{\infty} \gamma^{\frac{b}{b-p_n}} < \infty,$$

as claimed. □

1.4.2 Finitely additive measures

A digression aiming at preparing the ground for a description of the dual $(\ell^{p_n})^*$ in the case $p_+ = \sup_{n \in \mathbb{N}} p_n = \infty$ is in order. We start with the following standard definition.

Definition 1.4.2. *Let $X \neq \emptyset$ and $\mathcal{A} \subset 2^X$ be closed under complements and finite unions. A real valued function*

$$\mu : \mathcal{A} \longrightarrow \mathbb{R}$$

is said to be a finitely additive measure on $(X, \mathcal{A})$ if for any finite subset $\{M_1, ...M_n\} \subset \mathcal{A}$ it holds that

(i)

$$\mu\left(\bigcup_{k=1}^{N} M_k\right) = \sum_{k=1}^{N} \mu(M_k)$$

(ii)

$$\mu(\emptyset) = 0$$

(iii)

$$\sup_{M \in \mathcal{A}} |\mu(M)| < \infty.$$

The total variation of μ is (the finitely additive measure) given by:

$$|\mu|(A) =$$

$$\sup\left\{\sum_{j=1}^{k} |\mu(A_k)|,\ A_i \cap A_j = \emptyset \text{ if } i \neq j\ ,\ \mathcal{A} \ni A_k \subseteq A \text{ for } j = 1, 2, ..., k\right\}.$$

The finitely additive measure μ is said to be of bounded variation if $|\mu|(X) < \infty$. In that case the total variation $\|\mu\|$ of μ is given by

$$\|\mu\| = |\mu|(X).$$

Recall that we denote the subspace of ℓ^∞ consisting of all simple functions of the form

$$\varphi = \sum_{k=1}^{N} \varphi_k \mathbf{1}_{E_k}$$

by $\mathcal{S}_\infty$ (a word of caution: it is possible that the set $E_k \subseteq \mathbb{N}$ is infinite for some k). A finitely additive measure μ with $\|\mu\| < \infty$ defines naturally a linear functional $\gamma \in \mathcal{S}_\infty^*$, namely

$$\gamma\left(\sum_{k=1}^{N} a_k \mathbf{1}_{E_k}\right) = \sum_{k=1}^{N} a_k \mu(E_k)$$

for $a_j \in \mathbb{R}$, $j = 1, 2, ...N$. It follows immediately from this definition that

$$\sup_{u\in\mathcal{S}_\infty : \|u\|_{\ell^\infty} \leq 1} |\langle \gamma, u\rangle| \leq \|\mu\|.$$

The functional γ associated to μ in the above fashion can be continuously extended to ℓ^∞ on account of the density of $\mathcal{S}_\infty$ in ℓ^∞; specifically, if $(s_k) \subset \mathcal{S}_\infty$ is a Cauchy sequence in ℓ^∞, then so is $(\gamma(s_k))$, by virtue of the finiteness of $\|\mu\|$. Moreover if $(s_k) \subset \mathcal{S}_\infty$ and $s_k = \sum_{j=1}^{N_k} a_{k,j} \mathbf{1}_{E_{k,j}} \longrightarrow 0$ in ℓ^∞ as $k \to \infty$, then

$$\begin{aligned} |\gamma(s_k)| &= \left|\sum_{j=1}^{N_k} a_{k,j} \mu\left(E_{k,j}\right)\right| \\ &\leq \sup_{k,j} |a_{k,j}| \|\mu\| \to 0 \quad \text{as} \quad k \to \infty. \end{aligned} \tag{1.43}$$

Inequality (1.43) shows that if $\|s_k - a\|_{\ell^\infty} \to 0$ as $k \to \infty$, then $\lim_{k\to\infty} \gamma(s_k)$ exists and is independent of the sequence (s_k). Hence the natural extension (still denoted by γ) defined as

$$\begin{aligned} &\gamma : \ell^\infty \longrightarrow \mathbb{R} \\ &\gamma(a) = \lim \gamma(s_k) := \int_{\mathbb{N}} a \, d\mu \end{aligned}$$

is well defined, linear and bounded; in fact the preceding discussion shows that

$$\|\gamma\|_{(\ell^\infty)^*} \leq \|\mu\|. \tag{1.44}$$

Lemma 1.4.20. *Any finitely additive measure μ of finite total variation $\|\mu\|$ on $2^{\mathbb{N}}$ induces a naturally defined bounded, linear functional $\gamma \in (\ell^\infty)^*$ via (1.43). Moreover (1.44) holds. In addition, if μ vanishes on finite sets, then γ vanishes on $\mathbf{c}_0$.*

Proof. In view of the discussion preceding the statement of Lemma 1.4.20 it remains only to prove the last assertion. To this end, fix a finitely additive measure μ on $(\mathbb{N}, 2^{\mathbb{N}})$ that vanishes on finite sets and consider $\mathbf{x} \in \mathbf{c}_0$. Fix $\delta > 0$ and let $n \geq N \in \mathbb{N}$ imply $\sup_{\mathbb{N} \ni j \geq N} |x_j| < \delta$. Then

$$\begin{aligned}
|\gamma(\mathbf{x})| &= \left|\gamma(\mathbf{x}\mathbf{1}_{\{1\leq j\leq N\}} + \mathbf{x}\mathbf{1}_{\{j>N\}})\right| \leq \left|\gamma(\mathbf{x}\mathbf{1}_{\{1\leq j\leq N\}})\right| + \left|\gamma(\mathbf{x}\mathbf{1}_{\{j>N\}})\right| \\
&= \left|\sum_{j=1}^{N} x_j \gamma(\mathbf{1}_j)\right| + |\gamma(\mathbf{x}\mathbf{1}_{\{j>N\}})| \\
&\leq \sum_{j=1}^{N} |x_j||\mu(\{j\})| + \|\gamma\|_{(\ell^\infty)^*} \|\mathbf{x}\mathbf{1}_{\{j>N\}}\|_{\ell^\infty} \\
&\leq \|\gamma\|_{(\ell^\infty)^*} \delta
\end{aligned}$$

from which the claim is clear. □

Lemma 1.4.21. *If* $\Lambda \in (\ell^\infty)^*$ *vanishes on* $\mathbf{c}_0$*, then there exists a finitely additive measure* μ *on* $(\mathbb{N}, 2^{\mathbb{N}})$ *such that with finite total variation* $\|\mu\|$ *such that*

(i)

$$\|\mu\| \leq \|\Lambda\|_{(\ell^\infty)^*}$$

(ii) For any sequence $\mathbf{x} \in \ell^\infty$ *one has*

$$\Lambda(\mathbf{x}) = \int_{\mathbb{N}} \mathbf{x}\, d\mu. \tag{1.45}$$

(iii) μ *vanishes on finite sets.*

Proof. For $E \subset \mathbb{N}$ set $\mu(E) = \Lambda(\mathbf{1}_E)$. It is straightforward to show that, due to the linearity of Λ, μ is a finitely additive measure on $\mathbb{N}$. It remains to prove that the total variation $\|\mu\|$ is finite. To this end, consider a finite sequence of mutually disjoint subsets of $\mathbb{N}$, say (E_j), $1 \leq j \leq n$ and set

$$\xi_j = \mathbf{sgn}\Lambda(\mathbf{1}_{E_j}).$$

Then

$$\begin{aligned}
\sum_{j=1}^{n} |\mu(E_j)| &= \sum_{j=1}^{n} |\Lambda(\mathbf{1}_{E_j})| = \sum_{j=1}^{n} \xi_j \Lambda\left(\mathbf{1}_{E_j}\right) = \sum_{j=1}^{n} \Lambda\left(\xi_j \mathbf{1}_{E_j}\right) \\
&\leq \Lambda\left(\sum_{j=1}^{n} \xi_j \mathbf{1}_{E_j}\right) \leq \|\Lambda\|_{(\ell^\infty)^*} \left\|\sum_{j=1}^{n} \xi_j \mathbf{1}_{E_j}\right\|_{\ell^\infty} \\
&\leq \|\Lambda\|_{(\ell^\infty)^*}.
\end{aligned}$$

Equality (1.45) follows straight from the discussion preceding the statement of Lemma 1.4.20. □

Definition 1.4.3. *A bounded linear functional $x^* \in (\ell^{p_n})^*$ is said to be singular if it vanishes on $\ell^{p_n}_a$.*

Proposition 1.4.22. *In view of Lemma 1.4.16, it is easily seen that $x^* \in (\ell^{p_n})^*$ is singular if and only if $x^*(\mathbf{e_j}) = 0$ for all $j \in \mathbb{N}$.*

We now turn to the question of duality in the absence of the Δ_2 condition. More precisely, we describe the dual of ℓ^{p_n} under the assumption that the exponent sequence (p_n) is unbounded.
The first order of business is to underline the fact that if ℓ^{p_n} contains sequences of non-absolutely continuous norms, then ℓ^{p_n} is "locally" isomorphic to ℓ^∞. Specifically:

Lemma 1.4.23. *(See [76, 101]) Let $\ell^{p_n}_a \subsetneq \ell^{p_n}$. Then ℓ^{p_n} contains a subspace isomorphic with ℓ^∞.*

Proof. By assumption, one can select $\mathbf{a} = (a_n) \in \ell^{p_n} \setminus \ell^{p_n}_a$ with $\|(a_n)\|_{\ell^{p_n}} = 1$. Define the sequence $\mathbf{a^k}$ as

$$\mathbf{a^k} := \mathbf{a}\ \mathbf{1}_{\{n:n\geq k\}} = (a_n^k),$$

that is $a_n^k := a_n$ for $n \geq k$ and $a_n^k := 0$ for $n < k$, and set

$$\mathbf{a^{k,l}} := \mathbf{a}\ \mathbf{1}_{\{n:k\leq n\leq l\}} = (a_n^{k,l}). \tag{1.46}$$

In other words $a_n^{k,l} = a_n$ for $k \leq n \leq l-1$ and $a_n^{k,l} := 0$ for $n < k$ or $l \leq n$.
By assumption on $\mathbf{a}$, there exist $\delta > 0$ such that $\|\mathbf{a^k}\|_{\ell^{p_n}} > 2\delta$ for each $k \geq 1$. Observing that $\|\mathbf{a^{k,l}}\|_{\ell^{p_n}} \nearrow \|\mathbf{a^k}\|_{\ell^{p_n}}$ as $l \to \infty$ one quickly derived the existence of an increasing sequence (k_i) such that $k_i \to \infty$ as $i \to \infty$ and that $\|\mathbf{a^{k_i,k_{i+1}}}\|_{\ell^{p_n}} > \delta$ for each i.
Note that for $i \neq j$ the supports of $\mathbf{a^{k_i,k_{i+1}}}$ and $\mathbf{a^{k_i,k_{i+1}}}$ are disjoint.
Set

$$Z_{\mathbf{a}} := \left\{ g \in \ell^{p_n} : \text{ there exists } (c_i) \in \ell^\infty \text{ such that } g = \sum_{i=1}^\infty c_i \mathbf{a^{k_i,k_{i+1}}} \right\}. \tag{1.47}$$

Take $c = (c_i) \in \ell^\infty$ and $\varepsilon > 0$. There exists $i \in \mathbb{N}$ such that $\|c\|_{\ell^\infty} \leq |c_i| + \varepsilon$. Then

$$\begin{aligned}
\delta\|c\|_{\ell^\infty} &\leq \delta|c_i| + \delta\varepsilon \leq \|c_i \mathbf{a^{k_i,k_{i+1}}}\|_{\ell^{p_n}} + \delta\varepsilon \leq \left\| \sum_{m=1}^\infty c_m \mathbf{a^{k_m,k_{m+1}}} \right\|_{\ell^{p_n}} + \delta\varepsilon \\
&\leq \left\| \sum_{m=1}^\infty \|c\|_{\ell^\infty} \mathbf{a^{k_m,k_{m+1}}} \right\|_{\ell^{p_n}} + \delta\varepsilon \leq \|c\|_{\ell^\infty} \left\| \sum_{m=1}^\infty \mathbf{a^{k_m,k_{m+1}}} \right\|_{\ell^{p_n}} + \delta\varepsilon \\
&\leq \|c\|_{\ell^\infty} + \delta\varepsilon.
\end{aligned}$$

In all, the inequality:

$$\delta\|c\|_{\ell^\infty} \le \left\| \sum_{m=1}^{\infty} c_m \mathbf{a}^{\mathbf{k_m},\mathbf{k_{m+1}}} \right\|_{\ell^{p_n}} \le \|c\|_{\ell^\infty}$$

is derived. In other words, the mapping

$$J : \ell^\infty \longrightarrow \ell^{p_n} \tag{1.48}$$
$$J\left((c_m)\right) = \sum_{m=1}^{\infty} c_m \mathbf{a}^{\mathbf{k_m},\mathbf{k_{m+1}}}$$

is an isomorphism from ℓ^∞ into $J\left(\ell^\infty\right) \subset \ell^{p_n}$. □

The preceding lemma is key to the characterization of the dual space $(\ell^{p_n})^*$ in the absence of the Δ_2 condition. More specifically, we have the following result:

Theorem 1.4.24. *Let $(p_n) \subset (1, \infty)$ be a variable exponent and assume that $p_n \to \infty$ as $n \to \infty$.*
If $\Lambda \in (\ell^{p_n})^*$ and Λ vanishes on $\ell^{p_n}_a$ then for any $\mathbf{b} \in \ell^{p_n} \setminus \ell^{p_n}_a$ there exists $\tilde{\Lambda} \in (\ell^\infty)^*$ such that

(i) $\tilde{\Lambda}$ vanishes on $\mathbf{c}_0$

(ii) $\tilde{\Lambda}(\mathbf{1}_{\mathbb{N}}) = \Lambda(\mathbf{b})$.

Conversely, if $\Lambda \in (\ell^\infty)^*$ vanishes on $\mathbf{c}_0$, then given any fixed $b \in \ell^{p_n} \setminus \ell^{p_n}_a$, there exists $\gamma \in (\ell^{p_n})^*$ that fulfills the following:

(i) γ vanishes on $\ell^{p_n}_a$

(ii) $\gamma(\mathbf{1}_{\mathbb{N}}) = \tilde{\Lambda}(\mathbf{b})$

(iii) If $Z_{\mathbf{b}}$ is as in (1.47), then $\gamma|_{Z_{\mathbf{b}}} = \Lambda$.

Proof. To prove (i), let $\Lambda \in (\ell^{p_n})^*$ and fix $\mathbf{b} \in \ell^{p_n}$. Let $Z_{\mathbf{b}}$ denote the subspace of ℓ^{p_n} defined in (1.47); then $b \in Z_{\mathbf{b}} \sim \ell^\infty$. Any $\Lambda \in (\ell^{p_n})^*$ that vanishes on $\ell^{p_n}_a$ engenders a functional $\tilde{\Lambda} \in (\ell^\infty)^*$ via the isomorphism $J : Z_{\mathbf{b}} \longrightarrow \ell^\infty$ introduced in (1.48). Specifically

$$\tilde{\Lambda} = \Lambda \circ J^{-1}.$$

It is easy to see that

$$Z_{\mathbf{b}} \cap \ell^{p_n}_a = J^{-1}\left(\mathbf{c}_0\right), \tag{1.49}$$

as a consequence of which one readily obtains that $\tilde{\Lambda}$ vanishes on $\mathbf{c}_0$. Furthermore, by the very definition of J one has

$$\Lambda(\mathbf{b}) = \tilde{\Lambda}(\mathbf{1}_{\mathbb{N}}).$$

To prove the converse assertion, fix $\Lambda \in (\ell^\infty)^*$ and set

$$\tilde{\Lambda} \in (Z_{\mathbf{b}})^*$$

as $\tilde{\Lambda} = \Lambda \circ J$ and extending $\tilde{\Lambda}$ to ℓ^{p_n} via the Hahn-Banach Theorem (we continue to denote this Hahn-Banach extension by $\tilde{\Lambda}$).
In the light of Lemma 1.4.16, *(iii)*, one can write

$$\tilde{\Lambda} = \gamma_1 + \gamma_2,$$

where $\gamma_i \in (\ell^{p_n})^*$ for $i = 1, 2$ and γ_1 vanishes on $\ell_a^{p_n}$. In fact, for $x \in \ell^{p_n}$

$$\gamma_2(x) = \sum_1^\infty x_k \tilde{\Lambda}(\mathbf{e_k}).$$

The restrictions to $Z_{\mathbf{b}}$ of γ_1 and γ_2 satisfy

$$\tilde{\Lambda} = \gamma_1|_{Z_{\mathbf{b}}} + \gamma_2|_{Z_{\mathbf{b}}}.$$

By assumption and (1.49), $\tilde{\Lambda}$ and γ_1 both vanish on $J^{-1}(\mathbf{c}_0)$. Consequently, $\gamma_2|_{Z_{\mathbf{b}}}$ also vanishes on $J^{-1}(\mathbf{c}_0)$.
As was observed in the proof of Lemma 1.4.23, the assumption $\mathbf{b} \in \ell^{p_n} \setminus \ell_a^{p_n}$ and the choice of $\delta > 0$ leads to the existence of an increasing sequence $(k_j) \subset \mathbb{N}$ such that $\|\mathbf{b}^{\mathbf{k_j},\mathbf{k_{j+1}}}\|_{\ell^{p_n}} > \delta$ for each j (here $\mathbf{b}^{\mathbf{k_j},\mathbf{k_{j+1}}}$ is defined as in (1.46)).
We claim that for any sequence $\mathbf{x} = \sum_1^\infty c_m \mathbf{b}^{\mathbf{k_m},\mathbf{k_{m+1}}} \in Z_{\mathbf{b}}$ one has $\gamma_2(x) = 0$. Indeed, for each $m \in \mathbb{N}$, the sequence $(b_n^{k_m,k_{m+1}}) = \mathbf{b}^{\mathbf{k_n},\mathbf{k_{m+1}}} \in \ell_a^{p_n}$. Consequently, for each $N \in \mathbb{N}$ we have:

$$\begin{aligned}\gamma_2\left(\sum_{m=1}^N c_m \mathbf{b}^{\mathbf{k_m},\mathbf{k_{m+1}}}\right) &= \sum_{m=1}^N c_m \gamma_2(\mathbf{b}^{\mathbf{k_m},\mathbf{k_{m+1}}}) \\ &= 0.\end{aligned} \tag{1.50}$$

Writing $\mathbf{x} = (x_j)$ it is immediate by definition that:

$$\gamma_2(x) = \sum_1^\infty x_j \tilde{\Lambda}(\mathbf{e_j}) \tag{1.51}$$

and that for each k_m as above,

$$\sum_1^{k_{m+1}} x_j \tilde{\Lambda}(\mathbf{e_j}) = \sum_{j=1}^m c_j \sum_{k=1}^{k_j} b_k \tilde{\Lambda}(\mathbf{e_j}) = 0,$$

on account of (1.50). We conclude that there exists a strictly increasing sequence $K_1, K_2, , ...$ of natural numbers such that for each $k \in \mathbb{N}$ it holds

$$\sum_1^{K_N} x_j \tilde{\Lambda}(\mathbf{e_j}) = 0$$

and by virtue of (1.51), $\gamma_2(x) = 0$. In conclusion, $\gamma_1 \in (\ell^{p_n})^*$ vanishes on $\ell^{p_n}_a$, coincides with $\tilde{\Lambda}$ on $Z_{\mathbf{b}}$ and satisfies $\gamma_1(\mathbf{b}) = \Lambda(\mathbf{1}_{\mathbb{N}})$. Theorem 1.4.24 is thus proved.

□

1.4.3 Geometric properties of ℓ^{p_n}

In this section we investigate the modular geometric properties of the variable exponent sequence spaces ℓ^{p_n}. At this point, a digression is in order. It will be shown in a more general context (Corollary 2.6.2) that if the exponent sequence (p_n) satisfies $1 < p_-$ and $p_+ < \infty$ then the Banach space $(\ell^{p_n}, \|\cdot\|_{\ell^{p_n}})$ is uniformly convex. It is however desirable, at least in some mathematical situations to have a concept of "uniform convexity" associated directly to the modular. It is in this spirit that we undertake the discussion presented in this section. Surprisingly, some form of modular uniform convexity is evidenced even in the case when $p_- = 1$ or $p_+ = \infty$. In order to lay the groundwork for the main results, some technical lemmas need to be proved.

Lemma 1.4.25. *For any* $p : 1 \leq p \leq 2$ $q(p) = 2 + p(p-1) - 2^p \leq 0$, *in fact the strict inequality holds if* $1 < p < 2$.

Proof. Since $q'''(p) < 0$ and $q''(2) = 2 - 4\ln 2^2 > 0$, $q''(p) > 0$ for all $p \in [1,2]$ and q' is increasing on $[1,2]$. On the other hand, $q'(1) = 1 - 2\ln 2 < 0$ and $q'(2) = 3 - 4\ln 2 > 0$. It is clear then that there exists a unique $x_0 \in (1,2)$ such that q decreases on $(1, x_0)$ and increases on $(x_0, 2)$. Since $q(1) = q(2) = 0$, it is immediate that $q < 0$ on $(1,2)$, as claimed. □

Lemma 1.4.26. *If* $1 \leq p \leq 2$ *and* $h : [0,1] \to \mathbb{R}$ *be defined as*

$$h(s) = s^{p-1} - (s+1)^{p-1} + \frac{p(p-1)}{2},$$

then $h(s) \leq 0$ *on* $[0,1]$.

Proof. It is immediately seen that h is increasing and that $h(1) = 1 - 2^{p-1} + \frac{p(p-1)}{2} = \gamma(p) < 0$ on account of the previous lemma. Hence $h(s) < 0$ for any $x \in (0,1)$. □

Lemma 1.4.27. *Let* $a \in \mathbb{R}$, $b \in \mathbb{R}$ $0 < b < a$, $1 \leq p \leq 2$. *Then*

$$\left|\frac{a-b}{2}\right|^p + \frac{p(p-1)}{2}\left|\frac{a+b}{2}\right|^p \leq \frac{1}{2}\left(|a|^p + |b|^p\right).$$

Proof. Using the transformation $s = \frac{t-1}{t+1}$ one has $t \geq 1 \iff 0 \leq s < 1$ and thus, Lemma 1.4.26 implies that for $t \in [1,\infty)$ one has:

$$p(t+1)^{p-1}\left(\left(\frac{t-1}{t+1}\right)^{p-1}+\frac{p(p-1)}{2}-2^{p-1}\left(\frac{t}{t+1}\right)^{p-1}\right)$$
$$=p(t-1)^{p-1}+\frac{p^2(p-1)}{2}(t+1)^{p-1}-p2^{p-1}t^{p-1}$$
$$=\frac{d}{dt}\left((t-1)^p+\frac{p(p-1)}{2}(t+1)^p-2^{p-1}t^p\right)\le 0.$$

It follows that

$$w(t)=(t-1)^p+\frac{p(p-1)}{2}(t+1)^p-2^{p-1}t^p$$

is decreasing on $[1,\infty)$ and since $w(1)=\left(\frac{p(p-1)}{2}-1\right)2^{p-1}<0$ it is plain that $w(t)\le 0$ for $t\ge 1$.
Writing $t=\frac{a}{b}$, the lemma follows at once. □

Lemma 1.4.28. *If $1<p\le 2$ and consider $G:[0,1)\longrightarrow\mathbb{R}$*

$$G(s)=2\left(2+p(p-1)s^2-((1+s)^p+(1-s)^p)\right),$$

then $G(s)<0$ for any $s: 0\le s<1$.

Proof. Routine calculations reveal that $G'''(s)<0$, so that $G''(s)$ is decreasing. On the other hand, $G''(0)=0$, and as a consequence, G' is decreasing on $(0,1)$, which in conjunction with the fact that $G'(0)=0$ yields that G is decreasing on $(0,1)$. Finally, $G(0)=0$, which proves the claim. □

Lemma 1.4.29. *Let $0<b<a$ be real numbers and $1\le p\le 2$. Then*

$$\left|\frac{a+b}{2}\right|^p+\frac{p(p-1)}{2}\left|\frac{a-b}{|a|+|b|}\right|^{2-p}\left|\frac{a-b}{2}\right|^p\le\frac{1}{2}\left(|a|^p+|b|^p\right).$$

Proof. From Lemma 1.4.28 it follows that

$$\left(\frac{2}{1-s}\right)^2+\frac{p(p-1)}{2}\left(\frac{2s}{1-s}\right)^2-2^{p-1}\left(\left(\frac{1+s}{1-s}\right)^p+1\right)\left(\frac{2}{1-s}\right)^{2-p}<0$$

for $0\le s<1$. Using the transformation

$$s\longrightarrow\frac{t-1}{t+1}$$

the latter implies that for any $t\ge 1$,

$$(t+1)^2+\frac{p(p-1)}{2}(t-1)^2-2^{p-1}(t^p+1)(t+1)^{2-p}<0,$$

which is equivalent to

$$(t+1)^p + \frac{p(p-1)}{2}\frac{(t-1)^2}{(t+1)^{2-p}} - 2^{p-1}(t^p+1) < 0 \text{ on } [1,\infty);$$

in other words

$$(t+1)^p + \frac{p(p-1)}{2}\left|\frac{(t-1)}{(t+1)}\right|^{2-p}|t-1|^p - 2^{p-1}(t^p+1) < 0 \text{ on } [1,\infty);$$

the lemma follows by setting $t = \frac{a}{b}$ and multiplying the inequality times 2^{-p}. ☐

Theorem 1.4.30. *(Clarkson, [17]) For $p \geq 2$ and arbitrary real numbers a, b one has*

$$\left|\frac{a+b}{2}\right|^p + \left|\frac{a-b}{2}\right|^p = \frac{1}{2}\left(|a|^p + |b|^p\right). \tag{1.52}$$

Theorem 1.4.31. *(Sundaresan, [99]) If $a \in \mathbb{R}$, $b \in \mathbb{R}$, $|a| + |b| \neq 0$ and $1 < p \leq 2$, then*

$$\left|\frac{a+b}{2}\right|^p + \frac{p(p-1)}{2}\left|\frac{a-b}{|a|+|b|}\right|^{2-p}\left|\frac{a-b}{2}\right|^p \leq \frac{1}{2}\left(|a|^p + |b|^p\right). \tag{1.53}$$

Proof. Inequality (1.53) is obvious if either $a = 0$ or $b = 0$. To proceed with the proof in the remaining case, we distinguish the cases $ab > 0$ and $ab < 0$.

If a and b have different signs, there is no loss of generality by considering $a > 0$ and writing $b = -c$, $c > 0$. Inequality (1.53) is then, in this case, nothing but the conclusion of Lemma 1.4.27. If $ab > 0$, on the other hand and with no loss of generality we assume $a > b$, then inequality (1.53) is Lemma 1.4.29. ☐

Theorem 1.4.32. *([7]) Denote the space of all real sequences (a_n) by V and let $(p_j) \subset \mathbb{R}$ be a sequence with $1 < \inf_{j\geq 1} p_j$. Then, the modular ρ_{p_n} on V defined as*

$$\rho_{p_n} : V \longrightarrow [0,\infty]$$
$$\mathbf{a} = (a_n) \longrightarrow \rho_{p_n}(\mathbf{a}) = \sum_{j=1}^{\infty} \frac{|a_j|^{p_j}}{p_j}$$

satisfies the $UCC2$ condition.

Proof. We fix the sequence (p_n) and define ρ_{p_n} as in Proposition 1.24, Section 1.4. For a subset $A \subseteq \mathbb{N}$ we write

$$\rho^A_{p_n}((a_n)) = \sum_{n\in A} \frac{|a_n|^{p_n}}{p_n}.$$

Let $r > 0$, $\epsilon > 0$ and consider $\mathbf{u} = (u_n) \in \ell^{p_n}, \mathbf{v} = (v_n) \in \ell^{p_n}$ such that

$$\rho_{p_n}(\mathbf{u}) \leq r\,,\ \rho_{p_n}(\mathbf{v}) \leq r\,,\ \rho_{p_n}\left(\frac{\mathbf{u}-\mathbf{v}}{2}\right) \geq \epsilon r.$$

Notice that the convexity of ρ_{p_n} forces $\epsilon \leq 1$, for

$$r\epsilon \leq \rho_{p_n}\left(\frac{\mathbf{u}-\mathbf{v}}{2}\right) \leq r.$$

Let $A := \{n \in \mathbb{N} : p_n \geq 2\}$. By assumptions on $\mathbf{u}$ and $\mathbf{v}$ and on account of the definition of A one must necessarily have either

$$\rho^A_{p_n}\left(\frac{\mathbf{u}-\mathbf{v}}{2}\right) \geq \frac{r\epsilon}{2} \tag{1.54}$$

or

$$\rho^{\mathbb{N}\setminus A}_{p_n}\left(\frac{\mathbf{u}-\mathbf{v}}{2}\right) \geq \frac{r\epsilon}{2}. \tag{1.55}$$

If (1.54) holds, Clarkson's inequality (1.52) yields

$$\rho^A_{p_n}\left(\frac{\mathbf{u}-\mathbf{v}}{2}\right) + \rho^A_{p_n}\left(\frac{\mathbf{u}+\mathbf{v}}{2}\right) \leq \frac{1}{2}\left(\rho^A_{p_n}(\mathbf{u}) + \rho^A_{p_n}(\mathbf{v})\right);$$

it is clear then that

$$\rho^A_{p_n}\left(\frac{\mathbf{u}+\mathbf{v}}{2}\right) \leq \frac{1}{2}\left(\rho^A_{p_n}(\mathbf{u}) + \rho^A_{p_n}(\mathbf{v})\right) - \frac{r\epsilon}{2}.$$

Thus,

$$\begin{aligned}
\rho_{p_n}\left(\frac{\mathbf{u}+\mathbf{v}}{2}\right) &= \rho^A_{p_n}\left(\frac{\mathbf{u}+\mathbf{v}}{2}\right) + \rho^{\mathbb{N}\setminus A}_{p_n}\left(\frac{\mathbf{u}+\mathbf{v}}{2}\right) \\
&\leq \frac{1}{2}\left(\rho^A_{p_n}(\mathbf{u}) + \rho^A_{p_n}(\mathbf{v})\right) - \frac{r\epsilon}{2} + \frac{1}{2}\left(\rho^{\mathbb{N}\setminus A}_{p_n}(\mathbf{u}) + \rho^{\mathbb{N}\setminus A}_{p_n}(\mathbf{v})\right) \\
&= \frac{1}{2}\left(\rho_{p_n}(\mathbf{u}) + \rho_{p_n}(\mathbf{v})\right) - \frac{r\epsilon}{2} \\
&\leq r\left(1 - \frac{\epsilon}{2}\right).
\end{aligned}$$

In case (1.55) holds, we define

$$B := \left\{n \in \mathbb{N} \setminus A : |u_n - v_n| \leq \frac{\epsilon}{4}(|u_n| + |v_n|)\right\}.$$

It follows that

$$\begin{aligned}
\rho^B_{p_n}\left(\frac{\mathbf{u}-\mathbf{v}}{2}\right) &\leq \frac{\epsilon}{8}\left(\rho^B_{p_n}(\mathbf{u}) + \rho^B_{p_n}(\mathbf{v})\right) \\
&\leq \frac{\epsilon}{8}\left(\rho_{p_n}(\mathbf{u}) + \rho_{p_n}(\mathbf{v})\right) \leq \frac{r\epsilon}{8}.
\end{aligned}$$

The validity of (1.55) implies in particular that

$$\rho_{p_n}^{\mathbb{N}\setminus(A\cup B)}\left(\frac{u-v}{2}\right) = \rho_{p_n}^{\mathbb{N}\setminus A}\left(\frac{u-v}{2}\right) - \rho_{p_n}^{B}\left(\frac{u-v}{2}\right) \geq \frac{r\epsilon}{2} - \frac{r\epsilon}{4} = \frac{r\epsilon}{4}.$$

Next, as a direct consequence of inequality (1.53), if $n \in \mathbb{N}\setminus(A\cup B)$, one has

$$\begin{aligned}
&\left|\frac{u_n+v_n}{2}\right|^{p_n} + \left(\inf_{\mathbb{N}} p_n - 1\right)\frac{\epsilon}{8}\left|\frac{u_n-v_n}{2}\right|^{p_n} \\
&\leq \left|\frac{u_n+v_n}{2}\right|^{p_n} + \frac{p_n(p_n-1)}{2}\left(\frac{\epsilon}{4}\right)^{2-p_n}\left|\frac{u_n-v_n}{2}\right|^{p_n} \\
&\leq \left|\frac{u_n+v_n}{2}\right|^{p_n} + \frac{p_n(p_n-1)}{2}\left|\frac{u_n-v_n}{|u_n|+|v_n|}\right|^{2-p_n}\left|\frac{u_n-v_n}{2}\right|^{p_n} \\
&\leq \frac{1}{2}(|u_n|^{p_n} + |v_n|^{p_n}).
\end{aligned}$$

Consequently, integrating the preceding inequality over $\mathbb{N}\setminus(A\cup B)$ it is plain that

$$\begin{aligned}
&\rho_{p_n}^{\mathbb{N}\setminus(A\cup B)}\left(\frac{\mathbf{u}+\mathbf{v}}{2}\right) + \left(\inf_{\mathbb{N}} p_n - 1\right)\frac{\epsilon}{8}\rho_{p_n}^{\mathbb{N}\setminus(A\cup B)}\left(\frac{\mathbf{u}-\mathbf{v}}{2}\right) \\
&\leq \frac{1}{2}\left(\rho_{p_n}^{\mathbb{N}\setminus(A\cup B)}(\mathbf{u}) + \rho_{p_n}^{\mathbb{N}\setminus(A\cup B)}(\mathbf{v})\right)
\end{aligned}$$

and thus

$$\rho_{p_n}^{\mathbb{N}\setminus(A\cup B)}\left(\frac{\mathbf{u}+\mathbf{v}}{2}\right) \leq \frac{1}{2}\left(\rho_{p_n}^{\mathbb{N}\setminus(A\cup B)}(\mathbf{u}) + \rho_{p_n}^{\mathbb{N}\setminus(A\cup B)}(\mathbf{v})\right) - \left(\inf_{\mathbb{N}} p_n - 1\right)\frac{\epsilon^2}{32}r.$$

In fact, then,

$$\begin{aligned}
\rho_{p_n}\left(\frac{\mathbf{u}+\mathbf{v}}{2}\right) &= \rho_{p_n}^{(A\cup B)}\left(\frac{\mathbf{u}+\mathbf{v}}{2}\right) + \rho_{p_n}^{\mathbb{N}\setminus(A\cup B)}\left(\frac{\mathbf{u}+\mathbf{v}}{2}\right) \\
&\leq \frac{1}{2}\left(\rho_{p_n}^{(A\cup B)}(\mathbf{u}) + \rho_{p_n}^{(A\cup B)}(\mathbf{v})\right) \\
&\quad + \frac{1}{2}\left(\rho_{p_n}^{\mathbb{N}\setminus(A\cup B)}(\mathbf{u}) + \rho_{p_n}^{\mathbb{N}\setminus(A\cup B)}(\mathbf{v})\right) - \left(\inf_{\mathbb{N}} p_n - 1\right)\frac{\epsilon^2}{32}r \\
&\leq r - \left(\inf_{\mathbb{N}} p_n - 1\right)\frac{\epsilon^2}{32}r = r\left(1 - \left(\inf_{\mathbb{N}} p_n - 1\right)\frac{\epsilon^2}{32}\right).
\end{aligned}$$

In all, for any $r > 0$, $\epsilon > 0$ and arbitrary sequences $\mathbf{u}$ and $\mathbf{v}$ in $D(r,\epsilon)$ as specified in Definition 1.3.8, it holds that

$$1 - \frac{1}{r}\rho_p\left(\frac{\mathbf{u}+\mathbf{v}}{2}\right) \geq \min\left\{\frac{\epsilon}{2}, \left(\inf_{\mathbb{N}} p_n - 1\right)\frac{\epsilon^2}{32}\right\} > 0,$$

and ℓ^{p_n} is $UUC2$. □

Theorem 1.4.32 settles the question of the property $UCC2$ in the context of variable exponent sequence spaces. The next natural order of business is to consider the $UCED$ property for such spaces. The next theorem points toward this direction. We refer the reader to Definition 1.3.11.

Theorem 1.4.33. *Consider the vector space ℓ^{p_n}. The following statements are equivalent:*

(i) The set $\{n \in \mathbb{N} :\ p_n = 1\}$ has at most one element (property (AO)).

(ii) The modular ρ_{p_n} is (UUCED).

(iii) The modular ρ_{p_n} is (SC).

Proof. Note that (ii) easily implies (iii).

We next prove that (iii) implies (i). Assume that ρ_{p_n} is (SC) and (a_n) fails the condition (i). There must then exist $i, j \in \mathbb{N}$ such that $i \neq j$ and $p_i = p(j) = 1$.
Set $\mathbf{x} = (x_n)$, where $x_n = 0$ if $n \neq i$ and $x_i = 1$, and $\mathbf{y} = (y_n)$, where $y_n = 0$ if $n \neq j$ and $y_j = 1$. Then

$$\rho_{p_n}(\mathbf{x}) = \rho_{p_n}(\mathbf{y}) = \rho_{p_n}\left(\frac{\mathbf{x}+\mathbf{y}}{2}\right) = 1$$

and $\mathbf{x} \neq \mathbf{y}$. This will contradict our assumption that ρ_{p_n} is (SC). In order to complete the proof of Theorem 1.4.33, we only need to prove that (i) implies (ii). Assume that condition (i) holds; we will demonstrate that ρ is $(UUCED)$. To this end, let $\mathbf{z}_1 = (z_n^1), \mathbf{z}_2 = (z_n^2) \in \ell^{p_n}$ such that $\mathbf{z}_1 \neq \mathbf{z}_2$. Let $R > 0$ and $\mathbf{x} \in \ell^{p_n}$ such that

$$\rho_{p_n}(\mathbf{x} - \mathbf{z}_1) \leq R \ \ and \ \ \rho_{p_n}(\mathbf{x} - \mathbf{z}_2) \leq R.$$

Set $K = \{n \in \mathbb{N};\ z_n^1 \neq z_n^2\}$. It is immediate from the assumption $\mathbf{z}_1 \neq \mathbf{z}_2$ that K is not empty. We have $K = K_1 \cup K_2 \cup K_3$, where $K_1 = \{n \in K;\ p_n \geq 2\}$, $K_2 = \{n \in K;\ 1 < p_n < 2\}$ and $K_3 = \{n \in K;\ p_n = 1\}$. Since K is not empty, at least one of the subsets K_1, K_2 or K_3 is not empty. The proof requires an analysis of the three different scenarios.
<u>**First case:**</u> To start with, we assume that $K_1 \neq \emptyset$. In the light of Theorem 1.4.30 it is clear that

$$\left|x_i - \frac{z_i^1 + z_i^2}{2}\right|^{p_i} + \left|\frac{z_i^1 - z_i^2}{2}\right|^{p_i} \leq \frac{1}{2}\Big(|x_i - z_i^1|^{p_i} + |x_i - z_i^2|^{p_i}\Big),$$

for any $i \in K_1$. In addition:

$$\left|x_n - \frac{z_n^1 + z_n^2}{2}\right|^{p_n} \leq \frac{1}{2}\Big(|x_n - z_n^1|^{p_n} + |x_n - z_n^2|^{p_n}\Big),$$

for any $n \notin K_1$, which imply

$$\rho_{p_n}\left(x - \frac{\mathbf{z}_1 + \mathbf{z}_2}{2}\right) + \sum_{i \in K_1} \frac{1}{p_i} \left|\frac{z_i^1 - z_i^2}{2}\right|^{p_i} \leq \frac{\rho_{p_n}(\mathbf{x} - \mathbf{z}_1) + \rho_{p_n}(\mathbf{x} - \mathbf{z}_2)}{2} \leq R.$$

In this case, we take

$$\delta(\mathbf{z}_1, \mathbf{z}_2, R) = \frac{1}{R} \sum_{i \in K_1} \frac{1}{p_i} \left|\frac{z_i^1 - z_i^2}{2}\right|^{p_i} > 0.$$

__Second case:__ If K_2 is not empty, Theorem 1.4.30 yields

$$\left|x_i - \frac{z_i^1 + z_i^2}{2}\right|^{p_i} + A_i \leq \frac{1}{2}\Big(|x_i - z_i^1|^{p_i} + |x_i - z_i^2|^{p_i}\Big),$$

where

$$\begin{aligned} A_i = & \quad \frac{p_i(p_i-1)}{2} \left|\frac{z_i^1 - z_i^2}{|x_i - z_i^1| + |x_i - z_i^2|}\right|^{2-p_i} \left|\frac{z_i^1 - z_i^2}{2}\right|^{p_i} \\ = & \quad \frac{p_i(p_i-1)}{2^{1+p_i}} \frac{1}{\Big(|x_i - z_i^1| + |x_i - z_i^2|\Big)^{2-p_i}} |z_i^1 - z_i^2|^2, \end{aligned} \tag{1.56}$$

for any $i \in K_2$. On the other hand, the inequalities $\rho(\mathbf{x} - \mathbf{z}_1) \leq R$ and $\rho(\mathbf{x} - \mathbf{z}_2) \leq R$ combined yield:

$$\frac{1}{p_i}|x_i - z_i^1|^{p_i} \leq R \text{ and } \frac{1}{p_i}|x_i - z_i^2|^{p_i} \leq R,$$

as a consequence of which it is readily obtained that:

$$|x_i - z_i^1| + |x_i - z_i^2| \leq 2\Big(p_i\ R\Big)^{1/p_i} \leq 2\ \Big(2\ R\Big)^{1/p_i} = 2^{1+1/p_i}\ R^{1/p_i} \leq 2^2\ R^{1/p_i},$$

for any $i \in K_2$. Hence,

$$\Big(|x_i - z_i^1| + |x_i - z_i^2|\Big)^{2-p_i} \leq \Big(2^2\ R^{1/p_i}\Big)^{2-p_i} \leq 2^2\ R^{(2-p_i)/p_i};$$

it follows from here that:

$$\frac{1}{2^{1+p_i}} \frac{1}{\Big(|x_i - z_i^1| + |x_i - z_i^2|\Big)^{2-p_i}} \geq \frac{1}{2^{1+p_i}} \frac{1}{2^2\ R^{(2-p_i)/p_i}} \geq \frac{1}{2^5\ R^{(2-p_i)/p_i}}.$$

In all,

$$A_i \geq \frac{p_i(p_i-1)}{2^5\ R^{(2-p_i)/p_i}} |z_i^1 - z_i^2|^2,$$

for any $i \in K_2$. Therefore, we have:

$$\sum_{i\in K_2} \frac{(p_i-1)}{2^5\ R^{(2-p_i)/p_i}}\ |z_i^1 - z_i^2|^2 \le \sum_{i\in K_2} A_i.$$

It follows as a direct consequence of Theorem 1.53 that, if A_i is as defined in (1.56), then

$$\begin{aligned} &\rho_{p_n}^{K_2}\left(\mathbf{x} - \frac{\mathbf{z}_1+\mathbf{z}_2}{2}\right) + \sum_{i\in K_2} \frac{(p_i-1)}{2^5\ R^{(2-p_i)/p_i}}\ |z_i^1 - z_i^2|^2 \\ &\le \rho_{p_n}^{K_2}\left(\mathbf{x} - \frac{\mathbf{z}_1+\mathbf{z}_2}{2}\right) + \sum_{i\in K_2} A_i \\ &\le \frac{1}{2}\sum_{n\in K_2} (|x_{p_n} - z_{p_n}^1| + |x_{p_n} - z_{p_n}^2|). \end{aligned} \tag{1.57}$$

In conjunction with Theorem 1.4.30 and the equality

$$\rho_{p_n}\left(\mathbf{x} - \frac{\mathbf{z}_1+\mathbf{z}_2}{2}\right) = \rho_{p_n}^{K_1}\left(\mathbf{x} - \frac{\mathbf{z}_1+\mathbf{z}_2}{2}\right) + \rho_{p_n}^{K_2}\left(\mathbf{x} - \frac{\mathbf{z}_1+\mathbf{z}_2}{2}\right),$$

inequality (1.57) yields that

$$\begin{aligned} &\rho_{p_n}\left(\mathbf{x} - \frac{\mathbf{z}_1+\mathbf{z}_2}{2}\right) + \sum_{i\in K_2} \frac{(p_i-1)}{2^5\ R^{(2-p_i)/p_i}}\ |z_i^1 - z_i^2|^2 \\ &\le \frac{\rho_{p_n}(\mathbf{x}-\mathbf{z}_1) + \rho_{p_n}(\mathbf{x}-\mathbf{z}_2)}{2} \le R. \end{aligned}$$

One can therefore take

$$\delta(\mathbf{z}_1,\mathbf{z}_2,R) = \sum_{i\in K_2} \frac{(p_i-1)}{2^5\ R^{1+(2-p_i)/p_i}}\ |z_i^1 - z_i^2|^2 = \sum_{i\in K_2} \frac{(p_i-1)}{2^5\ R^{2/p_i}}\ |z_i^1 - z_i^2|^2 > 0.$$

Third case: We finally tackle the case in which $K_1 = K_2 = \emptyset$ and $K_3 \neq \emptyset$. Since condition (i) is fulfilled it follows that K_3 is a singleton; say $K_3 = \{i\}$. For the sake of notational simplicity let us write $K_3 = K$ and $\mathbb{N}\setminus K = K^c$. Our assumptions on x, z_1 and z_2 imply the identities

$$\rho_{p_n}^{K^c}(\mathbf{x}-\mathbf{z}_1) = \rho_{p_n}^{K^c}(\mathbf{x}-\mathbf{z}_2) = \rho_{p_n}^{K^c}(\mathbf{x}-(\mathbf{z}_1+\mathbf{z}_2)/2) := R(x); \tag{1.58}$$

moreover it is clear that for $j = 1, 2$

$$\rho_{p_n}^{K}(\mathbf{x}-\mathbf{z}_j) = |x_i - z_i^j| \le R - R(x).$$

On account of Theorem 1.4.30, we have

$$\left|x_i - \frac{z_i^1+z_i^2}{2}\right|^2 + \left|\frac{z_i^1-z_i^2}{2}\right|^2 = \frac{|x_i - z_i^1|^2 + |x_i - z_i^2|^2}{2} \le (R - R(x))^2$$

and from here one concludes that

$$\left| x_i - \frac{z_i^1 + z_i^2}{2} \right|^2 \leq (R - R(x))^2 \left(1 - \frac{1}{4(R - R(x))^2} |z_i^1 - z_i^2|^2 \right).$$

In other words, by the very definition of K:

$$\rho_{p_n}^K \left(\mathbf{x} - \frac{\mathbf{z}_1 + \mathbf{z}_2}{2} \right) \leq (R - R(x)) \left(1 - \frac{1}{4(R - R(x))^2} |z_i^1 - z_i^2|^2 \right)^{1/2}.$$

Hence, for $R(x)$ as defined in (1.58) it is easy to derive the inequality:

$$\begin{aligned} \rho_{p_n} \left(\mathbf{x} - \frac{\mathbf{z}_1 + \mathbf{z}_2}{2} \right) &\leq (R - R(x)) \left(1 - \frac{1}{4(R - R(x))^2} |z_i^1 - z_i^2|^2 \right)^{1/2} + R(x) \\ &= R \left(1 - \frac{1}{4(R - R(x))^2} |z_i^1 - z_i^2|^2 \right)^{1/2} \\ &+ R(x) \left(1 - \left(1 - \frac{1}{4(R - R(x))^2} |z_i^1 - z_i^2|^2 \right)^{1/2} \right) \\ &= R(1 - B) + R(x) B, \end{aligned} \tag{1.59}$$

where we have introduced the positive quantity

$$B = 1 - \sqrt{1 - \frac{1}{4(R - R(x))^2} |z_i^1 - z_i^2|^2}.$$

As it transpires from (1.59), we have

$$\rho_{p_n} \left(\mathbf{x} - \frac{\mathbf{z}_1 + \mathbf{z}_2}{2} \right) \leq R \left(1 - \frac{R - R(x)}{R} B \right).$$

On the other hand, note that

$$|z_i^1 - z_i^2| \leq |z_i^1 - x| + |x - z_i^2| \leq 2(R - R(x)) \leq 2R;$$

we also underline the elementary fact that the function

$$s \longrightarrow 1 - \left(1 - \frac{1}{4s^2} |z_i^1 - z_i^2|^2 \right)^{1/2}$$

is decreasing. It follows from the preceding considerations that

$$1 - \sqrt{1 - \frac{1}{4(R - R(x))^2} |z_i^1 - z_i^2|^2} \geq 1 - \sqrt{1 - \frac{1}{4R^2} |z_i^1 - z_i^2|^2},$$

which implies

$$\frac{R-R(x)}{R}B = \frac{R-R(x)}{R}\left(1-\sqrt{1-\frac{1}{4(R-R(x))^2}|z_i^1-z_i^2|^2}\right)$$
$$\geq \frac{|z_i^1-z_i^2|}{2R}\left(1-\sqrt{1-\frac{1}{4R^2}|z_i^1-z_i^2|^2}\right).$$

Setting $\delta(\mathbf{z}_1,\mathbf{z}_2,R) = \frac{|z_i^1-z_i^2|}{2R}\left(1-\sqrt{1-\frac{1}{4R^2}|z_i^1-z_i^2|^2}\right)$, it follows that

$$\rho_{p_n}\left(\mathbf{x}-\frac{\mathbf{z}_1+\mathbf{z}_2}{2}\right) \leq R(1-\delta(\mathbf{z}_1,\mathbf{z}_2,R)).$$

Note that $\delta(\mathbf{z}_1,\mathbf{z}_2,R) > 0$. In all three instances we have $\delta(\mathbf{z}_1,\mathbf{z}_2,R) \geq \delta(\mathbf{z}_1,\mathbf{z}_2,\bar{R})$, whenever $R \leq \bar{R}$, i.e., the modular ρ is $(UUCED)$ as claimed. □

1.4.4 Applications: Fixed point theorems on ℓ^{p_n} spaces

A remarkable fact about the above discussion is that the $UUC2$ property holds even if $\sup_{n\in\mathbb{N}} p_n = \infty$, that is, in the absence of the Δ_2 condition. This observation makes the $UUC2$ condition a valuable tool for dealing with certain applications that have been hitherto heavily Δ_2-dependent. We will present a few applications to problems arising in the theory of fixed points. For an exhaustive treatment of the interplay between the theory of modular spaces and the theory of fixed point of mappings we refer the reader to the monograph [55].

If the sequence of exponents (p_n) satisfies $1 < p_- = \inf_n p_n$ and $\sup_n p_n = p_+ < \infty$, there is no substantial difference between dealing with the norm and dealing with the modular in ℓ^{p_n}. However in the general case, Theorem 1.3.8 and the example following it show that modular convergence and norm convergence part ways. It is sometimes more convenient to deal with modular convergence. With this aim in mind we introduce some terminology before proceeding any further: A subset $W \in \ell^{p_n}$ will be called **ρ_{p_n}-bounded** if there exists a constant $C \geq 0$ such that the inequality

$$\rho_{p_n}(\mathbf{u}) \leq C$$

holds for any $\mathbf{u} \in W$. W is said to be **ρ_{p_n}-closed** if whenever

$$\mathbf{u}_n \overset{\rho_{p_n}}{\to} \mathbf{u}$$

one has $\mathbf{u} \in W$. The next observation is of particular importance in the sequel: Consider sequences $\mathbf{x}$, $\mathbf{y}$ and let $(\mathbf{y^n})$ be ρ_{p_n}-convergent to $\mathbf{y}$. Fatou's Lemma yields the following inequality

$$\rho_{p_n}(\mathbf{x}-\mathbf{y}) \leq \liminf_{n\to\infty} \rho_{p_n}(\mathbf{x}-\mathbf{y^n}).$$

For obvious reasons, the above is known as the Fatou property of the modular ρ_{p_n}.

Theorem 1.4.34. *([7]) Let $p : \mathbb{N} \longrightarrow (1, \infty)$; assume $\inf_{n \in \mathbb{N}} p_n > 1$. Let $W \subset \ell^{p_n}$ be convex and ρ_{p_n}-closed and $u \in \ell^{p_n}$ satisfy*

$$d_{\rho_{p_n}}(\mathbf{u}, W) = \inf \{\rho_{p_n}(\mathbf{u} - \mathbf{v}) : \mathbf{v} \in W\} < \infty.$$

Then there exists a unique $v_0 \in W$ for which

$$d_{\rho_{p_n}}(\mathbf{u}, W) = \rho_{p_n}(\mathbf{u} - \mathbf{v}_0).$$

Proof. One can clearly assume that $\mathbf{u} \notin W$, otherwise there is nothing to prove. Under this assumption, one must have $d(\mathbf{u}, W) > 0$, due to the ρ_{p_n}-closedness of W. Let $(\mathbf{v}_n) \subseteq W$ be such that

$$\rho_{p_n}(\mathbf{u} - \mathbf{v}_n) < d(\mathbf{u}, W) \left(1 + \frac{1}{n}\right).$$

Then the sequence $\left(\frac{\mathbf{v}_n}{2}\right)$ must be ρ_{p_n}-Cauchy, i.e. it must necessarily hold that $\rho_{p_n}(2^{-1}(\mathbf{v}_n - \mathbf{v}_m)) \to 0$ as $m, n \to \infty$. The latter follows by contradiction. Indeed, if otherwise, there would exist $\delta > 0$ and strictly increasing subsequences $(n_k)_{k \geq 1}$ and $(m_k)_{k \geq 1}$ with $n_k > m_k$ for every k, such that

$$\rho_{p_n}\left(\frac{\mathbf{v}_{n_k} - \mathbf{v}_{m_k}}{2}\right) \geq \delta \tag{1.60}$$

for each $k \in \mathbb{N}$. Since $n_k > m_k$, it holds that

$$\max\{\rho_{p_n}(\mathbf{u} - \mathbf{v}_{n_k}), \rho_{p_n}(\mathbf{u} - \mathbf{v}_{m_k})\} \leq d(\mathbf{u}, W) \left(1 + \frac{1}{m_k}\right) := r_k.$$

Together with (1.60) and in by virtue of Definitions (1.3.8) and (1.3.9) and Theorem 1.4.32, there exists $\eta > 0$

$$1 - \frac{1}{r}\rho_{p_n}\left(\mathbf{u} - \frac{(\mathbf{v}_{m_k} + \mathbf{v}_{n_k})}{2}\right) \geq \eta > 0,$$

for any $k \in \mathbb{N}$. Though not mentioned explicitly there, it is apparent from the proof of Theorem 1.4.32 that η is independent of r_k. Since W is convex by assumption, the last inequality above yields

$$\begin{aligned} d(\mathbf{u}, W) &\leq \rho_{p_n}\left(\mathbf{u} - \frac{(\mathbf{v}_{m_k} + \mathbf{v}_{n_k})}{2}\right) \leq r(1 - \eta) \\ &= d(\mathbf{u}, W) \left(1 + \frac{1}{m_k}\right)(1 - \eta). \end{aligned}$$

Letting k tend to ∞ one clearly reaches a contradiction: in conclusion, the sequence $\left(\frac{\mathbf{v}_n}{2}\right)$ is ρ_{p_n}-Cauchy, as claimed. Since ℓ^{p_n} is ρ_{p_n} complete, we define $\mathbf{v}$ as

$$\lim_{n\to\infty} \rho_{p_n}(\mathbf{v} - 2^{-1}\mathbf{v}_n) = 0.$$

Notice that

$$\rho_{p_n}\left(2\mathbf{v} - \left(\mathbf{v} + \frac{\mathbf{v}_k}{2}\right)\right) \to 0 \text{ as } k \to \infty;$$

and that for fixed $k \in \mathbb{N}$, $\left(\frac{\mathbf{v}_k+\mathbf{v}_n}{2}\right)_n$ converges to $\frac{\mathbf{v}_k}{2} + \mathbf{v}$. The convexity and ρ_{p_n}-closedness of W implies then that $\frac{\mathbf{v}_k}{2} + \mathbf{v} \in W$ for each k and invoking again the ρ_{p_n}-closedness of W we conclude that $2\mathbf{v} \in W$. On account of the Fatou's property for the modular ρ_{p_n} one concludes that

$$\begin{aligned}
d(\mathbf{u}, W) \le \rho_{p_n}(\mathbf{u} - 2\mathbf{v}) &\le \liminf_{k\to\infty} \rho_{p_n}\left(\mathbf{u} - \left(\mathbf{v} + \frac{\mathbf{v}_k}{2}\right)\right) \\
&\le \liminf_{n\to\infty} \liminf_{k\to\infty} \rho_{p_n}\left(\mathbf{u} - \left(\frac{\mathbf{v}_n + \mathbf{v}_k}{2}\right)\right) \\
&\le \liminf_{n\to\infty} \liminf_{k\to\infty} \frac{1}{2}\left(\rho_{p_n}(\mathbf{u} - \mathbf{v}_n) + \rho_{p_n}(\mathbf{u} - \mathbf{v}_k)\right) \\
&= d(\mathbf{u}, W).
\end{aligned}$$

It follows that

$$d(\mathbf{u}, W) = \rho_{p_n}(\mathbf{u} - 2\mathbf{v}).$$

If $\mathbf{w} \in W$ and $d(\mathbf{u}, W) = \rho_{p_n}(\mathbf{u} - \mathbf{w})$ it is therfore concluded that

$$d \le \rho_{p_n}\left(\mathbf{u} - \frac{2\mathbf{v} + \mathbf{w}}{2}\right) \le \frac{1}{2}\left(\rho_{p_n}(\mathbf{u} - 2\mathbf{v}) + \rho_{p_n}(\mathbf{u} - \mathbf{w})\right) = d.$$

Lemma 1.3.17 guarantees that ρ_{p_n} is strictly convex and hence that $\mathbf{w} = 2\mathbf{v}$, which yields the uniqueness statement. □

Aiming at presenting further applications of the $UUC2$ property for ℓ^{p_n} we state and prove the following :

Theorem 1.4.35. *([7]) Let $(C_n)_n$ be a non-increasing sequence of ρ_{p_n}-closed, convex, nonempty subsets of ℓ^{p_n} with*

$$\inf_{n\in\mathbb{N}} p_n > 1.$$

Assume that there exists $\mathbf{v} \in \ell^{p_n}$ such that $\sup_{n\ge1} d(\mathbf{v}, C_n) < \infty$. Then

$$\bigcap_{n=1}^{\infty} C_n \ne \emptyset.$$

Proof. It suffices to assume that for some $n_0 \in \mathbb{N}$ it holds $\mathbf{v} \notin C_{n_0}$, for otherwise there would be nothing to prove. From the ρ_{p_n}-closedness of C_{n_0} it is easily derived that $d(\mathbf{v}, C_{n_0}) > 0$. Since the sequence $(C_n)_n$ is non-increasing by assumption, the inequalities

$$\infty > \sup_{j\geq 1} d(\mathbf{v}, C_j) \geq d(\mathbf{v}, C_n) = \inf_{\mathbf{u}\in C_n} d(\mathbf{v}, \mathbf{u}) \geq \inf_{\mathbf{u}\in C_{n-1}} d(\mathbf{v}, \mathbf{u}) = d(\mathbf{v}, C_{n-1})$$

are clear for any $n > 1$. Thus, the sequence $d(\mathbf{v}, C_n)$ is non-decreasing and bounded. Let $L = \lim_{n\to\infty} d(\mathbf{v}, C_n) < \infty$; clearly $L > 0$. For each $n \in \mathbb{N}$ let $\mathbf{u}_n \in C_n$ be chosen so that $\rho_{p_n}(\mathbf{v} - \mathbf{u}_n) = d(\mathbf{v}, C_n)$ as in Theorem 1.4.34. It follows that the sequence $\left(\frac{\mathbf{u}_n}{2}\right)_n$ is ρ_{p_n}-Cauchy in ℓ^{p_n} and hence it ρ_{p_n}-converges to, say, $\mathbf{u}/2 \in \ell^{p_n}$. Fix $k \in \mathbb{N}$. Then the sequence $\left(\frac{\mathbf{u}_n}{2}\right)_{n\geq k}$ is contained in C_k and ρ_{p_n}-converges to $\frac{\mathbf{u}}{2}$, which implies that $\frac{\mathbf{u}}{2} \in C_k$, since C_k is ρ_p-closed. In conclusion,

$$\frac{\mathbf{u}}{2} \in \bigcap_{n=1}^{\infty} C_n,$$

i.e., $\bigcap_{n=1}^{\infty} C_n \neq \emptyset$, as claimed. □

Theorem 1.4.36. *([7]) Let $\inf_{n\in\mathbb{N}} p_n > 1$ and $\emptyset \neq C \subset \ell^{p_n}$ be a ρ_{p_n}-closed, ρ_{p_n}-bounded, convex set and $(C_i)_{i\in I} \subset 2^C$ be a family of subsets of C having the finite intersection property, i.e., such that for every finite subset $\{i_1, ...i_k\} \subset I$ it holds that $\bigcap_{j=1}^{k} C_{i_j} \neq \emptyset$. Then*

$$\bigcap_{i\in I} C_i \neq \emptyset.$$

Proof. C is ρ_{p_n}-bounded, it is therefore immediate that for any $\mathbf{u} \in C$ and $i \in I$,

$$d(\mathbf{u}, C_i) = \inf_{\mathbf{v}\in C_i} \rho_{p_n}(\mathbf{u} - \mathbf{v}) \leq \sup_{\mathbf{v}\in C} \rho_{p_n}(\mathbf{u} - \mathbf{v}) < \infty.$$

For any finite subset $A \subset I$ let

$$d_A = d\left(\mathbf{u}, \bigcap_{j\in A} C_j\right).$$

If A and B be finite subsets of I, $A \subseteq B$, then $\bigcap_{j\in B} C_j \subseteq \bigcap_{j\in A} C_j$ and consequently,

$$d\left(\mathbf{u}, \bigcap_{j\in A} C_j\right) = \inf_{\mathbf{v}\in \bigcap_{j\in A} C_j} \rho_{p_n}(\mathbf{u}, \mathbf{v}) \leq \inf_{\mathbf{v}\in \bigcap_{j\in B} C_j} \rho_{p_n}(\mathbf{u}, \mathbf{v}),$$

i.e., $d_A \leq d_B$. Write

$$d_I = \sup\left\{ d\left(\mathbf{u}, \bigcap_{i\in J} C_i\right) \quad J \subset I: \text{ and } \bigcap_{i\in J} C_i \neq \emptyset \right\}.$$

Let (A_n) be the sequence defined by

$$d_I - \frac{1}{n} < d_{A_n} \leq d_I,$$

write $B_n = \bigcup_{k=1}^{n} A_k$ and $J = \bigcup_{n=1}^{\infty} B_n$. It is clear then that for each $n \in \mathbb{N}$, the set $\bigcap_{i\in B_n} C_i$ is ρ_p-closed, convex and non-empty and that the sequence $\left(\bigcap_{i\in B_n} C_i\right)$ is non-increasing. Hence, Theorem 1.4.35 applies and we have

$$S = \bigcap_{i\in J} C_i \neq \emptyset.$$

By definition, for each $n \in \mathbb{N}$, it holds that

$$\bigcap_{i\in J} C_i \subseteq \bigcap_{i\in A_n} C_i,$$

and it follows that for each n one has

$$d_I - \frac{1}{n} < d_{A_n} \leq d(\mathbf{u}, S) \leq d_I.$$

Thus $d(\mathbf{u}, S) = d_I$. On account of Theorem 1.4.34, there exists a unique $\mathbf{z} \in S$ which satisfies $\rho_{p_n}(\mathbf{u} - \mathbf{z}) = d_I$ and therefore, for any index $i_0 \in I$ one has

$$S \supseteq S \cap C_{i_0} = \bigcap_{i\in J\cup\{i_0\}} C_i \neq \emptyset;$$

it is seen immediately that $d_I \leq d(\mathbf{u}, S) \leq d(\mathbf{u}, S \cap C_{i_0}) \leq d_I$. In all,

$$d(\mathbf{u}, S) = d(\mathbf{u}, S \cap C_{i_0})$$

and by Theorem 1.4.34 there exists a unique $\mathbf{w} \in S \cap C_{i_0}$ for which

$$\rho_{p_n}(\mathbf{u} - \mathbf{w}) = d(\mathbf{u}, S \cap C_{i_0}) = d_I.$$

In particular, $\mathbf{w} \in S$, thus, invoking the uniqueness part of Theorem 1.4.34, one must necessarily have $\mathbf{w} = \mathbf{z}$. Since i_0 is arbitrary, it is concluded that $\mathbf{z} \in \bigcap_{i\in I} C_i$ and hence, the latter intersection is non-empty, as claimed. □

The following theorem is another consequence of the $UUC2$ property for ℓ^{p_n}.

Theorem 1.4.37. *([7]) Let* $\inf_{n\in\mathbb{N}} p_n > 1$ *and* $\emptyset \neq C \subset \ell^{p_n}$ *be a convex,* ρ_{p_n}*-closed,* ρ_{p_n} *bounded and assume that* C *is not a singleton (i.e.,* C *at least two distinct points). Then there exists* $\mathbf{u} \in C$ *for which*

$$\sup_{v\in C} \rho_{p_n}(\mathbf{u}-\mathbf{v}) < \operatorname{diam}(C),$$

where as usual $\operatorname{diam}(C) = \sup_{\mathbf{a},\mathbf{b}\in C} \rho_{p_n}(\mathbf{a}-\mathbf{b})$ *stands for the* ρ_{p_n}*-diameter of* C.

The property established in Theorem 1.4.37 is commonly referred to as the ρ_{p_n}**-normal structure property**. Theorem 1.4.37 can thus be rephrased as asserting that if $\inf_{n\in\mathbb{N}} p_n > 1$, then ℓ^{p_n} has ρ_{p_n}**-normal structure.**

Proof. The assumptions imply that $\delta(C) > 0$ and that there exist two distinct points $\mathbf{u} \in C$, $\mathbf{v} \in C$, $\mathbf{u} \neq \mathbf{v}$. For any $\mathbf{w} \in C$, invoking the $UUC2$ property it follows at once that, for δ as in the definition of $UUC2$, (1.3.8),

$$\begin{aligned}\rho_{p_n}\left(\frac{\mathbf{u}+\mathbf{v}}{2}-\mathbf{w}\right) &= \rho_{p_n}\left(\frac{\mathbf{u}-\mathbf{w}+\mathbf{v}-\mathbf{w}}{2}\right)\\ &\leq \operatorname{diam}(C)\left(1-\delta\left(\operatorname{diam}(C), \frac{\epsilon}{\operatorname{diam}(C)}\right)\right).\end{aligned}$$

The arbitrariness of $\mathbf{w}$ in concert with the convexity of C yields the claim. □

Theorem 1.4.38. *([7]) If* $\inf_{n\in\mathbb{N}} p_n > 1$, $\emptyset \neq C \subset \ell^{p_n}$ *is convex,* ρ_{p_n}*-closed and* ρ_{p_n}*-bounded, then any map*

$$T : C \longrightarrow C$$

for which the bound

$$\rho_{p_n}(T(\mathbf{u}) - T(\mathbf{v})) \leq \rho_{p_n}(\mathbf{u}-\mathbf{v})$$

holds for any $\mathbf{u} \in C, \mathbf{v} \in C$ *(i.e.,* T *is nonexpansive), has a fixed point, that is to say, there exists* $w \in C$ *such that*

$$T(\mathbf{w}) = \mathbf{w}.$$

Proof. It is obvious that the theorem is true if C is a singleton. Thus, it can be assumed that the cardinality of C is at least 2. Let

$$\mathcal{F} = \{\emptyset \neq K \subset C : K \text{ is } \rho_{p_n}\text{–closed and } T(K) \subseteq K\}.$$

Since $C \in \mathcal{F}$, $\mathcal{F} \neq \emptyset$. Moreover, $\mathcal{F}$ is partially ordered by the order relation

$$X \leq Y \iff Y \subseteq X.$$

If $\mathcal{G}$ is a totally ordered subfamily of $\mathcal{F}$, then $\mathcal{G}$ possesses the finite intersection property and on account of Theorem 1.4.36 it follows that

$$\bigcap_{X \in \mathcal{G}} X \neq \emptyset;$$

this clearly implies that $\bigcap_{X \in \mathcal{G}} X \in \mathcal{F}$ and therefore that $\bigcap_{X \in \mathcal{G}} X$ is an upper bound for $\mathcal{G}$.

Zorn's Lemma yields the existence of a maximal element $X_0 \in \mathcal{F}$. We set about to prove that X_0 contains exactly one point. Denote the intersection of all ρ_{p_n}-closed, convex subsets of C that contain $T(X_0)$ by $\overline{T(X_0)}$. In particular, since $X_0 \in \mathcal{F}$,

$$\overline{T(X_0)} \subseteq X_0.$$

On the other hand, $\overline{T(X_0)} \in \mathcal{F}$, for it is convex, ρ_{p_n}-closed and

$$T\left(\overline{T(X_0)}\right) \subseteq T(X_0) \subseteq \overline{T(X_0)}.$$

As a consequence of the maximality of X_0 with respect to the indicated inclusion, one has

$$\overline{T(X_0)} = X_0. \tag{1.61}$$

Theorem 1.4.37 yields the existence of an element $\mathbf{x}_0 \in X_0$ such that

$$r_0 = \sup_{u \in X_0} \rho_{p_n}(\mathbf{x}_0 - \mathbf{u}) < \operatorname{diam}(X_0). \tag{1.62}$$

Let $B_{\rho_{p_n}}(\mathbf{a}, s)$ denote the ρ_{p_n}-ball of radius s centered at $\mathbf{a}$; we remark the obvious fact that the convexity and the Fatou property of the modular ρ_{p_n} imply that $B_{\rho_{p_n}}(\mathbf{a}, s)$ is ρ_{p_n}-closed and convex. Set

$$M = \bigcap_{\mathbf{v} \in X_0} B_{\rho_{p_n}}(\mathbf{v}, r_0) \cap X_0 = \left\{ \mathbf{u} \in X_0 : \sup_{v \in X_0} \rho_{p_n}(\mathbf{u} - \mathbf{v}) \leq r_0 \right\};$$

then M is ρ_{p_n}-closed and convex and $M \subset X_0$. Moreover, if $\mathbf{x} \in M$, then for any $\mathbf{v} \in X_0$

$$\rho_{p_n}\left(T(\mathbf{x}) - T(\mathbf{v})\right) \leq \rho_{p_n}\left(\mathbf{x} - \mathbf{v}\right) \leq r_0.$$

In other words, if $\mathbf{v} \in X_0$, $\rho_{p_n}(T(\mathbf{v}) - T(\mathbf{x})) \leq r_0$, i.e., $T(X_0) \subseteq B_{\rho_{p_n}}(T(\mathbf{x}), r_0)$. By definition of $\overline{T(X_0)}$ it is plain that:

$$\overline{T(X_0)} \subseteq B_{\rho_{p_n}}(T(\mathbf{x}), r_0);$$

and from (1.61) it follows that

$$X_0 = \overline{T(X_0)} \subseteq B_{\rho_{p_n}}(T(\mathbf{x}), r_0);$$

that is, for any $v \in X_0$, $\rho_{p_n}(T(\mathbf{x})-v) \leq r_0$, i.e., $T(\mathbf{x}) \in B_{\rho_{p_n}}(v, r_0)$. It follows from the definition of M that

$$T(M) \subseteq M,$$

so that $M \in \mathcal{F}$. and since $M \subseteq X_0$ and X_0 is maximal, one has *a fortiori*:

$$X_0 = M.$$

By definition, then, if $\mathbf{w} \in X_0$,

$$\rho_{p_n}(\mathbf{w} - \mathbf{x}_0) \leq r_0;$$

this forces the inequality $\mathrm{diam}(X_0) \leq r_0$, which contradicts (1.62) unless $\mathrm{diam}(X_0) = 0$. Hence, $\mathrm{diam}(X_0) = 0$ and $X_0 = \{\mathbf{a}\}$ is a singleton. Since also $T(X_0) \subseteq X_0$, necessarily

$$T(\mathbf{a}) = \mathbf{a}$$

we conclude that T has a fixed point, as claimed. □

From the modular uniform convexity proved in [7] the authors were able to derive an intersection property similar to that of weak-compactness in Banach spaces. Since, in general, the new uniform convexity discussed in the previous section does not imply any good intersection property, we will need the following definition.

Definition 1.4.4. *Let C be a nonempty ρ_{p_n}-closed and ρ_{p_n}-bounded convex subset of ℓ^{p_n}. We will say that C satisfies the property (R), if for any decreasing sequence $\{C_n\}_{n\geq 1}$ of ρ_{p_n}-closed convex nonempty subsets of C, we have*

$$\bigcap_{n\geq 1} C_n \neq \emptyset.$$

The following technical proposition will be useful throughout.

Proposition 1.4.39. *Let C be a nonempty ρ_{p_n}-closed and ρ_{p_n}-bounded convex subset of ℓ^{p_n}.*

(i) Assume that C satisfies the property (R) and let K be a nonempty ρ_{p_n}-closed convex subset of C. Then K is ρ_{p_n}-proximinal in C, i.e. for any $\mathbf{x} \in C$, the set $P_{\rho_{p_n},K}(\mathbf{x}) = \{\mathbf{y} \in C;\ \rho_{p_n}(\mathbf{x}-\mathbf{y}) = \inf_{\mathbf{z}\in K} \rho_{p_n}(\mathbf{x}-\mathbf{z})\}$ is not empty. Moreover if ρ_{p_n} is (SC) (Definition 1.3.11), then K is a Čebyšev subset, i.e. $P_{\rho_{p_n},K}(\mathbf{x})$ is a singleton for any $\mathbf{x} \in C$.

(ii) Assume that C satisfies the property (R) and that ρ_{p_n} is (SC) (Definition 1.3.11). Then for any family $\{C_\alpha\}_{\alpha\in\Gamma}$ of ρ_{p_n}-closed convex nonempty subsets of C such that $\bigcap_{\alpha\in\Gamma_f} C_\alpha \neq \emptyset$ for any finite subset $\Gamma_f \subset \Gamma$, we have $\bigcap_{\alpha\in\Gamma} C_\alpha \neq \emptyset$.

(iii) Assume that C satisfies the property (R) and ρ_{p_n} is (UUCED). Let K be a nonempty ρ_{p_n}-closed convex subset of C. Then K has a unique ρ_{p_n}-Čebyšev center $\mathbf{x} \in K$, i.e.,

$$\sup\{\rho_{p_n}(\mathbf{x}-\mathbf{y});\ \mathbf{y} \in K\} = \inf_{z \in K} \Big(\sup\{\rho_{p_n}(\mathbf{z}-\mathbf{y});\ \mathbf{y} \in K\}\Big).$$

Proof. Assume that C satisfies the property (R). Let K be a nonempty ρ_{p_n}-closed convex subset of C. For $\mathbf{x} \in C$, we have

$$d_{\rho_{p_n}}(\mathbf{x}, K) = \inf\{\rho_{p_n}(\mathbf{x}-\mathbf{y});\ \mathbf{y} \in K\} < \infty,$$

since C is ρ_{p_n}-bounded. Moreover, from the equality

$$P_{\rho_{p_n}, K}(\mathbf{x}) = \bigcap_{n \geq 1} B_{\rho_{p_n}}\Big(\mathbf{x}, d_{\rho_{p_n}}(\mathbf{x}, K) + 1/n\Big) \cap K,$$

where $B_{\rho_{p_n}}(\mathbf{x}, r)$ is the ρ_{p_n}-ball centered at $\mathbf{x}$ with radius r, it follows that the property (R) implies that $P_{\rho_{p_n}, K}(\mathbf{x})$ is not empty. It is clear that if ρ_{p_n} is (SC), then $P_{\rho_{p_n}, K}(\mathbf{x})$ must be reduced to a single point, which completes the proof of (i).

In order to prove (ii), assume that C satisfies the property (R) and that ρ_{p_n} is (SC). Let $\{C_\alpha\}_{\alpha \in \Gamma}$ be a family of ρ_{p_n}-closed convex nonempty subsets of C such that $\bigcap_{\alpha \in \Gamma_f} C_\alpha$ is not empty, for any finite subset $\Gamma_f \subset \Gamma$. Let $\mathbf{x} \in C$. Then $\sup_{\alpha \in \Gamma} d_{\rho_{p_n}}(\mathbf{x}, C_\alpha) < \infty$ holds since C is ρ_{p_n}-bounded. For any subset $F \subset \Gamma$, set $d_F = d_{\rho_{p_n}}(\mathbf{x}, \bigcap_{\alpha \in F} C_\alpha)$. Note that if $F_1 \subset F_2 \subset \Gamma$ are finite subsets, then $d_{F_1} \leq d_{F_2}$. Set

$$d_\Gamma = \sup\Big\{d_{\rho_{p_n}}\Big(\mathbf{x}, \bigcap_{\alpha \in J} C_\alpha\Big),\ J \subset \Gamma \textit{ such that } \bigcap_{\alpha \in J} C_\alpha \neq \emptyset\Big\}.$$

For any $n \geq 1$, there exists a subset $F_n \subset \Gamma$ such that $d_\Gamma - 1/n < d_{F_n} \leq d_\Gamma$. Set $F_n^* = F_1 \cup \cdots \cup F_n$, for $n \geq 1$. Then $\Big\{\bigcap_{\alpha \in F_n^*} C_\alpha\Big\}$ is a decreasing sequence of nonempty ρ_{p_n}-closed convex subsets of C. The property (R) satisfied by C implies that for

$$J = \bigcup_{n \geq 1} F_n^* = \bigcup_{n \geq 1} F_n$$

it holds that

$$\bigcap_{\alpha \in J} C_\alpha \neq \emptyset.$$

Set $K = \bigcap_{\alpha \in J} C_\alpha$. Note that $d_{\rho_{p_n}}(x, K) = d_\Gamma$ because, for any $n \geq 1$,

$$d_\Gamma - 1/n < d_{F_n} \leq d_{\rho_{p_n}}(\mathbf{x}, K) \leq d_\Gamma.$$

Because of (i), there exists a unique $\mathbf{y} \in K$ for which

$$\rho_{p_n}(\mathbf{x} - \mathbf{y}) = d_{\rho_{p_n}}(\mathbf{x}, K) = d_\Gamma.$$

For fixed $\alpha_0 \in \Gamma$, one has:

$$K \cap C_{\alpha_0} = \bigcap_{\alpha \in J \cup \{\alpha_0\}} C_\alpha \neq \emptyset$$

because of the same argument based on the property (R). It follows then that

$$d_{\rho_{p_n}}(\mathbf{x}, K) \leq d_{\rho_{p_n}}(\mathbf{x}, K \cap C_{\alpha_0}) \leq d_\Gamma.$$

Hence $d_{\rho_{p_n}}(\mathbf{x}, K \cap C_{\alpha_0}) = d_{\rho_{p_n}}(\mathbf{x}, K) = d_\Gamma$ which implies $\mathbf{y} \in K \cap C_{\alpha_0}$. Therefore, we have $\mathbf{y} \in \bigcap_{\alpha \in \Gamma} C_\alpha$; in turn this proves that

$$\bigcap_{\alpha \in \Gamma} \{C_\alpha\} \neq \emptyset.$$

In order to prove (iii), assume that C satisfies the property (R) and ρ_{p_n} is $(UUCED)$. Let K be a nonempty ρ_{p_n}-closed convex subset of C. Set $r_{\rho_{p_n}}(\mathbf{x}, K) = \sup_{\mathbf{y} \in K} \rho_{p_n}(\mathbf{x} - \mathbf{y})$, for any $\mathbf{x} \in K$, and

$$R_{\rho_{p_n}}(K) = \inf \{r_{\rho_{p_n}}(\mathbf{x}, K); \mathbf{x} \in K\}.$$

All the above numbers are finite since C is ρ_{p_n}-bounded. Note that the set

$$K_n = \{\mathbf{x} \in K;\ r_{\rho_{p_n}}(\mathbf{x}, K) \leq R_{\rho_{p_n}}(K) + 1/n\},$$

is non-empty for any $n \geq 1$ and that

$$K_n = \bigcap_{\mathbf{y} \in K} B_{\rho_{p_n}}\Big(\mathbf{y}, R_{\rho_{p_n}}(K) + 1/n\Big) \cap K,\ n \geq 1,$$

which shows that (K_n) is a decreasing sequence of ρ_{p_n}-closed convex nonempty subsets of K. The property (R) implies that $\bigcap_{n \geq 1} K_n$ is nonempty. Clearly, any $\mathbf{x} \in \bigcap_{n \geq 1} K_n$ will satisfy $r_{\rho_{p_n}}(\mathbf{x}, K) = R_{\rho_{p_n}}(K)$. We next set out to show that $\bigcap_{n \geq 1} K_n$ is reduced to a single point. Assume that there exists $\mathbf{z} \in K$ such that $\mathbf{z} \neq \mathbf{x}$ and $r_{\rho_{p_n}}(\mathbf{z}, K) = R_{\rho_{p_n}}(K)$. Since

$$\rho_{p_n}(\mathbf{x} - \mathbf{z}) \leq r_{\rho_{p_n}}(\mathbf{x}, K) = R_{\rho_{p_n}}(K)$$

and $\mathbf{x} \neq \mathbf{z}$, we conclude that $R_{\rho_{p_n}}(K) > 0$. From the $(UUCED)$ property of ρ_{p_n} one derives the existence of $\delta = \delta(\mathbf{x}, \mathbf{z}, R_{\rho_{p_n}}(K)) > 0$ such that

$$\begin{cases} \rho_{p_n}(\mathbf{y} - \mathbf{x}) & \leq & R_{\rho_{p_n}}(K) \\ \rho_{p_n}(\mathbf{y} - \mathbf{z}) & \leq & R_{\rho_{p_n}}(K) \end{cases} \implies \rho_{p_n}\left(\mathbf{y} - \frac{\mathbf{x} + \mathbf{z}}{2}\right) \leq R_{\rho_{p_n}}(K)(1 - \delta),$$

for any $\mathbf{y} \in K$, which implies

$$R_{\rho_{p_n}}(K) \leq r_{\rho_{p_n}}\left(\frac{\mathbf{x}+\mathbf{z}}{2}, K\right) \leq R_{\rho_{p_n}}(K)(1-\delta).$$

This contradiction finishes the proof of (*iii*), which in turn completes the proof of Proposition 1.4.39. □

We aim at utilizing the above ideas for proving an analogue to Kirk's fixed point theorem in ℓ^{p_n}. The following definition will be needed to that effect:

Definition 1.4.5. *Consider a nonempty set $C \subset \ell^{p_n}$ and a mapping $T : C \to C$. T is said to be ρ_{p_n}-Lipschitzian if for some constant $K \geq 0$ one has*

$$\rho_{p_n}(T(\mathbf{x}) - T(\mathbf{y})) \leq K\ \rho_{p_n}(\mathbf{x}-\mathbf{y}), \quad \textit{for any } \mathbf{x}, \mathbf{y} \in C.$$

In particular T is called ρ_{p_n}-nonexpansive if $K = 1$ and $\mathbf{x} \in C$ is called a fixed point of T if $T(\mathbf{x}) = \mathbf{x}$. The collection of all fixed points of T will be denoted by $Fix(T)$.

The following modular version of Kirk's fixed point theorem is an improvement from the one discovered in [7] since it does not require that $\inf\limits_{n\in\mathbb{N}} p_n > 1$.

Theorem 1.4.40. *Assume that ρ_{p_n} satisfies the condition (AO) (Theorem 1.4.33). Let C be a nonempty ρ_{p_n}-closed convex ρ_{p_n}-bounded subset of ℓ^{p_n} which has the property (R). Let $T : C \to C$ be a ρ-nonexpansive mapping. Then $Fix(T)$ is a nonempty ρ_{p_n}-closed convex subset of C.*

Proof. Let $\emptyset \neq C \subset \ell^{p_n}$ be ρ_{p_n}-closed, convex and ρ-bounded and consider $T : C \to C$ to be a ρ_{p_n}-nonexpansive mapping. Assume that C is not a singleton: it is clear that no generality is lost with this assumption.
Consider the family

$$\mathcal{F} = \{K \subset C; K \neq \emptyset,\ K\ \rho_{p_n}\text{-closed, convex and } T(K) \subset K\}.$$

The family $\mathcal{F}$ is not empty since $C \in \mathcal{F}$. Our assumption on the sequence (p_n) implies that ρ_{p_n} is $(UUCED)$. Zorn's Lemma in concert with Proposition 1.4.39 yields the existence of a minimal element of $\mathcal{F}$, which we denote by K_0. We contend that K_0 consists of exactly one point. For otherwise we set $co(T(K_0))$ to be the intersection of all ρ_{p_n}-closed, convex subsets of C which contain $T(K_0)$. In particular, $co(T(K_0)) \subset K_0$ since $T(K_0) \subset K_0$ and we readily conclude that

$$T\Big(co(T(K_0))\Big) \subset T(K_0) \subset co(T(K_0)).$$

The minimality of K_0 yields $K_0 = co(T(K_0))$. Next, set $\mathbf{x} \in K_0$ to be the unique ρ_{p_n}-Čebyšev center of K_0, that is:

$$r_{\rho_{p_n}} x(\mathbf{x}, K_0) = \sup\{\rho_{p_n}(\mathbf{x}-\mathbf{y});\ \mathbf{y} \in K_0\} = \inf_{\mathbf{z}\in K_0} \Big(\sup\{\rho_{p_n}(\mathbf{z}-\mathbf{y});\ \mathbf{y} \in K_0\}\Big),$$

which exists according to *(iii)* of Proposition 1.4.39. Since T is ρ_{p_n}-nonexpansive and $K_0 \subset B_{\rho_{p_n}}(\mathbf{x}, r_{\rho_{p_n}}(\mathbf{x}, K_0))$, it is concluded that

$$T(K_0) \subset B_{\rho_{p_n}}(T(\mathbf{x}), r_{\rho_{p_n}}(\mathbf{x}, K_0)).$$

Since all ρ_{p_n}-balls are ρ_{p_n}-closed and convex, it is easily obtained that

$$K_0 = co(T(K_0)) \subset B_{\rho_{p_n}}(T(\mathbf{x}), r_{\rho_{p_n}}(\mathbf{x}, K_0)).$$

One concludes, therefore, that

$$\begin{aligned} r_{\rho_{p_n}}(T(\mathbf{x}), K_0) \le r_{\rho_{p_n}}(\mathbf{x}, K_0) &= \sup\{\rho_{p_n}(\mathbf{x}-\mathbf{y});\ \mathbf{y} \in K_0\} \\ &= \inf_{\mathbf{z}\in K_0} \Big(\sup\{\rho_{p_n}(\mathbf{z}-\mathbf{y});\ \mathbf{y} \in K_0\}\Big), \end{aligned}$$

i.e., $T(x)$ is also ρ_{p_n}-Čebyšev center of K_0. Therefore it must *a fortiori* hold that

$$T(\mathbf{x}) = \mathbf{x}. \tag{1.63}$$

Assertion (1.63) implies $K_0 = \{\mathbf{x}\}$; this is a contradiction to our assumption that K_0 contains more than one point.

Hence any minimal element of $\mathcal{F}$ is reduced to one point: Consequently, $Fix(T)$ is not empty. Since T is ρ_{p_n}-nonexpansive, $Fix(T)$ is ρ_{p_n}-closed. In fact, the set $Fix(T)$ is also convex. Indeed, set $\mathbf{z}_1, \mathbf{z}_2 \in Fix(T)$ with $\mathbf{z}_1 \neq \mathbf{z}_2$. Let $\alpha \in [0,1]$. Then:

$$\begin{aligned} \rho_{p_n}(\mathbf{z}_i - T(\alpha\ \mathbf{z}_1 + (1-\alpha)\mathbf{z}_2)) &= \rho_{p_n}(T(\mathbf{z}_i) - T(\alpha\ \mathbf{z}_1 + (1-\alpha)\mathbf{z}_2)) \\ &\le \rho_{p_n}(\mathbf{z}_i - (\alpha\ \mathbf{z}_1 + (1-\alpha)\mathbf{z}_2)), \end{aligned}$$

for $i = 1, 2$. Since ρ_{p_n} is $(UUCED)$, then ρ_{p_n} is (SC). It is therefore concluded that

$$T(\alpha\mathbf{z}_1 + (1-\alpha)\mathbf{z}_2) = \alpha\ \mathbf{z}_1 + (1-\alpha)\mathbf{z}_2,$$

i.e., $\alpha\ \mathbf{z}_1 + (1-\alpha)\mathbf{z}_2 \in Fix(T)$, which completes the proof of Theorem 1.4.40. □

As a consequence of the properties of the fixed point set proved in Theorem 1.4.40 we present the following common fixed point result.

Theorem 1.4.41. *Consider the vector space ℓ^{p_n}. Assume that ρ_{p_n} satisfies the condition (AO). Let C be a nonempty ρ-closed convex ρ_{p_n}-bounded subset of ℓ^{p_n} which satisfies the property (R). Let $\{T_\alpha\}_{\alpha\in\Gamma}$ be a commutative family of ρ_{p_n}-nonexpansive self-mappings defined on C. Then $\bigcap_{\alpha\in\Gamma} Fix(T_\alpha)$ is a nonempty ρ_{p_n}-closed convex subset of C.*

Proof. Let $S, T : C \to C$ be two commutative ρ_{p_n}-nonexpansive mappings. Theorem 1.4.40 implies that $Fix(T)$ is a nonempty ρ_{p_n}-closed convex subset of C. Since S and T commute, then we have $S(Fix(T)) \subset Fix(T)$. A further application of Theorem 1.4.40 yields that the restriction of S to $Fix(T)$ has a fixed point, i.e.,

$$Fix(T) \cap Fix(S) \neq \emptyset.$$

It follows from this argument that for any finite subset Γ_f of Γ, $\bigcap\limits_{\alpha \in \Gamma_f} Fix(T_\alpha)$ is a nonempty ρ_{p_n}-closed convex subset of C. Since C satisfies the property (R), we conclude that $\bigcap\limits_{\alpha \in \Gamma} Fix(T_\alpha) \neq \emptyset$. The fact that this intersection is ρ_{p_n}-closed and convex follows easily. □

1.4.5 Further remarks

The (modular) spaces ℓ^{p_n} were used by W. Orlicz in 1931 [85] to clarify some questions on lacunary series that had been studied by Banach [8].
To get at the bottom of the problem in point we recall the well-known fact that if

$$\frac{a_0}{2} + \sum_{n=1}^{\infty} (a_n \cos nx + b_n \sin nx)$$

is the Fourier series of $f \in L^2((0, 2\pi))$, then

$$\sum_{n=1}^{\infty} \left(|a_n|^2 + |b_n|^2\right) < \infty. \tag{1.64}$$

By virtue of the Riesz-Fischer theorem [111, Theorem 1.1], the power 2 in (1.64) is optimal, that is, given any $\delta > 0$, there must exist $g \in L^2(0, 2\pi)$ such that the sequence of its Fourier coefficients is not in $\ell^{2-\delta}$. In this regard, Carleman [16] constructed a continuous function g on $[0, 2\pi]$, the numerical series of whose Fourier coefficients satisfies

$$\sum_{n=1}^{\infty} \left(|a_n|^{2-\delta} + |b_n|^{2-\delta}\right) = \infty$$

for any $\delta > 0$. Thus the power 2 cannot be lowered even under the continuity assumption on f.
A special class of trigonometric series has appealed the attention of mathematicians since the nineteenth century, namely, the class of lacunary series. As early as 1872, the famous example of a continuous, nowhere differentiable function provided by Weierstrass involved a series of this type. Far from attempting any exhaustive treatment of this deep topic, we present some isolated facts as an example of how, almost inadvertently, spaces with variables exponents start creeping into the center stage of mathematics through this theme. Specifically:

Definition 1.4.6. *The trigonometric series*

$$\frac{a_0}{2} + \sum_{n=1}^{\infty} (a_n \cos \alpha_n x + b_n \sin \alpha_n x) \tag{1.65}$$

is said to be lacunary if for each $n \in \mathbb{N}$,

$$\frac{\alpha_{n+1}}{\alpha_n} > q > 1.$$

As far back as 1872, the famous example of a continuous, nowhere differentiable function provided by Weierstrass involved a series of this type. It is well known that the Fourier series of an integrable function might diverge pointwise almost everywhere. However, Kolmogorov proved that a lacunary Fourier series converges almost everywhere [59].
The following two theorems due to Sidon [97, 98] and Zygmund [110], respectively.

Theorem 1.4.42. *(Sidon, 1927) If the series (1.65) is the Fourier series of a measurable, bounded function, then*

$$\sum_{n=1}^{\infty} (|a_n| + |b_n|) < \infty.$$

Theorem 1.4.43. *(Zygmund, 1930) If the series (1.65) is the Fourier series of a function $f \in L^1((0, 2\pi))$, then in fact $f \in L^p((0, 2\pi))$ for every $p : p \in (0, \infty)$ and in particular*

$$\sum_{n=1}^{\infty} \left(|a_n|^2 + |b_n|^2\right) < \infty.$$

Enlightened by the preceding two theorems and influenced by the aforementioned work by Carleman, in 1930, Banach elaborated on the work done by Zygmund in [110]. More specifically, he undertook the task of proving the following:

Theorem 1.4.44. *(Banach, 1930) There exist a sequence (ε_n) with $0 < \varepsilon_n < 1$ and $\lim_{n \to \infty} \varepsilon_n = 0$ and a continuous function $f \in C([0, 2\pi])$ with lacunary Fourier series (1.65) for which*

$$\sum_{n=1}^{\infty} \left(|a_n|^{2-\epsilon_n} + |b_n|^{2-\epsilon_n}\right) = \infty. \tag{1.66}$$

As a consequence of the preceding theorem one has the following:

Theorem 1.4.45. *(Orlicz, 1931) There exist sequences (a_n), (b_n) such that*

$$\sum_{n=0}^{\infty} \left(|a_n|^{\frac{2-\epsilon_n}{1-\epsilon_n}} + |a_n|^{\frac{2-\epsilon_n}{1-\epsilon_n}}\right) < \infty$$

and that the series

$$\sum_{n=1}^{\infty}(a_n \cos nx + b_n \sin nx) \tag{1.67}$$

is not the Fourier series of any function.

Proof. Let a'_n, b'_n denote the Fourier coefficients of the continuous function u given in Theorem 1.4.44. On account of Theorem 1.4.45 and the divergence statement (1.66), there must exist sequences (a_n), (b_n) with

$$\sum_{n=1}^{\infty}\left(|a_n|^{\frac{2-\epsilon_n}{1-\epsilon_n}} + |a_n|^{\frac{2-\epsilon_n}{1-\epsilon_n}}\right) < \infty$$

for which the series

$$\sum_{n=1}^{\infty}(a_n a'_n + b_n b'_n) \tag{1.68}$$

diverges. Theorem 1.4.43, though, guarantees the convergence of the series (1.68) for the Fourier coefficients $(a_n), (b_n)$ of any integrable function. It follows immediately that the series (1.67) cannot be the Fourier series of any integrable function. □

1.5 Forerunners of the Musielak-Orlicz class: Orlicz spaces, $L^{p(x)}$ spaces

In what follows we present two examples of Banach function spaces intended to serve as the transition from the general theory to the idea of Musielak-Orlicz spaces. Focusing attention on these particular cases is not only consistent with the chronological development of the subject but it is also pedagogically adequate. The first such instance is given by the Orlicz spaces. Since Orlicz spaces are in turn a particular case of the Musielak-Orlicz spaces, which will be analyzed in detail in the sequel, our interest in Orlicz spaces is only historical, as an example of the mathematical structures introduced in the preceding sections. It is impossible for the authors to provide an exhaustive account of the extremely vast literature devoted to spaces of Orlicz type. For a quite comprehensive account, the interested reader might start by referring to [91, 92] and the bibliography therein.
Let $(\Omega, \mathcal{A}, \mu)$ be a measure space, assume μ is σ-finite and complete and denote the class of all real-valued, $\mathcal{A}$-measurable functions on Ω by $\mathcal{M}(\Omega)$.

Definition 1.5.1. *A function*

$$\varphi : \Omega \longrightarrow [0, \infty)$$

is said to be an Orlicz function if

(i) $\varphi(0) = 0$ *and* $\varphi(t) > 0$ *if* $t > 0$,

(ii) $\lim_{t\to\infty} \varphi(t) = \infty, \ \lim_{t\to 0^+} \varphi(t) = 0,$

(iii) φ *is nondecreasing and convex,*

(iv) φ *is continuous.*

The definition of Orlicz function is more or less stringent than ours depending on the authors. For an exhaustive treatment of this topic, which is here presented only as an example, we refer the reader to [2, 11, 62, 91, 92]. In this setting, the functional defined on $V = \mathcal{M}(\Omega)$ by

$$\mathcal{M}(\Omega) \ni v \longrightarrow \rho_\varphi(v) = \int_\Omega \varphi(|v(x)|)\, dx$$

is a convex, left-continuous modular. The Orlicz class is the collection of all measurable functions $v \in \mathcal{M}(\Omega)$ such that

$$\rho_\varphi(v) < \infty.$$

The vector space V_{ρ_φ} consists, in this example of those $v \in \mathcal{M}(\Omega)$ such that, for some $\lambda > 0$

$$\int_\Omega \varphi(\lambda|v(x)|)\, dx < \infty$$

and the Luxemburg norm on V_{ρ_φ} is given by

$$\|v\|_{\rho_\varphi} = \inf\left\{\lambda > 0 : \int_\Omega \varphi\left(\lambda^{-1}|v(x)|\right)\, dx \le 1\right\};$$

the Banach space $(V_{\rho_\varphi}, \|v\|_{\rho_\varphi})$ is called the Orlicz space corresponding to φ and is usually denoted by $L^\varphi(\Omega)$. Examples include the classical L^p-spaces for $1 \le p < \infty$ ($\varphi(t) = t^p$) and the space $L^\xi(\Omega)$ with $\xi(t) = e^{t^{\frac{n}{n-1}}} - 1, n = 2, 3, 4...$ encountered naturally in the formulation of one of the classical Sobolev embedding theorems [42, Theoremn 7.15].

We select yet another example before embarking in the abstract theory of Musielak-Orlicz spaces. Our motivation to present this paradigmatic case separately stems in part from its intrinsic importance in current applications and from the fact that, though in a primitive stage, this is the second example of Musielak-Orlicz space ever mentioned in the literature [85].
Specifically let $\mathcal{P}(\Omega)$ stand for the class of all measurable, extended real-valued function

$$p : \Omega \longrightarrow [1, \infty].$$

It is easy to see that

$$\varphi : \Omega \times [0,\infty) \longrightarrow [0,\infty]$$
$$\varphi(x,t) = t^{p(x)}$$

is also measurable. For each such p define the sets:

$$\Omega_0 = \{x \in \Omega : 1 < p(x) < \infty\}$$
$$\Omega_1 = \{x \in \Omega : p(x) = 1\}$$
$$\Omega_\infty = \{x \in \Omega : p(x) = \infty\}$$

and set:

$$p_- = \operatorname*{ess\,inf}_{x\in\Omega_0} p(x) \text{ and } p_+ = \operatorname*{ess\,sup}_{x\in\Omega_0} p(x) \text{ if } |\Omega_0| > 0,$$

$$p_- = p_+ = 1 \text{ otherwise.}$$

Theorem 1.5.1. *The function*

$$\rho_p : \mathcal{M}(\Omega) \longrightarrow [0,\infty]$$
$$\rho_p(u) = \int_{\Omega_0\cup\Omega_1} |u(x)| d\mu + \operatorname*{ess\,sup}_{x\in\Omega_\infty} |u(x)|$$

defines a convex, continuous modular on $\mathcal{M}(\Omega)$.

Proof. Properties $(i)-(iii)$ of Definition 1.3.1 have to be checked. It is obvious that $\rho_p(u) = 0 \iff u = 0\, a.e.$ and that for $|\lambda| = 1$ and measurable u

$$\rho_p(\lambda u) = \int_{\Omega\setminus\Omega_\infty} |\lambda u(x)| d\mu + \operatorname*{ess\,sup}_{x\in\Omega_\infty} |\lambda u| = \rho_p(u).$$

Convexity and continuity follow at once. □

Notice that the corresponding Luxemburg norm defined on $\mathcal{M}(\Omega)_{\rho_p}$ (i.e., on the space of all Borel-measurable functions v on Ω for which $\rho_p(\lambda v) < \infty$ for some $\lambda > 0$, see Definition 1.3.2) is given by

$$\|u\|_p = \inf\left\{\lambda > 0 : \rho_p(\lambda^{-1}u) \le 1\right\}.$$

In particular, if $\Omega \subseteq \mathbb{R}^n$, $\mathcal{A}$ is the σ-algebra of Borel subsets of Ω and μ is the Lebesgue measure, $L^p(\Omega)$ is referred to as the Lebesgue space with variable exponent. If $p(x)$ is constant in Ω, the space $L^p(\Omega)$ coincides with the classical Lebesgue space $L^p(\Omega)$ and the Luxemburg norm is identical to the Lebesgue norm

$$u \longrightarrow \|u\|_p = \left(\int_\Omega |u(x)|^p \, dx\right)^{\frac{1}{p}} = (\rho_p(u))^{\frac{1}{p}}.$$

Proposition 1.5.2. *Let $p \in \mathcal{P}(\Omega)$. Then $\|u\|_p \le 1$ if and only if $\rho_p(u) \le 1$; moreover, if $\|u\|_p \le 1$, then $\rho_p(u) \le \|u\|_p$. If $u \in L^p(\Omega)\backslash\{0\}$ and*

$$\operatorname*{ess\,sup}_{\{x:u(x)\neq 0\}} p(x) < \infty,$$

then

$$\rho_p\left(u/\|u\|_p\right) = \int_{\Omega_0\cup\Omega_1} \left(\frac{|u(x)|}{\|u\|_p}\right)^{p(x)} dx + \operatorname*{ess\,sup}_{x\in\Omega_\infty} \frac{|u(x)|}{\|u\|_p} = 1.$$

Proof. The first assertion is contained in Proposition 1.3.2, Section 1.3, the second in Corollary 1.3.4 of Section 1.3. Now suppose that $u \in L^p(\Omega)\backslash\{0\}$ and that

$$\operatorname*{ess\,sup}_{\{x:u(x)\neq 0\}} p(x) < \infty,$$

assume that $K := \rho_p\left(u/\|u\|_p\right) < 1$ and let $\lambda \in \left(0, \|u\|_p\right)$ be such that $\left(\|u\|_p/\lambda\right)^a K \le 1$. Then

$$\begin{aligned}
\rho_p(u/\lambda) &= \int_{\Omega_0\cup\Omega_1} \left(\frac{|u(x)|}{\lambda}\right)^{p(x)} dx + \operatorname*{ess\,sup}_{x\in\Omega_\infty} \frac{|u(x)|}{\lambda} \\
&= \left(\|u\|_p/\lambda\right)^a \left(\lambda/\|u\|_p\right)^a \left(\int_{\Omega_0\cup\Omega_1} \left(\frac{|u(x)|}{\lambda}\right)^{p(x)} dx + \operatorname*{ess\,sup}_{x\in\Omega_\infty} \frac{|u(x)|}{\lambda} \right) \\
&\le \left(\|u\|_p/\lambda\right)^a \left(\int_{\Omega_0\cup\Omega_1} \left(\lambda/\|u\|_p\right)^a \left(\frac{|u(x)|}{\lambda}\right)^{p(x)} dx + \operatorname*{ess\,sup}_{x\in\Omega_\infty} \frac{|u(x)|}{\|u\|_p} \right) \\
&\le \left(\|u\|_p/\lambda\right)^a \left(\int_{\Omega_0\cup\Omega_1} \left(\frac{|u(x)|}{\|u\|_p}\right)^{p(x)} dx + \operatorname*{ess\,sup}_{x\in\Omega_\infty} \frac{|u(x)|}{\|u\|_p} \right) \le 1.
\end{aligned}$$

Since $\lambda < \|u\|_p$, we have a contradiction, and so $K \ge 1$. Let (λ_k) be a decreasing sequence with limit $\|u\|_p$. Then

$$\rho_p(u/\lambda_k) = \int_{\Omega_0\cup\Omega_1} \left(\frac{|u(x)|}{\lambda_k}\right)^{p(x)} dx + \operatorname*{ess\,sup}_{x\in\Omega_\infty} \frac{|u(x)|}{\lambda_k} \le 1,$$

so that the integrals $\int_{\Omega_0\cup\Omega_1} \left(|u(x)|/\lambda_k\right)^{p(x)} dx$ are bounded above. Since $|u(x)|/\lambda_k \nearrow |u(x)|/\|u\|_p$, it follows from monotone convergence that

$$K = \int_{\Omega_0\cup\Omega_1} \left(\frac{|u(x)|}{\|u\|_p}\right)^{p(x)} dx + \operatorname*{ess\,sup}_{x\in\Omega_\infty} \frac{|u(x)|}{\|u\|_p} \le 1.$$

Thus $K = 1$ and the proof is complete. □

As a matter of fact $L^p(\Omega)$ has a richer structure than that or a mere modular space: we next show that it is a Banach function space.

Theorem 1.5.3. *If $\Omega \subseteq \mathbb{R}^n$, $\mathcal{A}$ is the σ-algebra of Borel subsets of Ω, μ is the Lebesgue measure and $p \in \mathcal{P}(\Omega)$, then $L^p(\Omega)$ is a Banach function space. In particular, $L^p(\Omega)$ is a Banach space.*

Proof. Properties (i), (ii) and (iv) of a Banach function space are clearly satisfied. For (iii), let $0 \le u_k \nearrow u$ a.e.: then $\left(\|u_k\|_p\right)$ is non-decreasing. If $\|u\|_p < \infty$, suppose that there exists $\lambda > 0$ such that $\|u_k\|_p \nearrow \lambda < \|u\|_p$. Then

$$\begin{aligned}
1 &\ge \int\limits_{\Omega_0 \cup \Omega_1} \left|\frac{u_k(x)}{\|u_k\|_p}\right|^{p(x)} dx + \operatorname*{ess\,sup}_{x\in\Omega_\infty} \left|\frac{u_k(x)}{\|u_k\|_p}\right| \\
&\ge \int\limits_{\Omega_0 \cup \Omega_1} \left|\frac{u_k(x)}{\lambda}\right|^{p(x)} dx + \operatorname*{ess\,sup}_{x\in\Omega_\infty} \left|\frac{u_k(x)}{\lambda}\right| \\
&\nearrow \int\limits_{\Omega_0 \cup \Omega_1} \left|\frac{u(x)}{\lambda}\right|^{p(x)} dx + \operatorname*{ess\,sup}_{x\in\Omega_\infty} \left|\frac{u(x)}{\lambda}\right| > 1,
\end{aligned}$$

and we have a contradiction. The case in which $\|u\|_p = \infty$ is handled analogously.
In order to address Property (v), let $E \subset \Omega$ be such that $|E| < \infty$, let $u \in X$ and put

$$M = \{x \in E \cap (\Omega_0 \cup \Omega_1) : |u(x)| < 1\}$$

and

$$N = \{x \in E \cap (\Omega_0 \cup \Omega_1) : |u(x)| \ge 1\}.$$

Then:

$$\begin{aligned}
&\frac{1}{\|u\|_p}\left(\int\limits_{E\cap(\Omega_0\cup\Omega_1)} |u(x)|\, dx + \operatorname*{ess\,sup}_{x\in E\cap\Omega_\infty} |u(x)|\right) \\
&\quad = \int\limits_M \frac{|u(x)|}{\|u\|_p} dx + \int\limits_N \frac{|u(x)|}{\|u\|_p} dx + \operatorname*{ess\,sup}_{x\in E\cap\Omega_\infty} \frac{|u(x)|}{\|u\|_p} \\
&\quad \le |M| + \int\limits_N \frac{|u(x)|^{p(x)}}{\|u\|_p^{p(x)}} dx + \operatorname*{ess\,sup}_{x\in E\cap\Omega_\infty} \frac{|u(x)|}{\|u\|_p} \\
&\quad \le |M| + 1,
\end{aligned}$$

□

Sharper relationships between the modular and the norm can be obtained under stricter conditions on p.

Proposition 1.5.4. *Let $p \in \mathcal{P}(\Omega)$ be such that $1 < p_- \leq p_+ < \infty$. Then for all $u \in L^p(\Omega)$,*

$$\min\left\{\|u\|_p^{p_-}, \|u\|_p^{p_+}\right\} \leq \rho_p(u) \leq \max\left\{\|u\|_p^{p_-}, \|u\|_p^{p_+}\right\}.$$

Proof. First note that by Proposition 1.5.2, $\rho_p(u/\|u\|_p) = 1$. Thus if $\|u\|_p = \lambda > 1$,

$$\lambda^{-p_+}\rho_p(u) \leq \rho_p(u/\lambda) = 1 \leq \lambda^{-p_-}\rho_p(u),$$

and so

$$\|u\|_p^{p_-} \leq \rho_p(u) \leq \|u\|_p^{p_+}.$$

The case $\|u\|_p < 1$ is handled in a similar manner. □

Theorem 1.5.5. *If $p_+ < \infty$, then the modular ρ_p satisfies the Δ_2 condition.*

Proof. Let $f \in L^p(\Omega)$. Clearly $2^{p(x)} \leq 2^{p_+} < \infty$ in Ω, which yields

$$\begin{aligned}\rho_p(2u) &= \int\limits_{\Omega_1\cup\Omega_0} |2u|^p d\mu + \operatorname*{ess\ sup}_{x\in\Omega_\infty} |2u| \\ &\leq 2^{p_+}\rho_p(u).\end{aligned}$$

□

Corollary 1.5.6. *If $p_+ < \infty$, then ρ_p modular convergence is equivalent to $\|\cdot\|_p$ convergence in $L^p(\Omega)$.*

Proof. The corollary follows immediately from Theorem 1.5.5 and Corollary 1.3.9. □

Spaces with variable exponent also resemble classical Lebesgue spaces with regard to Lebesgue points. We recall that if p is a constant and $1 \leq p < \infty$, then given any $v \in L^p(\mathbb{R}^n)$, a point $x \in \mathbb{R}^n$ such that

$$\lim_{r\to 0} |B(x,r)|^{-1} \int\limits_{B(x,r)} |v(y) - v(x)|^p \, dy = 0$$

is called a Lebesgue point of v. It is a very familiar fact (see [36, Corollary 1, p. 44]), that almost every point of $\mathbb{R}^n$ is a Lebesgue point of v. By a simple adaptation of the Corollary just mentioned it can be shown that if $p \in \mathcal{P}(\mathbb{R}^n)$, $p_+ < \infty$ and $v \in L^p(\mathbb{R}^n)$, then

$$\lim_{r\to 0} |B(x,r)|^{-1} \int\limits_{B(x,r)} |v(y) - v(x)|^{p(y)} \, dy = 0$$

for a.e. $x \in \mathbb{R}^n$. We finish this section with a useful technical lemma reminiscent of a similar property in the classical setting.

Proposition 1.5.7. *Suppose* $p \in \mathcal{P}(\Omega)$ *is such that* $p_+ < \infty$. *Then* $\rho_p(u_k) \to 0$ *if and only if* $\|u_k\|_p \to 0$; *moreover, if* $\|u_k\|_p \to 0$, *then* $u_k \to 0$ *in measure.*

Proof. Suppose that $\rho_p(u_k) \to 0$, let $\varepsilon \in (0,1)$ and let $N \in \mathbb{N}$ be such that $\rho_p(u_k) < \varepsilon$ if $k > N$. Then

$$\begin{aligned}
&\rho_p\left(\rho_p\left(u_k\right)^{-1/p_+} u_k\right) \\
&\quad \leq \rho_p\left(u_k\right)^{-1} \int\limits_{\Omega\setminus\Omega_\infty} |u_k(x)|^{p(x)}\, dx + \rho_p\left(u_k\right)^{-1/p_+} \|u_k\|_{\infty,\Omega_\infty} \\
&\quad \leq \rho_p\left(u_k\right)^{-1} \rho_p\left(u_k\right) = 1.
\end{aligned}$$

Thus

$$\|u_k\|_p \leq \rho_p\left(u_k\right)^{1/p_+} < \varepsilon^{1/p^*},$$

so that $\|u_k\|_p \to 0$. That $\|u_k\|_p \to 0$ implies $\rho_p(u_k) \to 0$ is clear.

Finally, suppose that $\|u_k\|_p \to 0$ and u_k does not converge to 0 in measure. Then there exist $\varepsilon > 0$, $\delta \in (0,1)$ and a subsequence $(u_{k_l})_{l\in\mathbb{N}}$ such that for all $l \in \mathbb{N}$,

$$\inf |\{x \in \Omega : |u_{k_l}(x)| > \varepsilon\}| \geq \delta.$$

Thus $\rho_p\left(u_{k_l}\right) \geq \delta\varepsilon^{p_+}$ for all $l \in \mathbb{N}$. But by the first part of the proposition, $\rho_p(u_{k_l}) \to 0$: contradiction. □

The variable exponent Lebesgue spaces constitute one important example of Banach functions spaces; their ubiquity in various areas of theoretical and applied mathematics [24, 60] justifies their careful study. We will continue to direct our efforts to the study of these spaces in the light of a more general mathematical framework, namely the theory of Musielak-Orlicz spaces which in particular, encompasses the variable exponent Lebesgue scale.

Further properties of the variable exponent Lebesgue scale $L^p(\Omega)$ will be investigated in the general context of Musielak-Orlicz spaces in Chapter 2.

Chapter 2

Musielak-Orlicz spaces

2.1 Introduction, definition and examples 91
2.2 Embeddings between Musielak-Orlicz spaces 101
 2.2.1 The Δ_2-condition 110
 2.2.2 Absolute continuity of the norm 113
2.3 Separability 115
2.4 Duality of Musielak-Orlicz spaces 118
 2.4.1 Conjugate Musielak-Orlicz functions 118
 2.4.2 Dual of $L^{\varphi}(\Omega)$ 120
2.5 Density of regular functions 124
2.6 Uniform convexity of Musielak-Orlicz spaces 128
2.7 Carathéodory functions and Nemytskii operators on Musielak-Orlicz spaces 132
2.8 Further properties of variable exponent spaces 137
 2.8.1 Duality maps on spaces of variable integrability 150
2.9 The Matuszewska-Orlicz index of a Musielak-Orlicz space 151
2.10 Historical notes 155

2.1 Introduction, definition and examples

The modular spaces ℓ^{p_n} introduced in Section 1.4 and the space $L^p(\Omega)$ defined in Section 1.5 are but particular instances of the same mathematical concept. The key observation to study their commonalities rather than emphasizing their differences is that the modulars ρ_{p_n} and ρ_p introduced in Section 1.4 and in Section 1.5, respectively entail a higher degree of complexity than that inherent to the mere definition of modular given in Section 1.3. Specifically, from now on we will focus on a detailed exploration of the intrinsic measure-theoretic character of both ρ_{p_n} and ρ_p. To the effect of precisely describing the specialness of the particular class of modulars we have in mind, some terminology must be introduced.

Definition 2.1.1. *For a given measure space $(\Omega, \mathcal{A}, \mu)$ we say that φ is a Musielak-Orlicz (MO if no confusion arises) function on $(\Omega, \mathcal{A}, \mu)$ (or on Ω if no confusion arises from such simplification) as the collection of all functions*

$$\varphi : \Omega \times [0, \infty) \longrightarrow [0, \infty)$$

such that

(i) $\varphi(\cdot, t)$ *is measurable for each* $t \in [0, \infty)$,

(ii) For a.e. $x \in \Omega$, $\varphi(x, \cdot)$ *is non-decreasing, convex, continuous and* $\varphi(x, 0) = 0$.

(iii) $\varphi(x, t) > 0$ *for* $t > 0$,

(iv) $\varphi(x, t) \longrightarrow \infty$ *as* $t \longrightarrow \infty$.

For example if $\Omega \subseteq \mathbb{R}^n$, $\mathcal{A}$ is the Borel σ-algebra of subsets of Ω, μ is the Lebesgue measure and p is a measurable function

$$p : \Omega \to [1, \infty),$$

then $\varphi(x, t) = t^{p(x)}$ is a Musielak-Orlicz function on Ω in this setting.

In particular the function φ in Definition 2.1.1 is said to be an Orlicz function if it is independent of $x \in \Omega$.

Definition 2.1.2. *The Musielak-Orlicz function* φ *is said to be locally integrable if for any* $t > 0$ *and any subset* $\mathcal{W} \subseteq \Omega$ *with* $\mu(\mathcal{W}) < \infty$ *one has*

$$\int_{\mathcal{W}} \varphi(x, t)\, dx < \infty.$$

Remark 2.1.1. The condition introduced in Definition 2.1.2 is referred to as local integrability of φ. It is emphatically not an artificial technical condition. As we shall see (for example, in Corollary 3.1.2 of Chapter 3), local integrability of φ is required to have a meaningful definition of the Sobolev spaces of Musielak-Orlicz type.

Definition 2.1.3. *If* $(\Omega, \mathcal{A}, \mu)$ *is a measure space, a simple function is a (finite) linear combination of characteristic functions of sets of finite measure. The class of simple functions will be denoted by* S.

Example 2.1.1. If $\Omega = \mathbb{N}$, μ is the counting measure, $\mathcal{A} = 2^{\mathbb{N}}$ and $p : \mathbb{N} \longrightarrow \infty$, the function

$$\varphi : \mathbb{N} \times [0, \infty) \longrightarrow [0, \infty)$$
$$\varphi(n, t) = t^{p(n)}$$

is a proper, locally integrable Musielak-Orlicz function on $\mathbb{N}$.
On the other hand, it is easy to produce a Musielak-Orlicz function that is not locally integrable: Consider the Lebesgue measure on the interval $(0, 1)$ and for any $x \in (0, 1)$ let n_x be the unique natural number for which $x \in (2^{-(n_x+1)}, 2^{-(n_x)}]$. Define p as

$$p : (0, 1) \to [1, \infty)$$
$$p(x) = n_x + 1$$

and define the Musielak-Orlicz function ψ by

$$\phi(x,t) = t^{p(x)}.$$

Then

$$\int_0^1 \phi(x,3)dx = \int_0^1 3^{p(x)}dx = \sum_{n=0}^{\infty} \int_{2^{-(n+1)}}^{2^{-n}} 3^{n+1}dx$$
$$= \sum_{n=0}^{\infty} \left(\frac{3}{2}\right)^{n+1} = \infty.$$

Let the elementary fact be observed at this point that if

$$u : \Omega \longrightarrow [-\infty, \infty]$$

is a $\mathcal{A}$-measurable function on Ω, then so is

$$x \longrightarrow \varphi(x, |u(x)|).$$

Let $\mathcal{M}(\Omega)$ stand for the set of all extended real valued $\mathcal{A}$-measurable functions defined on Ω.

Theorem 2.1.2. *The functional*

$$\rho_\varphi : \mathcal{M}(\Omega) \longrightarrow [0,\infty] \tag{2.1}$$
$$\rho_\varphi(u) = \int_\Omega \varphi(x, |u(x)|)\, dx$$

is a convex, left-continuous modular.

Proof. If $\rho_\varphi(u) = 0$ then necessarily $\varphi(x,|u(x)|) = 0$ *a.e.*; by virtue of (*ii*) in Definition 2.1.4, $u(x) = 0$ *a.e.* in Ω.
If $|\alpha| = 1$. $\varphi(x, |\alpha u(x)|) = \varphi(x, |u(x)|)$ for any $u \in \mathcal{M}(\Omega)$. Convexity follows immediately from the assumed convexity of φ. If $\lambda \to \lambda_0$, $u \in L^\varphi(\Omega)$ it is plain, on account of the monotonicity and continuity of φ and the monotone convergence theorem that

$$\rho_\varphi(\lambda u) = \int_\Omega \varphi(x, \lambda|u(x)|)dx \to \int_\Omega \varphi(x, \lambda_0|u(x)|)dx = \rho_\varphi(\lambda_0 u).$$

Let the fact be noticed that the above reasoning holds even if one assumes

$$\rho_\varphi(\lambda_0 u) = \infty.$$

□

Definition 2.1.4. *Let φ be an Musielak-Orlicz function on Ω. When the σ-algebra $\mathcal{A}$ and the measure μ are clear from the context, which will be the case throughout this monograph, we set*

$$L^{\varphi}(\Omega) = \left\{ u \in \mathcal{M} : \int_{\Omega} \varphi(x, \lambda |u(x)|)\, dx < \infty \text{ for some } \lambda > 0 \right\}.$$

It is easy to see that under the prescribed hypothesis $L^{\varphi}(\Omega)$ is a vector space.
When endowed with the Luxemburg norm (Proposition 1.3.2)

$$\|u\|_{\varphi} = \inf \left\{ \lambda > 0 : \int_{\Omega} \varphi\left(x, \frac{|u(x)|}{\lambda}\right) dx \leq 1 \right\}$$

$L^{\varphi}(\Omega)$ becomes a normed space (in fact, a Banach space if μ is σ-finite, as will be shown in Theorem 2.1.6), and will be called the Musielak-Orlicz space generated by φ (MO space for short, if there is no room for confusion).

Lemma 2.1.3. *Let φ be an Musielak-Orlicz function on the measure space $(\Omega, \mathcal{A}, \mu)$, $(u_k) \subseteq \mathcal{M}(\Omega)$ and $u \in \mathcal{M}(\Omega)$. Then:*

(i) If $u_k \to u$ a.e. in Ω, then $\rho_{\varphi}(u) \leq \liminf_{k\to\infty} \rho_{\varphi}(u_k)$.

(ii) If $0 \leq u_k \nearrow u$ a.e. in Ω, then $\rho_{\varphi}(u) = \lim_{k\to\infty} \rho_{\varphi}(u_k)$.

Proof. The proof relies on the fact that *a.e.* $x \in \Omega$ the function

$$t \longrightarrow \varphi(x, t)$$

is lower semicontinuous. For, if $\varepsilon > 0$, then by virtue of the monotonicity of φ,

$$\varphi(t_0 - \varepsilon) \leq \inf_{x, t_0 - \varepsilon < t < t_0 + \varepsilon} \varphi(x, t)$$

and left-continuity implies $\varphi(x, t_0) \leq \liminf_{t\to t_0} \varphi(x, t)$ *a.e.*.
Statement (i) follows from Fatou's Lemma and from the lower semicontinuity of φ :

$$\begin{aligned}
\rho_{\varphi}(u) &= \int_{\Omega} \varphi(x, |u(x)|)\, d\mu = \int_{\Omega} \varphi(x, \liminf_{k\to\infty} |u_k(x)|)\, d\mu \\
&\leq \int_{\Omega} \liminf_{k\to\infty} \varphi(x, |u_k(x)|)\, d\mu \\
&\leq \liminf_{k\to\infty} \int_{\Omega} \varphi(x, |u_k(x)|)\, d\mu.
\end{aligned}$$

Statement (ii) follows similarly from the monotonicity of φ and the Monotone Convergence Theorem. □

Example 2.1.2. To the authors' best knowledge, the variable exponent Lebesgue spaces $L^{p}(\Omega)$ introduced in Section 1.5 constitute the first historical example of generalized Musielak-Orlicz spaces [85].

Let $(\Omega, \mathcal{A}, \mu)$ be a measure space with a σ-finite measure μ; consider a measurable function

$$p : \Omega \longrightarrow [1, \infty)$$

and set

$$\begin{aligned} \varphi_p &: \Omega \times [0, \infty) \longrightarrow [0, \infty] \\ \varphi_p(x, t) &= t^{p(x)}. \end{aligned}$$

Then φ_p is a Musielak-Orlicz function and the resulting normed space

$$\left(L^{\varphi_p}(\Omega), \|\cdot\|_{\varphi_p}\right) = (L^{p}(\Omega), \|\cdot\|_{p})$$

was proved in Theorem 1.5.3 to be a Banach space.

Example 2.1.3. We single out one fundamental case. If φ is an Orlicz function (i.e., $\varphi(x,t) = \varphi(t)$, see Definition 1.5.1) the corresponding space $L^{\varphi}(\Omega)$ is called the φ-Orlicz space. The Orlicz modular in this case is given by

$$u \longrightarrow \rho_{\varphi}(u) = \int_{\Omega} \varphi\left(|u(x)|\right) d\mu.$$

The history of Orlicz spaces originated in the second decade of the twentieth century (see Section 2.10) with the works of Young, Orlicz and Birnbaum,s among others. Orlicz spaces have been exhaustively treated in the literature (see for example [91, 92] and the references therein).
The idea of Orlicz spaces was in fact inspired by one of its most important particular cases, namely the Lebesgue spaces L^{p}, introduced in 1910 by Friedrich Riesz [93]. Specifically, if $\mathcal{B}$ is the Borel σ-algebra of subsets of a Borel-measurable set $\Omega \subset \mathbb{R}^n$, $1 \le p < \infty$ and

$$\begin{aligned} \varphi &: \Omega \longrightarrow [0, \infty) \\ \varphi(t) &= t^{p}, \end{aligned}$$

then the Orlicz space $L^{\varphi}(\Omega)$ is the classical Lebesgue space $L^{p}(\Omega)$.

Turning back to the general theory, it is important to underline the fact that in general, $L^{\varphi}(\Omega)$ does not contain the class of simple functions.

Example 2.1.4. For $x \in (0,1)$ let n_x be the unique natural number satisfying $x \in 2^{-(n_x+1)} < x \le 2^{-n_x}$ and let ϕ be the Musielak-Orlicz function on $(0,1)$ defined by

$$\phi(x,t) = t2^{2n_x}.$$

For any $\lambda > 0$ let k_λ be the smallest natural number such that $\lambda > 2^{-k_\lambda}$. Then

$$\int_0^1 \phi(x,\lambda)dx \geq \int_0^1 \phi(x, 2^{-n_\lambda})dx = \sum_{n=0}^{\infty} \int_{2^{-(n+1)}}^{2^{-n}} \phi(x, 2^{-k_\lambda})dx$$
$$= \sum_{n=0}^{\infty} 2^{-(n+1)} 2^{2n-k_\lambda} = \infty.$$

It is apparent that $\mathbf{1}_{(0,1)} \notin L^\phi((0,1))$. If φ is locally integrable Musielak-Orlicz function it easily seen that $L^\varphi(\Omega)$ contains the class S of simple functions. However, it is not hard to verify that in Example 2.1.1, $L^\phi(\Omega)$ contains the class S in spite of ϕ not being locally integrable.

Definition 2.1.5. *Let $(\Omega, \mathcal{A}, \mu)$ be a measure space. The Musielak-Orlicz function φ (or the space $L^\varphi(\Omega)$) is said to be proper iff:*

(i) $L^\varphi(\Omega)$ contains the class of simple functions.

(ii) If $F \in \mathcal{A}$ and $\mu(F) < \infty$, then there exists a positive constant $c(F)$ such that for any $u \in L^\varphi(\Omega)$ it holds that

$$\int_\Omega \mathbf{1}_F(x)u(x)d\mu \leq c(F)\|u\|_\varphi.$$

Example 2.1.5. The Musielak-Orlicz space ℓ^{p_n} already introduced in Section 1.4 is easily seen to be proper.

Example 2.1.6. Let $\Omega \subseteq \mathbb{R}^n$, $\mathcal{A}$ the Borel σ-algebra of subsets of Ω, μ be the Lebesgue measure and p a measurable function on Ω with $1 \leq p < \infty$. Then it is easy to verify that the Musielak-Orlicz function

$$\begin{aligned} &\varphi : \Omega \times [0,\infty) \longrightarrow [0,\infty) \\ &\varphi(x,t) = t^{p(x)} \end{aligned} \tag{2.2}$$

is locally integrable if $p_+ < \infty$ but not if $p_+ = \infty$. It follows that φ is proper in the former case and not proper in the latter, for if $w = \sum_{k=1}^{M} s_k \mathbf{1}_{E_k}$ is a simple function on Ω then, if $p_+ < \infty$ it is clear that

$$\rho_\varphi(w) = \sum_{k=1}^{M} \int_{E_k} s_k^{p(x)} dx < \infty,$$

so that the class of simple functions is contained in $L^\varphi(\Omega)$.

For the remaining part of this work, the notation

$$\rho_p := \rho_\varphi \text{ and } L^p(\Omega) := L^\varphi(\Omega)$$

will be used in the particular case of φ given by (2.2). The corresponding norm will be denoted by $\|\cdot\|_{L^p(\Omega)}$ or $\|\cdot\|_p$ if there is no room for confusion or ambiguity.

Invoking Theorem 1.5.3 and Axiom (v) in the definition of Banach function space it is concluded that $L^{t^p}(\Omega) = L^p(\Omega)$ is proper under the assumption $p_+ < \infty$.

It was shown in Theorem 1.5.3 that $L^p(\Omega)$ is a Banach function space. If the Musielak-Orlicz function φ is proper, the theorem extends to the more general case of $L^\varphi(\Omega)$. More precisely:

Theorem 2.1.4. *Let $(\Omega, \mathcal{A}, \mu)$ be a measure space and φ be a proper Musielak-Orlicz function on Ω. Then, when endowed with the Luxemburg norm $\|\cdot\|_\varphi$, $L^\varphi(\Omega)$ is a Banach function space.*

Proof. Axioms $(i)-(v)$ of Definition 1.2.1 need to be checked. It is a straightforward matter to verify that the functional

$$u \longrightarrow \|\cdot\|_\varphi$$

has the properties of a norm on $\mathcal{M}(\Omega)$. Axiom (i) follows simply by observing that if $u \in L^\varphi(\Omega)$ then for some $\lambda > 0$ $\int_\Omega \varphi(x, \lambda|u(x)|)\, d\mu = \gamma < \infty$, then either $\gamma \leq 1$ or

$$\int_\Omega \varphi(x, \gamma^{-1}\lambda|u(x)|)\, d\mu \leq \gamma \int_\Omega \varphi(x, \lambda|u(x)|)\, d\mu < \infty;$$

from both cases one derives $\|u\|_\varphi < \infty$.
Axiom (ii) follows automatically from Definition 2.1.4. Properties (iv) and (v) are being assumed in the statement of the theorem. To prove property (iii) it suffices to consider the case $\|u\|_\varphi \neq 0$. We observe first that if $0 \leq u_k \nearrow u$ pointwise *a.e.* in Ω, then the continuity and the monotonicity of φ yield

$$\varphi\left(x, \frac{u_k(x)}{\|u\|_\varphi}\right) d\mu \nearrow \varphi\left(x, \frac{u(x)}{\|u\|_\varphi}\right) \quad a.e.,$$

from which we conclude

$$\int_\Omega \varphi\left(x, \frac{u_k(x)}{\|u\|_\varphi}\right) d\mu \nearrow \int_\Omega \varphi\left(x, \frac{u(x)}{\|u\|_\varphi}\right) d\mu \leq 1,$$

by virtue of the monotone convergence theorem, Theorem 2.1.2 and Proposition 1.3.2, (iii). Hence

$$\|u_k\|_\varphi \leq \|u\|_\varphi.$$

As a consequence of the monotonicity of φ, if $\rho_\varphi\left(x, \frac{u_{k+1}}{\lambda}\right) \leq 1$ for $\lambda > 0$, then also $\rho_\varphi\left(x, \frac{u_k}{\lambda}\right) \leq 1$, which readily yields

$$\|u_k\|_\varphi \leq \|u_{k+1}\|_\varphi.$$

Now, given $\delta > 0$ select λ_δ such that $\|u\|_\varphi - \delta < \lambda_\delta < \|u\|_\varphi$ and

$$\int_\Omega \varphi\left(x, \frac{u(x)}{\lambda_\delta}\right) > 1,$$

and since as $k \to \infty$,

$$\int_\Omega \varphi\left(x, \frac{u_k(x)}{\lambda_\delta}\right) \nearrow \int_\Omega \varphi\left(x, \frac{u(x)}{\lambda_\delta}\right),$$

for sufficiently large k one has

$$\int_\Omega \varphi\left(x, \frac{u_k(x)}{\lambda_\delta}\right) > 1 \geq \int_\Omega \varphi\left(x, \frac{u_k(x)}{\|u_k\|_\varphi}\right),$$

monotonicity yields

$$\|u\|_\varphi - \delta < \lambda_\delta < \|u_k\|_\varphi \leq \|u\|_\varphi$$

for k large enough. In all,

$$\|u_k\|_\varphi \nearrow \|u\|_\varphi.$$

□

Corollary 2.1.5. *Since every Banach function space is a Banach space by Theorem 1.2.3, the preceding result implies that the Musielak-Orlicz space $L^\varphi(\Omega)$ is a Banach space if φ is a proper Musielak-Orlicz function.*

Next, we present the proof of the completeness of $L^\varphi(\Omega)$ without the restriction imposed on φ in Corollary 2.1.5.

Theorem 2.1.6. *Let φ be as in Definition 2.1.4. If μ is σ-finite, $L^\varphi(\Omega)$ is a Banach space.*

Proof. Let (Ω_k) be a sequence of pairwise disjoint, measurable sets, $\mu(\Omega_k) < \infty$ for any $k \in \mathbb{N}$ such that

$$\Omega = \bigcup_{k=1}^{\infty} \Omega_k.$$

Write

$$\mathcal{A}_k = \mathcal{A} \cap 2^{\Omega_k}.$$

We remark in passing that for every $k \in \mathbb{N}$, $(\Omega_k, \mathcal{A}_k, \mu_k)$, where μ_k is the standard restriction of μ to $\mathcal{A}_k$, is a measure space itself. Further, we observe the obvious fact that for any $k \in \mathbb{N}$ and an arbitrary $\lambda > 0$, μ_k is absolutely continuous with respect to the measure η_k^λ defined on $\mathcal{A}_k$ by

$$\eta_k^\lambda(Y) = \int\limits_Y \varphi(x, r)\, d\mu_k.$$

Accordingly, since $\mu_k(\Omega_k) < \infty$, for any $\epsilon > 0$ there exists $\delta > 0$ such that for any $A \in \mathcal{A}_k$ one has

$$\eta_k^\lambda(A) < \delta \Rightarrow \mu_k(A) < \epsilon. \tag{2.3}$$

Fix $\lambda > 0$ and for $\epsilon > 0$, let $\delta : 0 < \delta < 1$ be selected to satisfy (2.3) for $k = 1$ and $r = \lambda$ and consider a (norm) Cauchy sequence $(u_j) \subset L^\varphi(\Omega)$. Fix $N > 0$ such that $\|u_n - u_m\|_\varphi < \delta$ holds for $m, n \geq N$, let λ be an arbitrary positive number and for fixed $m, n \geq N$ set

$$Y_\lambda = \{x \in \Omega_1 : |u_n(x) - u_m(x)| \geq \lambda\}.$$

Then

$$\begin{aligned}
\eta_1^\lambda(Y_\lambda) &= \int\limits_{Y_\lambda} \varphi(x, \lambda)\, d\mu_1 \\
&\leq \int_{Y_\lambda} \varphi(x, |u_n(x) - u_m(x)|)\, d\mu_1 \leq \rho_\varphi\left(|u_n - u_m|\right) \\
&= \rho_\varphi\left(\frac{\|u_n - u_m\|_\varphi}{\|u_n - u_m\|_\varphi}|u_n - u_m|\right) \leq \|u_n - u_m\|_\varphi < \delta.
\end{aligned}$$

Consequently $\mu_1(Y_\lambda) < \epsilon$ and the sequence of restrictions $(u_n|_{\Omega_1})$ is Cauchy in measure, hence μ_1-convergent in measure to an $\mathcal{A}_1$-measurable function v_1 defined in Ω_1. Thus, there exists a subsequence $(u_{n_j}^{(1)})_j$ that converges $\mu_1 - a.e.$ to v_1. Assume now that for $1 < k \in \mathbb{N}$ an $\mathcal{A}_i$-measurable function v_i defined on Ω_i is given for each $1 \leq i \leq k$ and a subsequence

$$(u_{n_j}^{(k)})_j$$

of the original sequence (u_n) has been found such that

$$(u_{n_j}^{(k)}(x))_j \to \sum_{i=1}^{k} v_i(x)\mathbf{1}_{\Omega_i(x)} \quad \mu - a.e. \text{ in } \bigcup_{k=1}^{k} \Omega_i.$$

Almost verbatim, the preceding argument implies the existence of a $\mathcal{A}_{k+1}$-measurable function u_{k+1} and a subsequence $(u_{n_j}^{(k+1)})_j$ of (u_n) that converges μ-a.e. to $\sum_{i=1}^{k+1} v_i \mathbf{1}_{\Omega_i}$.

It is obvious that the function u defined on Ω as

$$u = \sum_{k=1}^{\infty} v_k(x)\mathbf{1}_{\Omega_k}$$

is $\mathcal{A}$- measurable, that the original sequence (u_n) converges to u in measure and that $(u^k_{n_k})$ converges to u in measure in Ω.
For arbitrary $\lambda > 0$, $0 < \epsilon < 1$, let N_0 be chosen so large that $k, j \geq N_0$ implies $\|\lambda(u_k - u^{(j)}_{n_j})\|_\varphi < \epsilon$. Then

$$\begin{aligned}
\rho_\varphi(\lambda(u_k - u)) &= \int_\Omega \varphi(x, \lambda|u_k(x) - u(x)|)\, d\mu \\
&\leq \int_\Omega \lim_{j\to\infty} \varphi(x, \lambda|u_k(x) - u^{(j)}_{n_j}(x)|)\, d\mu \\
&\leq \liminf_{j\to\infty} \int_\Omega \varphi(x, \lambda|u_k(x) - u^{(j)}_{n_j}(x)|)\, d\mu \\
&\leq \liminf_{j\to\infty} \rho_\varphi(\lambda(u_k - u^{(j)}_{n_j})) \\
&= \|\lambda(u_k - u^{(j)}_{n_j})\|_\varphi \rho_\varphi\left(\frac{\lambda(u_k - u^{(j)}_{n_j})}{\|\lambda(u_k - u^{(j)}_{n_j})\|_\varphi}\right) \\
&= \|\lambda(u_k - u^{(j)}_{n_j})\|_\varphi < \epsilon.
\end{aligned}$$

By virtue of Theorem 1.3.8 the arbitrariness of λ yields

$$u_k \to u \text{ in } L^\varphi(\Omega) \text{ as } k \to \infty.$$

□

Theorem 2.1.6 can be extended to the case in which the Musielak-Orlicz function φ satisfies the condition

$$\varphi(x, t) \geq 0$$

a.e. in Ω, see [24]. Other variants of the theme are possible, such as the consideration of a different norm [82].

We emphasize the following corollary to Theorem 2.1.6:

Corollary 2.1.7. *If $(u_n) \subset L^\varphi(\Omega)$ and*

$$u_n \overset{L^\varphi(\Omega)}{\longrightarrow} u,$$

then there exists a subsequence of (u_n) that converges to u $\mu - a.e.$ in Ω.

Corollary 2.1.8. *If $(\Omega, \mathcal{A}, \mu)$ is σ-finite, in particular, if $\Omega \subseteq \mathbb{R}^n$ is a Borel set and the underlying measure is the Lebesgue's measure, then, for $1 \leq p_- \leq p_+ < \infty$, $L^p(\Omega)$ is a Banach space.*

2.2 Embeddings between Musielak-Orlicz spaces

In this section we shall discuss necessary and sufficient conditions on the Musielak-Orlicz functions φ and ξ for the validity of the embedding $L^\varphi \subset L^\xi$. The main result in this section is Theorem 2.2.1, which was proved in [82]. We follow the proof in [24].

Definition 2.2.1. *Let φ and ξ be Musielak-Orlicz functions on the same measure space $(\Omega, \mathcal{A}, \mu)$. We use the notation*

$$\varphi \precsim \xi$$

if there exists a positive constant B and $0 \leq h \in L^1(\Omega)$ such that

$$\varphi(x,t) \leq \xi(x, Bt) + h(t) \tag{2.4}$$

for all $t \in [0, \infty)$ and a.e $x \in \Omega$.

Theorem 2.2.1. *Let φ and ξ be Musielak-Orlicz functions on the measure space $(\Omega, \mathcal{M}, \mu)$. Assume that μ is non-atomic. Then the embedding*

$$E_{\xi\varphi} : L^\xi(\Omega) \hookrightarrow L^\varphi(\Omega) \tag{2.5}$$

is continuous if and only if $\varphi \precsim \xi$.

Proof. Assume first that $\varphi \precsim \xi$; the set-theoretical inclusion $L^\xi(\Omega) \subseteq L^\varphi(\Omega)$ is obvious and it is also clear from (2.4) that if $\lambda > 0$ and $\rho_\xi(\lambda^{-1}u) \leq 1$ then

$$\rho_\varphi(\lambda^{-1}B^{-1}u) \leq 1 + \|h\|_1.$$

Convexity quickly yields

$$\rho_\varphi\left((\lambda(1 + \|h\|_1)B)^{-1}u\right) \leq 1,$$

whence $\|u\|_\varphi \leq (1+\|h\|_1)B\|u\|_\xi$. This settles the sufficiency of condition (2.4).

For the necessity of condition (2.4), let $E_{\xi,\varphi}$ be continuous and set

$$\alpha(x,t) = \begin{cases} \varphi(x,t) - \xi(x, \|E_{\xi,\varphi}\|t) & \text{if } \xi(x,t) < \infty \\ \infty & \text{otherwise.} \end{cases}$$

Set $\mathbb{Q} \cap [0, \infty) = (r_k)_{k \in \mathbb{N}}$ with $r_1 = 0$ and

$$h_k(x) := \max_{1 \leq j \leq k} \alpha(x, r_j).$$

Clearly, each h_k is measurable and for each $x \in \Omega$,

$$h_k(x) \geq \alpha(x, r_1) = \alpha(x, 0) = 0.$$

Moreover, for $x \in \Omega$ the sequence $(h_k(x))_{k\in\mathbb{N}}$ is nondecreasing. Since ξ and φ are Musielak-Orliczs-functions, α is left-continuous, from which it is easily derived that

$$h(x) = \sup_{k\in\mathbb{N}} h_k(x) = \sup_{t\geq 0} \alpha(x,t)$$

and that a.e. $x \in \Omega$ it holds the inequality

$$\varphi(x,t) \leq \xi(x, \|E_{\xi,\varphi}\|t) + h(x).$$

Next, we set about to prove that $h \in L^1(\Omega)$ and that $\|h\|_{L^1(\Omega)} \leq 1$. We first show that

$$|\{x \in \Omega : h(x) = \infty\}| = 0.$$

A contradiction will be reached by assuming the contrary. If

$$|I| = |\{x : h(x) = \infty\}| > 0,$$

we set

$$V_k := \left\{x \in I : \alpha(x, r_k) \geq \frac{2}{|I|}\right\}$$

$$\text{and } W_{k+1} = V_{k+1} \setminus \left(\bigcup_{j=1}^{k} V_j\right).$$

Clearly

$$I = \bigcup_{k=2}^{\infty} W_k = \bigcup_{k=1}^{\infty} V_k.$$

Let

$$f := \sum_{k=2}^{\infty} r_k \mathbf{1}_{W_k}.$$

Notice that the condition $\xi(x, |f(x)|) < \infty$ holds *a.e.* $x \in I$. Indeed, $\xi(x, |f(x)|) = \xi(x, r_k)$ for some $k \in \mathbb{N}$ and

$$\alpha(x, r_k) > 0,$$

which by definition of α implies $\xi(x, |f(x)|) < \infty$.

Next, we observe that for $x \in I = \bigcup_{k=2}^{\infty} W_k$

$$\infty = h(x) = \sup_{t\geq 0} \alpha(x,t),$$

from which one immediately derives the existence of $k \in \mathbb{N}$ for which

$$\alpha(x, r_k) > \frac{2}{|I|}.$$

It follows straight from the definition of α and of W_k that for any $x \in I$,

$$\varphi(x, |f(x)|) \geq \xi\left(x, |f(x)| \|E_{\xi,\varphi}\|\right) + \frac{2}{|I|}. \tag{2.6}$$

In all, one concludes

$$\int_I \varphi(x, |f(x)|)\, dx \geq \int_I \xi\left(x, |f(x)| \|E_{\xi,\varphi}\|\right) dx + 2. \tag{2.7}$$

On the other hand the assumed continuity of the embedding $E_{\xi,\varphi}$ and Proposition 1.3.2 yield

$$\int_\Omega \xi(x, |f(x)| \|E_{\xi,\varphi}\|)\, dx \leq 1 \Leftrightarrow \|f\| E_{\xi,\varphi}\| \|_{L^\xi(\Omega)} \leq 1.$$

On account of the assumed boundedness of the embedding (2.5) the preceding statement yields

$$\|f\|_{L^\varphi(\Omega)} \leq 1.$$

Therefore, $\int_\Omega \varphi(x, |f(x)|)\, dx \leq 1$, which together with (2.7) allows us to conclude that

$$\int_\Omega \xi(x, |f(x)| \|E_{\xi,\varphi}\|)\, dx > 1.$$

Since ξ is finite *a.e.* and Lebesgue's measure is non-atomic, there exists a measurable set $M \subseteq I$ with

$$\int_I \xi(x, \|E_{\xi,\varphi}\| f(x) \mathbf{1}_M(x))\, dx = 1.$$

In particular, the last inequality implies that

$$\int_\Omega \varphi(x, |f(x)| \mathbf{1}_M(x))\, dx \leq 1. \tag{2.8}$$

Notice that $|M| > 0$. Integrating inequality (2.6) on M

$$\int_I \varphi(x, |f(x)| \mathbf{1}_M(x))\, dx \geq 1 + 2\frac{|M|}{|I|} > 1,$$

which contradicts (2.8).
Hence h is *a.e.* finite. Next we prove that in fact,

$$\|h\|_1 \leq 1.$$

Suppose, otherwise that is, assume that for some ϵ one has

$$\int_\Omega h(z)\, dz > \epsilon > 1. \tag{2.9}$$

In this case we set

$$V_k := \left\{ x \in I : \alpha(x, r_k) \geq \frac{2}{1+\epsilon} h(x) \right\}$$

$$W_{k+1} = V_{k+1} \setminus \left(\bigcup_{j=1}^{k} V_j \right).$$

Let f be as in the preceding argument. We claim that the inequality

$$\xi(x, |f(x)|) < \infty$$

holds *a.e.* $x \in \Omega$. Indeed, for $x \in \Omega \setminus \bigcup_{k=2}^{\infty} W_k$, one clearly has $f(x) = 0$ and it follows at once that $\xi(x, |f(x)|) = 0$. Otherwise, for any $k \in \mathbb{N}$, $k > 1$ one has $\alpha(x, r_k) > 0$, thus, $\xi(x, |f(x)|) = \xi(x, r_k) < \infty$, as claimed. By definition of α and W_k, it holds the inequality

$$\varphi(x, |f(x)|) \geq \xi\left(y, |f(x)| \|E_{\xi,\varphi}\|\right) + \frac{2}{1+\epsilon} h(x). \tag{2.10}$$

Indeed, (2.10) is obvious if $x \in W_k$ for $k \geq 2$; on the other hand if $x \in \Omega \setminus \bigcup_{k=2}^{\infty} W_k$ then clearly

$$\alpha(x, r_k) \leq \frac{2}{1+\epsilon} h(x),$$

which automatically extends to

$$\alpha(x, t) \leq \frac{2}{1+\epsilon} h(x) = \frac{2}{1+\epsilon} \sup_{s \geq 0} \alpha(x, s)$$

for any $t \in [0, \infty)$. In turn, the latter is only possible if $h(x) = 0$. Hence (2.10) holds trivially.

Integrating inequality (2.10) in Ω it follows at once that

$$\begin{aligned} \int_{\Omega} \varphi(x, |f(x)|)\, d\mu &\geq \int_{\Omega} \xi\left(x, |f(x)| \|E_{\xi,\varphi}\|\right) d\mu + \frac{2}{1+\epsilon} \int_{\Omega} h(x)\, d\mu \\ &\geq \frac{2}{1+\epsilon} \int_{\Omega} h(x)\, d\mu \geq \frac{2\epsilon}{1+\epsilon} > 1. \end{aligned} \tag{2.11}$$

Inequality (2.11) together with the continuity of $E_{\xi,\varphi}$ forces

$$\int_{\Omega} \xi(x, |f(x)| \|E_{\xi,\varphi}\|)\, d\mu > 1,$$

for otherwise it would be concluded from the continuity of $\|E\|_{\xi,\varphi}$ that

$$\|f\|_{L^{\varphi}(\Omega)} \leq 1,$$

a contradiction to (2.11) by virtue of Proposition 1.3.2. As in the preceding argument, it is concluded that there exists a measurable set $M \subset \Omega$ for which

$$\int_{\Omega} \xi(x, \|E_{\xi,\varphi}\| |f(x)| \mathbf{1}_M(x))\, d\mu = 1. \tag{2.12}$$

Invoking again Proposition 1.3.2 and the continuity of $E_{\xi,\varphi}$, the last equality yields

$$\int_M \varphi(x, |f(x)|)\, d\mu \leq 1. \tag{2.13}$$

Notice that

$$M \bigcap \{z : f(z) > 0\} = M \bigcap \bigcup_{k=2}^{\infty} W_k = M \bigcap \{z : h(z) > 0\},$$

which in concert with equality (2.12) yields

$$\left| M \bigcap \{z : h(z) > 0\} \right| > 0$$

and thus

$$\int_M h(z)\, d\mu > 0. \tag{2.14}$$

In all, assumption (2.9) yields inequalities (2.10) and (2.14). Integrating inequality (2.10) on M:

$$\begin{aligned}\int_M \varphi(x, |f(x)|)\, d\mu &\geq \int_M \xi\left(x, |f(x)| \|E_{\xi,\varphi}\|\right) d\mu + \frac{2}{1+\epsilon} \int_M h(x)\, d\mu \\ &\geq 1 + \frac{2}{1+\epsilon} \int_M h(x)\, d\mu > 1,\end{aligned}$$

which contradicts (2.13). This contradiction stems from assumption (2.9). Hence

$$\int_{\Omega} h(x)\, dx \leq 1,$$

as claimed. The proof of Theorem 2.2.1 is thus complete. □

Example 2.2.1. To illustrate the preceding results, we consider the case of embeddings among the variable exponent Lebesgue spaces. It is well known that when Ω has finite measure, the classical Lebesgue spaces are ordered: if $p < q$, then $L^q(\Omega) \hookrightarrow L^p(\Omega)$. That the same holds for spaces with variable exponent was first shown by Kováčik and Rákosník in Theorem 2.8 of [60]. Their argument is given below. Recall that in this example we fix a domain $\Omega \subseteq \mathbb{R}^n$ and consider the measure space $(\Omega, \mathcal{B}, \mu)$, where $\mathcal{B}$ is the Borel σ-algebra of subsets of Ω and μ stands for the Lebesgue measure. We continue to use the notation of Section 1.5.

Theorem 2.2.2. *Let $|\Omega| < \infty$ and suppose that $p, q \in \mathcal{P}(\Omega)$. Then $L^q(\Omega)$ is continuously embedded in $L^p(\Omega)$ if and only if $p(x) \leq q(x)$ a.e. in Ω. When the embedding id exists,*

$$\|id\| \leq 1 + |\Omega|.$$

Proof. First suppose that $p(x) \leq q(x)$ a.e. in Ω. Note that $\Omega^p_\infty \subset \Omega^q_\infty$. Let u belong to the closed unit ball of $L^q(\Omega)$. Then by Proposition 1.5.2,

$$\rho_q(u) = \int_{\Omega \backslash \Omega^q_\infty} |u|^q \, dx + \operatorname{ess\,sup}_{\Omega^q_\infty} |u| \leq 1.$$

Thus $|u(x)| \leq 1$ a.e. in Ω^q_∞. Hence

$$\begin{aligned}\rho_p(u) &\leq |\{x \in \Omega \backslash \Omega^q_\infty : |u(x)| \leq 1\}| \\ &\quad + \int_{\Omega \backslash \Omega^q_\infty} |u|^q \, dx + |\Omega^q_\infty \backslash \Omega^p_\infty| + \operatorname{ess\,sup}_{\Omega^q_\infty} |u(x)| \\ &\leq |\Omega| + \rho_q(u) \leq |\Omega| + 1.\end{aligned}$$

As ρ_p is convex,

$$\rho_p\left(u/\left(|\Omega| + 1\right)\right) \leq \left(|\Omega| + 1\right)^{-1} \rho_p(u) \leq 1,$$

from which we see that $\|\mathrm{id}\| \leq 1 + |\Omega|$.

Now suppose that it is not the case that $p(x) \leq q(x)$ a.e. in Ω. Then there is a subset Ω^* of Ω, with positive measure, such that $p(x) > q(x)$ for all $x \in \Omega^*$. First assume that $|\Omega^p_\infty \cap \Omega^*| > 0$. Then there exist $A \subset \Omega^p_\infty \cap \Omega^*$, with $0 < |A| < \infty$, and $r \in (1, \infty)$ such that $1 \leq q(x) \leq r < \infty = p(x)$ for all $x \in A$. There are pairwise disjoint sets A_k such that $A = \cup_{k=1}^\infty A_k$ and $|A_k| = 2^{-k}|A|$ for all $k \in \mathbb{N}$; define $u = \sum_{k=1}^{\infty} (3/2)^r \mathbf{1}_{A_k}$. For this function u we have $\|u\|_p \geq \|u_{\mathbf{1}_A}\|_\infty = \infty$, so that $u \notin L^p(\Omega)$. However,

$$\begin{aligned}\rho_q(u) &= \int_A |u|^q \, dx = \sum_{k=1}^{\infty} \int_{A_k} (3/2)^{kq/r} dx \\ &\leq \sum_{k=1}^{\infty} (3/2)^k |A_k| = |A| \sum_{k=1}^{\infty} (3/4)^k < \infty,\end{aligned}$$

and hence $u \in L^{q}(\Omega)$.

On the other hand, if $|\Omega_{\infty}^{p} \cap \Omega^{*}| = 0$, then $1 \leq q(x) < p(x) < \infty$ for a.e. $x \in \Omega^{*}$ and there is a set $A \subset \Omega^{*}$, with $0 < |A| < \infty$, and numbers $a > 0, r \in (1, \infty)$ such that $q(x) + a \leq p(x) \leq r$ whenever $x \in A$. Choose sets A_k as before and define $v(x) = \sum_{k=1}^{\infty} (2^k k^{-2})^{1/q(x)}$ $(x \in \Omega)$. Then

$$\rho_q(v) = \sum_{k=1}^{\infty} 2^k k^{-2} |A_k| = |A| \sum_{k=1}^{\infty} k^{-2} < \infty,$$

which shows that $v \in L^{q}(\Omega)$. But since for all $\lambda \in (0, 1]$,

$$\begin{aligned}\rho_p(\lambda v) &\geq \lambda^r \sum_{k=1}^{\infty} \int_{A_k} (2^k k^{-2})^{p/q} dx \geq \lambda^r \sum_{k=1}^{\infty} (2^k k^{-2})^{1+a/r} |A_k| \\ &= \lambda^r |A| \sum_{k=1}^{\infty} 2^{ak/r} k^{-2(1+a/r)} = \infty,\end{aligned}$$

we see that $v \notin L^{p}(\Omega)$. □

Corollary 2.2.3. *Suppose that $p \in \mathcal{P}(\Omega)$ and let (u_k) be a sequence in $L^{p}(\Omega)$ that converges to u in $L^{p}(\Omega)$. Then there is a subsequence of (u_k) that converges pointwise a.e. in Ω to u.*

Proof. First suppose that $|\Omega| < \infty$. By Theorem 2.2.2, $L^{p}(\Omega) \hookrightarrow L^{1}(\Omega)$ and so $\|u - u_k\|_1 \to 0$. As the result is true for $L^{1}(\Omega)$, there is nothing more to do. If $|\Omega| = \infty$, apply what has already been proved to $L^{p}(\Omega \cap B(0, m))$ $(m \in \mathbb{N})$ and use diagonalization. □

Corollary 2.2.4. *Let $p \in \mathcal{P}(\Omega)$ be such that $p_+ < \infty$ and let $u, u_k \in L^{p}(\Omega)$ $(k \in \mathbb{N})$. Then $u_k \to u$ in $L^{p}(\Omega)$ if and only if*

$$(u_k) \text{ converges to } u \text{ in measure on } \Omega \text{ and } \rho_p(u_k) \to \rho_p(u).$$

Proof. Suppose that $u_k \to u$ in $L^{p}(\Omega)$. By Proposition 1.5.7, (u_k) converges to u in measure on Ω and $\rho_p(u_k - u) \to 0$. By Corollary 2.2.3, there is a subsequence $(u_{l(k)})_{k \in \mathbb{N}}$ of (u_k) that converges pointwise a.e. on Ω to u. The inequality

$$|u_k(x)|^{p(x)} \leq 2^{p_+ - 1} \left(|u_k(x) - u(x)|^{p(x)} + |u(x)|^{p(x)} \right)$$

shows that given $\varepsilon > 0$, there exist $\delta > 0$ and $k_0 \in \mathbb{N}$ such that for all $B \subset \Omega$ with $|B| < \delta$ and all $k \geq k_0$,

$$\int_B |u_k|^p \, dx \leq \varepsilon.$$

Hence by Vitali's theorem (see [63, 2.1.4]),

$$\lim_{k\to\infty} \rho_p(u_{l(k)}) = \rho_p(u).$$

In fact, $\lim_{k\to\infty} \rho_p(u_k) = \rho_p(u)$; for if not, there would be a subsequence of $(\rho_p(u_k))$ whose elements would be uniformly bounded away from $\rho_p(u)$, and an application of the preceding argument to this subsequence would give a contradiction.

For the converse, suppose that (u_k) converges to f in measure on Ω and $\rho_p(u_k) \to \rho_p(u)$. Now use the inequality

$$|u_k(x) - u(x)|^{p(x)} \leq 2^{p_+-1} \left(|u_k(x)|^{p(x)} + |u(x)|^{p(x)} \right),$$

and argue as in the first part to obtain $\rho_p(u_k - u) \to 0$ and hence $\|u_k - u\|_p \to 0$. □

We now show (following [34]) that for embedding maps, greater precision in norm estimation can be obtained when p and q are near to one another in the sense that for some $\varepsilon \in (0,1)$,

$$p(x) \leq q(x) \leq p(x) + \varepsilon \text{ a.e. in } \Omega. \tag{2.15}$$

Lemma 2.2.5. *Suppose that* $|\Omega| < \infty$ *and* p, q *satisfy (2.15). If* $u \in \mathcal{M}(\Omega)$ *and* $\rho_q(u) \leq 1$, *then*

$$\rho_p(u) \leq \varepsilon |\Omega| + \varepsilon^{-\varepsilon}.$$

Proof. Put

$$\Omega^1 = \{x \in \Omega : |u(x)| < \varepsilon\}, \Omega^2 = \{x \in \Omega : \varepsilon \leq |u(x)| \leq 1\},$$
$$\Omega^3 = \{x \in \Omega : 1 < |u(x)|\}.$$

Then

$$\rho_p(u) = \sum_{j=1}^{3} \int_{\Omega^j} |u|^p \, dx = \sum_{j=1}^{3} A_j.$$

It is clear that

$$A_1 \leq \int_{\Omega^1} \varepsilon^p dx \leq \int_{\Omega^1} \varepsilon dx \leq \varepsilon |\Omega|, \; A_3 \leq \int_{\Omega^3} |u|^q \, dx.$$

On Ω^2,

$$\varepsilon^\varepsilon \leq \varepsilon^{q(x)-p(x)} \leq |u(x)|^{q(x)-p(x)} \leq 1,$$

and so

$$1 \leq |u(x)|^{p(x)-q(x)} \leq \varepsilon^{-\varepsilon}.$$

Hence

$$A_2 = \int_{\Omega^2} |u|^q |u(x)|^{p-q}\, dx \leq \varepsilon^{-\varepsilon} \int_{\Omega^2} |u|^q\, dx.$$

Thus

$$\begin{aligned}
\rho_p(u) &\leq \varepsilon |\Omega| + \varepsilon^{-\varepsilon} \int_{\Omega^2} |u|^q\, dx + \int_{\Omega^3} |u|^q\, dx \\
&\leq \varepsilon |\Omega| + \varepsilon^{-\varepsilon} \left(\int_{\Omega^2} |u|^q\, dx + \int_{\Omega^3} |u|^q\, dx \right) \\
&\leq \varepsilon |\Omega| + \varepsilon^{-\varepsilon} \int_{\Omega} |u|^q\, dx \leq \varepsilon |\Omega| + \varepsilon^{-\varepsilon}.
\end{aligned}$$

□

Lemma 2.2.6. *Suppose that $|\Omega| < \infty$ and p, q satisfy (2.15). Then*

$$\|id\|_{L^q \to L^p} \leq \varepsilon |\Omega| + \varepsilon^{-\varepsilon}.$$

Proof. Clearly $K := \varepsilon |\Omega| + \varepsilon^{-\varepsilon} > 1$. For each u such that $\rho_q(u) \leq 1$ we have, by Lemma 2.2.5,

$$\rho_p(u/K) \leq K^{-1} \rho_p(u) \leq \left(\varepsilon |\Omega| + \varepsilon^{-\varepsilon} \right) / K = 1,$$

and the result follows. □

The required estimate from below is established in two stages.

Lemma 2.2.7. *Suppose that $1 \leq |\Omega| < \infty$ and p, q satisfy (2.15). Then*

$$\|id\|_{L^q \to L^p} \geq 1.$$

Proof. Let $v(x) = |\Omega|^{-1/q(x)}$ $(x \in \Omega)$. Then $\rho_q(v) = 1$. Since

$$|\Omega|^{-p(x)/q(x)} \geq |\Omega|^{-1}$$

we have, for each $\lambda \in (0, 1)$,

$$\rho_p(v/\lambda) = \int_{\Omega} \frac{|\Omega|^{-p/q}}{\lambda^p} dx \geq \int_{\Omega} \frac{|\Omega|^{-1}}{\lambda^p} dx \geq \int_{\Omega} \frac{|\Omega|^{-1}}{\lambda} dx = \lambda^{-1} > 1.$$

Hence $\|\mathrm{id}\| \geq \lambda$ for each $\lambda \in (0, 1)$, which shows that $\|\mathrm{id}\| \geq 1$. □

Lemma 2.2.8. *Suppose that $0 < |\Omega| < 1$ and p, q satisfy (2.15). Then*

$$\|id\|_{L^q \to L^p} \geq |\Omega|^{\varepsilon}.$$

Proof. Let v be as in the last lemma. Since

$$|\Omega|^{1-p(x)/q(x)} \geq |\Omega|^{\varepsilon/q(x)} \geq |\Omega|^{\varepsilon},$$

we see that

$$\rho_p(v) = \int_\Omega |\Omega|^{-p/q}\, dx = |\Omega|^{-1} \int_\Omega |\Omega|^{1-p/q}\, dx \geq |\Omega|^{\varepsilon}.$$

Thus for each positive $\lambda < |\Omega|^{\varepsilon}$,

$$\begin{aligned}\rho_p(v/\lambda) &> \int_\Omega \left|\frac{v}{|\Omega|^{\varepsilon}}\right|^p dx = \int_\Omega \left|\frac{v}{|\Omega|^{\varepsilon/p}}\right|^p dx \\ &= |\Omega|^{-\varepsilon} \int_\Omega |v|^p\, dx \geq |\Omega|^{-\varepsilon} |\Omega|^{\varepsilon} = 1.\end{aligned}$$

Hence $\|\mathrm{id}\| \geq \lambda$ for each positive $\lambda < |\Omega|^{\varepsilon}$, and the result follows. □

From these lemmas we immediately have the following theorem and corollary.

Theorem 2.2.9. *Suppose that $0 < |\Omega| < \infty$ and p, q satisfy (2.15). Then the norm of the embedding id of $L^q(\Omega)$ in $L^p(\Omega)$ satisfies*

$$\min(1, |\Omega|^{\varepsilon}) \leq \|id\|_{L^q \to L^p} \leq \varepsilon |\Omega| + \varepsilon^{\varepsilon}.$$

Corollary 2.2.10. *Suppose that $0 < |\Omega| < \infty$ and $p \in \mathcal{P}(\Omega)$ is such that $1 < p_- \leq p_+ < \infty$; suppose that for each $k \in \mathbb{N}$, $q_k \in \mathcal{P}(\Omega)$ and $\varepsilon_k > 0$ are such that $\lim_{k\to\infty} \varepsilon_k = 0$ and*

$$p(x) \leq q_k(x) \leq p(x) + \varepsilon_k \quad (x \in \Omega, k \in \mathbb{N}).$$

Let id_k be the natural embedding of $L^{q_k}(\Omega)$ in $L^p(\Omega)$. Then

$$\lim_{k\to\infty} \|id_k\|_{L^{q_k} \to L^p} \| = 1.$$

2.2.1 The Δ_2-condition

It is evident from the theory presented so far that the Δ_2 condition plays an essential role in the treatment of modular spaces. In this section a more detailed analysis is undertaken.

Definition 2.2.2. *An Musielak-Orlicz function φ is said to fulfill the Δ_2 condition if there exist constants $C > 1$ and $S_0 > 0$ such that, uniformly in Ω, the following inequality holds:*

$$\varphi(x, 2s) \leq C\varphi(x, s) \text{ for all } s \geq S_0,\ x \in \Omega. \tag{2.16}$$

Lemma 2.2.11. *Let φ be a Musielak-Orlicz function. If there exists a natural number $K > 1$ and constants $C > 1$ and $S_0 > 0$ such that, uniformly in Ω, it holds that*

$$\varphi(x, Ks) \leq C\varphi(x,s) \text{ for all } s \geq S_0\,,\ x \in \Omega, \tag{2.17}$$

then φ satisfies the Δ_2 condition.

Proof. If $K > 1$, then $K^j > 2$ for some $j \in \mathbb{N}$. Inequality (2.17) implies thus that uniformly in Ω one has

$$\varphi(x, 2s) \leq \varphi(x, K^j s) \leq C^j \varphi(x,s) \text{ for all } s \geq S_0.$$

□

Example 2.2.2. Given a measure space $(\Omega, \mathcal{A}, \mu)$ and $1 \leq p < \infty$ the Lebesgue space $L^p(\Omega)$ is the Musielak-Orlicz space arising from the consideration of the Musielak-Orlicz function

$$\begin{aligned}&\varphi : \Omega \times [0,\infty) \longrightarrow [0,\infty)\\ &\varphi(t) = t^p.\end{aligned}$$

It is routine to verify the validity of the Δ_2 condition for $L^p(\Omega)$.
On the other hand, setting

$$\begin{aligned}&\varphi : \Omega \times [0,\infty) \longrightarrow [0,\infty)\\ &\varphi(t) = \begin{cases} t & \text{if} \quad 0 \leq t \leq 1 \\ e^{t-1} & \text{if} \quad t > 1 \end{cases}\end{aligned}$$

a simple calculation reveals that φ does not satisfy the Δ_2 condition.

We emphasize the fact that if φ is a Musielak-Orlicz function on a measure space $(\Omega, \mathcal{A}, \mu)$ and φ satisfies the Δ_2 condition, then

$$\left\{ u \in \mathcal{M}(\Omega) : \int_\Omega \varphi(x, |u(x)|)dx < \infty \right\} = L^\varphi(\Omega). \tag{2.18}$$

Proposition 2.2.12. *Let φ be a Musielak-Orlicz function on a measure space $(\Omega, \mathcal{A}, \mu)$ and assume that φ satisfies the Δ_2 condition (2.16). Then, there exist constants $\alpha = \alpha(\varphi, C) \geq 0$ and $\beta = \beta(\varphi, C) \geq 0$ such that for $0 \leq t \leq 1$ and all $s \geq S_0$,*

$$C^{-1} t^\alpha \varphi(x,s) \leq \varphi(x, ts) \leq t\varphi(x,s)$$

and that for $1 < t$ and all $s \geq S_0$,

$$t\varphi(x,s) \leq \varphi(x,ts) \leq Ct^\beta \varphi(x,s). \tag{2.19}$$

Proof. For any $s \geq S_0$, $x \in \Omega$ and $0 < t < 1$, let $2^n \leq \frac{1}{t} < 2^{n+1}$, $n \in \mathbb{N}_0$.

$$\begin{aligned}\varphi(x,s) = \varphi\left(x, t\frac{1}{t}s\right) &\leq C^{n+1}\varphi(x,ts) \leq CC^{\frac{-\ln t}{\ln C}}\varphi(x,ts)\\ &\leq C\left(\frac{1}{t}\right)^{\frac{\ln C}{\ln 2}}\varphi(x,ts);\end{aligned}$$

the right-hand inequality follows from convexity.
The left-hand side inequality for $t \geq 1$ follows from the convexity assumption on φ; the remaining inequality is proved as the one above:

$$\begin{aligned}\varphi(x,ts) &\leq \varphi(x, 2^{[\frac{\ln t}{\ln K}]+1}s) \leq CC^{[\frac{\ln t}{\ln K}]}\varphi(x,s)\\ &\leq Ct^{\frac{\ln C}{\ln 2}}\varphi(x,s).\end{aligned}$$

□

Lemma 2.2.13. *If φ satisfies the Δ_2 condition, then the modular ρ_φ given in (2.1) is continuous.*

Proof. Fix $u \in L^\varphi(\Omega)$, hence, for some $\gamma > 0$ it holds

$$\int_\Omega \varphi(x, \gamma|u(x)|)\, dx < \infty.$$

If $(\lambda_j) \subset (0,\infty)$ converges to λ, the monotonicity of φ yields

$$\varphi(x, \lambda_j|u(x)|) \leq \varphi\left(x, \frac{3}{2}\lambda|u(x)|\right) \leq \varphi(x, \gamma|u(x)|)$$

if $\frac{3}{2}\lambda \leq \gamma$. The Δ_2 condition comes into play when observing that for $\frac{3}{2}\lambda > \gamma$, inequality (2.19) implies that for $|u(x)| \geq S_0$ one has

$$\varphi\left(x, \frac{3}{2}\lambda|u(x)|\right) = \varphi\left(x, \frac{3}{2\gamma}\lambda\gamma|u(x)|\right) \leq C\left(\frac{3\lambda}{2\gamma}\right)^\alpha \varphi(x, \gamma|u(x)|).$$

A straightforward application of Lebesgue's dominated convergence theorem now yields

$$\begin{aligned}\rho_\varphi(\lambda_j u) = &\int_{|u|\leq S_0} \varphi(x, \lambda_j|u(x)|)\, dx + \int_{|u|\geq S_0} \varphi(x, \lambda_j|u(x)|)\, dx\\ &\longrightarrow \rho_\varphi(\lambda u) \text{ as } j \longrightarrow \infty\end{aligned}$$

i.e., the continuity claim. □

Lemma 2.2.14. *If φ is an Musielak-Orlicz function satisfying the Δ_2 condition, then for any $0 \neq u \in L^{\varphi}(\Omega)$, one has*

$$\int_{\Omega} \varphi\left(x, \frac{|u(x)|}{\|u\|_{\varphi}}\right) dx = 1.$$

Proof. Take a sequence $(\lambda_j) \subset (0, \infty)$ convergent to $\|u\|_{\varphi}$ with

$$\int_{\Omega} \varphi\left(x, \frac{|u(x)|}{\lambda_j}\right) \leq 1.$$

By virtue of Lemma 2.2.13, one must necessarily have

$$\int_{\Omega} \varphi\left(x, \frac{|u(x)|}{\|u\|_{\varphi}}\right) dx \leq 1, \tag{2.20}$$

with the same token it is apparent that the assumption of strict inequality in (2.20) leads to a contradiction. □

2.2.2 Absolute continuity of the norm

Recall from Section 1.2.2, if X be a Banach function space, two subspaces of X are to be distinguished, namely:

(*i*) The set of all elements of X with absolutely, continuous norm, denoted by X_a;

(*ii*) The closure of the set of all bounded functions supported in sets of finite measure, denoted by X_b.

In the next theorem we characterize the space $L^{\varphi}_a(\Omega)$.

Theorem 2.2.15. *Let φ be a Musielak-Orlicz function on the σ-finite measure space $(\Omega, \mathcal{A}, \mu)$. Then*

$$L^{\varphi}_a(\Omega) = \{u \in L^{\varphi}(\Omega) : \rho_{\varphi}(\lambda u) < \infty \text{ for all } \lambda > 0\} \tag{2.21}$$

Proof. Let (Ω_k) be a sequence of mutually disjoint open sets with $\mu(\Omega_k) < \infty$, whose union is Ω. If $u \in L^{\varphi}(\Omega)$ and $\rho_{\varphi}(\lambda u) = \infty$ for $\lambda > 0$ then

$$\sum_{k=1}^{\infty} \rho_{\varphi}(u\mathbf{1}_{\Omega_k}) = \sum_{k=1}^{\infty} \int_{\Omega_k} \varphi(x, \lambda|u(x)|)\, d\mu = \infty.$$

Let $N_1 \in \mathbb{N}$ satisfy

$$\sum_{n=1}^{N_1} \rho_{\varphi}(\lambda u\mathbf{1}_{\Omega_k}) > 1;$$

there exists then $N_2 > N_1$ with

$$\sum_{n=N_1+1}^{N_2} \rho_\varphi\left(\lambda u \mathbf{1}_{\Omega_k}\right) > 1;$$

in this fashion one obtains a strictly increasing sequence $(N_k) \subset \mathbb{N}_0$ satisfying $N_0 = 0$ and

$$\sum_{n=N_k+1}^{N_{k+1}} \rho_\varphi\left(\lambda u \mathbf{1}_{\Omega_k}\right) > 1$$

for $k \geq 0$. Set $\tilde{\Omega}_k = \bigcup\limits_{j=N_k+1}^{N_{k+1}} \Omega_j$ for $k > 1$; it is clear that the sequence $(\tilde{\Omega}_k)$ consists of mutually disjoint sets and that $\mathbf{1}_{\tilde{\Omega}_k} \to \emptyset$ as $k \to \infty$. By construction

$$\rho_\varphi\left(\lambda u \mathbf{1}_{\tilde{\Omega}_k}\right) > 1$$

and it follows straight from the definition of the Luxemburg norm that $\|u\mathbf{1}_{\tilde{\Omega}_k}\|_\varphi > \lambda^{-1}$, whence it is clear that $u \notin L_a^\varphi(\Omega)$. Thus,

$$L_a^\varphi(\Omega) \subseteq \{u \in L^\varphi(\Omega) : \rho_\varphi(\lambda u) < \infty \text{ for all } \lambda > 0\}.$$

Conversely, for any u in the right-hand side set of (2.21), any sequence (E_k) of measurable sets such that $\mathbf{1}_{E_k} \to \emptyset$ and any $\varepsilon > 0$, one has

$$\lim_{k\to\infty} \rho_\varphi\left(\varepsilon^{-1} u \mathbf{1}_{E_k}\right) = 0$$

on account of the dominated convergence theorem; it is readily derived from here that $\|u\mathbf{1}_{E_k}\|_\varphi \longrightarrow 0$ as $k \to \infty$. □

Corollary 2.2.16. *If a Musielak-Orlicz function φ on the σ-finite measure space $(\Omega, \mathcal{A}, \mu)$ satisfies the Δ_2 condition, then*

$$L_a^\varphi(\Omega) = L^\varphi(\Omega) = L_b^\varphi(\Omega).$$

Conversely, if φ is an Musielak-Orlicz function on $(\Omega, \mathcal{A}, \mu)$ and $L_a^\varphi(\Omega) = L^\varphi(\Omega)$, then there exist $C > 0$ and $0 \leq h \in L^1(\Omega)$ such that for each $t > 0$:

$$\varphi(x, 2t) \leq C\varphi(x, t) + h(t) \ \text{a.e.}\, x \in \Omega. \tag{2.22}$$

Proof. The first assertion follows immediately from the fact that the Δ_2 condition for φ implies that $\rho_\varphi(\lambda u) < \infty$ for any $u \in L^\varphi(\Omega)$ and any $\lambda > 0$, from Theorem 2.2.15 and Lemma 1.2.16.
For the remaining claim, observe that setting $\xi(x, t) = \varphi(x, 2t)$, the assumption and Theorem 2.2.15 yield the continuity of the embedding

$$E : L^\varphi(\Omega) \hookrightarrow L^\xi(\Omega)$$

and on account of Theorem 2.2.1 the latter implies (2.22). □

2.3 Separability

The study of separability of Musielak-Orlicz spaces is tied to both the structure of the underlying measure and of the Musielak-Orlicz function. Throughout this section we will consider the measure μ to be separable and the Musielak-Orlicz function φ on the measure space $(X, \mathcal{A}, \mu)$ to be locally integrable, that is (Definition 2.1.2) for any $t \geq 0$ and any $W \in \mathcal{A}$ with $\mu(W) < \infty$, it holds that

$$\int_W \varphi(x,t)\,d\mu < \infty. \tag{2.23}$$

Remark
We observe the useful fact that if φ is locally integrable and (B_k) is a sequence of measurable sets for which

$$\mu(B_k) \to 0 \text{ as } k \to \infty,$$

then necessarily, for any $\lambda > 0$, it holds that

$$\rho_\varphi(\lambda \mathbf{1}_{B_k}) = \int_{B_k} \varphi(x,\lambda)\,d\mu \to 0 \text{ as } k \to \infty.$$

For otherwise, there would exist $\delta > 0$ and a sequence $(B_k) \subset \mathcal{A}$ with

$$\mu(B_k) < 2^{-k} \text{ and } \rho_\varphi(\lambda \mathbf{1}_{B_k}) > \delta.$$

Then if $D_j = \bigcup_{k \geq j} B_k$ for $j \in \mathbb{N}$, $\mu(D_1) = 1$ and pointwise,

$$\mathbf{1}_{D_1} \geq \mathbf{1}_{D_j} \to 0 \text{ as } j \to \infty.$$

On account of Lebesgue's theorem the latter would imply

$$\int_{B_j} \varphi(x,\lambda)\,d\mu \leq \int_{D_j} \varphi(x,\lambda)\,d\mu \to 0 \text{ as } j \to \infty,$$

a contradiction.
Next, we recall the standard notation

$$A \triangle B = (A \setminus B) \bigcup (B \setminus A)$$

to denote the symmetric difference between the sets A and B.

Definition 2.3.1. *The measure space $(X, \mathcal{A}, \mu)$ is said to be separable if there exists a sequence $(E_k) \subset \mathcal{A}$, with $\mu(E_k) < \infty$ for all $k \in \mathbb{N}$, such that for any $E \in \mathcal{A}$ with $\mu(E) < \infty$ and any $\epsilon > 0$, there exists $k_0 \in \mathbb{N}$ with*

$$\mu(E \triangle E_{k_0}) < \epsilon.$$

If $\Omega \subseteq \mathbb{R}^n$ is Lebesgue measurable, then the (restriction of the) Lebesgue measure to Ω is separable.
Recall that $\mathcal{S}_0$ stands for the class of simple functions of the type

$$w = \sum_{k=0}^{J} w_k \mathbf{1}_{A_k} \text{ with } \mu(A_k) < \infty \text{ and } w_k \neq w_j \text{ for } k \neq j.$$

Lemma 2.3.1. *Given a separable measure space $(\Omega, \mathcal{A}, \mu)$ as in Definition 2.3.1 and a locally integrable Musielak-Orlicz function φ on Ω, the (countable) set $\mathcal{C} \subset \mathcal{S}_0$ consisting of simple functions of the type*

$$s(t) = \sum_{k=0}^{M} s_k \mathbf{1}_{E_k}, \quad s_k \in \mathbb{Q}, 1 \leq k \leq M$$

is $\|\cdot\|_\varphi$-dense in the class $\mathcal{S}_0$.

Proof. Let $\varepsilon > 0$ and $u = \sum_{k=1}^{J} u_k \mathbf{1}_{A_k} \in \mathcal{S}_0$ and set $\max\{s_1, ..., s_J\} := 2M$. On account of the separability of μ, the local integrability of φ and the Remark following Definition 2.23 there is a finite collection $(F_k)_{k=1}^{J}$ of the sequence (E_n) given in Definition 2.3.1 such that for $1 \leq k \leq J$ one has the inequality

$$\int_{F_k \triangle A_k} \varphi(x, 4\lambda JM)\, d\mu < \varepsilon/2.$$

Invoking again the local integrability of φ it is immediate that if $\lambda > 0$ is arbitrary and $A = \bigcup_{i=1}^{J} A_i$, then pointwise in Ω,

$$L^1(\Omega, \mu) \ni \varphi(x, \mathbf{1}_A(x)) \geq \varphi(x, 2\lambda r \mathbf{1}_A(x)) \to 0 \text{ as } r \to 0.$$

By virtue of Lebesgue's theorem there exists $\delta > 0$ such that

$$\int_{\Omega} \varphi(x, 2\lambda\delta \mathbf{1}_A(x)) < \varepsilon/2.$$

Then one can select rational number $r_1, ..., r_J$ with $|r_j - u_j| < \delta$ and $|r_j| \leq 2M$ for $j = 1, 2, ..., J$. Let

$$v = \sum_{j=1}^{J} r_k \mathbf{1}_{F_k}.$$

Clearly

$$|u - v| = \left| \sum_{j=1}^{J} (u_j - r_j) \mathbf{1}_{A_j} \right| + \sum_{j=1}^{J} |r_j| |\mathbf{1}_{A_j \setminus F_j}| \leq \delta \mathbf{1}_A + 2M \sum_{j=1}^{J} |\mathbf{1}_{A_j \setminus F_j}|.$$

In all, for any $\lambda > 0$

$$\begin{aligned}\rho_\varphi(\lambda(u-v)) &= \frac{1}{2}\int_A \varphi(x, 2\lambda\delta)\, d\mu + \frac{1}{2}\int_\Omega \varphi\left(x, 4M\lambda \sum_{j=1}^{J} |\mathbf{1}_{A_j\setminus F_j}|\right) d\mu \\ &\le \frac{\varepsilon}{2} + \frac{1}{2J}\sum_{j=1}^{J}\int_\Omega \varphi\left(x, 4MJ\lambda|\mathbf{1}_{A_j\setminus F_j}|\right) d\mu \\ &\le \frac{1}{2} + \frac{1}{2J}\sum_{j=1}^{J}\frac{\varepsilon}{2} = \varepsilon.\end{aligned}$$

The arbitrariness of λ implies the claim. □

Theorem 2.3.2. *Let $(X, \mathcal{A}, \mu)$ be a separable measure space and*

$$\varphi : \Omega \times [0, \infty) \longrightarrow [0, \infty)$$

be locally integrable. If $L^\varphi(\Omega) = L^\varphi_a(\Omega)$, then $L^\varphi(\Omega)$ is separable.

Proof. We notice first that as an immediate consequence of condition (2.23), the class $\mathcal{S}$ of simple functions of the type

$$w = \sum_{k=0}^{J} w_k \mathbf{1}_{E_k}, \ \ \mu(E_k) < \infty$$

is contained in $L^\varphi(\Omega)$. It is well known that for an $\mathcal{A}$-measurable function $u \ge 0$ in Ω there exists a non-decreasing sequence of simple non-negative functions (u_n) which converges pointwise to u in Ω. In particular, if $u \in \mathcal{M}(\Omega)$, $n \in \mathbb{N}$

$$\int_{\{x:|u(x)|\ge 2^n\}} \varphi(x, 2^n)\, d\mu \le \int_\Omega \varphi(x, |u(x)|)\, d\mu$$

the sequence can be chosen to be of the type $\mathcal{S}$. Select two nondecreasing sequences in $\mathcal{S}$, (a_n) and (b_n) converging pointwise to $u^+ = \max\{u, 0\}$ and $u^- = \max\{-u, 0\}$. Hence, pointwise

$$u_n = a_n - b_n \to u^+ - u^- = u \ \text{ as } n \to \infty.$$

In particular, if $u \in L^\varphi(\Omega)$, for any $x \in \Omega$

$$\begin{aligned}\varphi(x, \lambda(u_n(x) - u(x))) &\le \varphi(x, \lambda(a_n(x) - u^+(x) - b_n(x) + u^-(x))) \\ &\le \varphi(x, \lambda(2(u^+(x) + u^-(x))) \\ &\le \varphi(x, 2\lambda|u(x)|) < \infty,\end{aligned}$$

on account of Theorem 2.2.15. Thus, for any $\lambda > 0$ and by direct application of Lebesgue's dominated convergence theorem it follows that

$$\rho_\varphi(\lambda(u_n - u)) = \int_\Omega \varphi(x, \lambda(u_n(x) - u(x)))\, d\mu \to 0 \text{ as } n \to \infty.$$

Thus (Theorem 1.3.8) u_n converges to u in $L^\varphi(\Omega)$. Separability is implied by the preceding result in conjunction with Lemma 2.3.1. □

The remark following Definition 2.3.1 in concert with the preceding theorem yields

Corollary 2.3.3. *Let $\Omega \subseteq \mathbb{R}^n$ be open, consider the measure space $(\Omega, \mathcal{B}, \mu)$ where $\mathcal{B}$ is the Borel σ-algebra and μ is the Lebesgue measure. If φ is a locally integrable Musielak-Orlicz function on Ω and $L^\varphi(\Omega) = L^\varphi_a(\Omega)$, then $L^\varphi(\Omega)$ is separable.*

Remark 2.3.4. Corollary 2.3.3 will be obtained by different means in Theorem 2.5.6.

Corollary 2.3.5. *If $\Omega \subseteq \mathbb{R}^n$ is a Borel set, μ is the Lebesgue's measure on $\mathbb{R}^n$ and $p : \Omega \longrightarrow [1, \infty)$ is Borel measurable with $p_+ < \infty$, then the variable exponent Lebesgue space $L^{p}(\Omega)$ is separable.*

Proof. If $p_+ < \infty$, then the Musielak-Orlicz function $t^{p(x)}$ satisfies the Δ_2 condition (Theorem 1.5.5) which in turn implies that $L^p(\Omega)$ has absolutely continuous norm. Corollary 2.3.3 yields the desired result. □

2.4 Duality of Musielak-Orlicz spaces

Spaces of Musielak-Orlicz type are modular spaces; it is thus in order to relate the study of their duality to the setting of Section 1.3. The following discussion will facilitate the connection we wish to establish.

2.4.1 Conjugate Musielak-Orlicz functions

Fix an Musielak-Orlicz function φ on Ω. We observe the fact that for each fixed $x \in \Omega$ the functional

$$\rho_x(r) = \begin{cases} \varphi(x, r) & \text{if} \quad r \geq 0 \\ \varphi(x, -r) & \text{if} \quad r < 0 \end{cases}$$

is a modular on $\mathbb{R}$. The conjugate function φ^* on $\Omega \times [0, \infty)$ is defined as

$$\varphi^*(x,t) = \sup_{s \geq 0} \{st - \varphi(x,s)\}. \tag{2.24}$$

It is a matter of routine to verify that φ^* is also a Musielak-Orlicz function on the prescribed measure space. Moreover, on account of the observation made at the beginning of this section and Theorem 1.3.14 it is clear that $\varphi^{**} = \varphi$ holds for any Musielak-Orlicz function φ. In particular, for any $x \in \Omega$, Young's inequality is valid:

$$ts \leq \varphi(x,t) + \varphi^*(x,s). \tag{2.25}$$

The preceding inequality was first observed by Young in a much less general context (see Section 2.10).

Theorem 2.4.1. *Let $(\Omega, \mathcal{A}, \mu)$ be a measure space, φ a Musielak-Orlicz function on Ω. Then for any $u \in L^{\varphi}(\Omega)$ and $v \in L^{\varphi^*}(\Omega)$, the following generalized version of Hölder's inequality holds:*

$$\int_{\Omega} |uv| \, d\mu \leq 2\|u\|_{\varphi}\|v\|_{\varphi^*}. \tag{2.26}$$

Proof. The claim is trivially true if either $\|u\|_{\varphi} = 0$ or $\|v\|_{\varphi^*} = 0$. Otherwise, Young's inequality yields

$$\begin{aligned}\int_{\Omega} \frac{|u(x)|}{\|u\|_{\varphi}} \frac{|v(x)|}{\|v\|_{\varphi^*}} \, d\mu &\leq \int_{\Omega} \varphi\left(x, \frac{|u(x)|}{\|u\|_{\varphi}}\right) d\mu + \int_{\Omega} \varphi^*\left(x, \frac{|v(x)|}{\|v\|_{\varphi*}}\right) d\mu \\ &\leq 2.\end{aligned}$$

The latter inequality proves the claim. □

Example 2.4.1. If $\mathcal{A}$ is the σ-algebra of Borel measurable subsets of $\mathbb{R}^n$, $\Omega \in \mathcal{A}$ and μ is the Lebesgue measure, it is a matter of routine to check that for $p : \Omega \to (1, \infty)$, the Musielak-Orlicz functions $\varphi(x,t) = \frac{t^{p(x)}}{p(x)}$ and $\varphi^*(x,t) = \frac{t^{q(x)}}{q(x)}$ are conjugate if and only if $\frac{1}{p(x)} + \frac{1}{q(x)} = 1$ in Ω.

In particular, if p is constant on Ω, φ and φ^* fall into the special category of Orlicz functions, since they are independent of the spatial variable x. In this case Young's inequality takes the form

$$ts \leq \frac{1}{p}t^p + \frac{1}{q}t^q$$

and the well-known stronger version of Hölder's inequality holds, namely

$$\int_{\Omega} |u(x)v(x)| \leq \|u\|_p\|v\|_q.$$

However, if the exponents p and q in the preceding example are allowed to depend on the x, it is easy to verify that the constant 2 in (2.26) cannot be avoided. Considering the importance of this particular case in the applications, we state Hölder's inequality for variable exponent spaces separately:

Theorem 2.4.2. *In the notation of Example 2.1.2, let p and q be real-valued, Borel measurable functions on Ω with $1 \le p < \infty$ and $1 \le q < \infty$, a.e. in Ω and such that*

$$\frac{1}{p(x)} + \frac{1}{q(x)} = 1 \ \textit{a.e. in}\ \Omega.$$

Then for any functions $u \in L^{p}(\Omega)$ and $v \in L^{q}(\Omega)$, the following generalized Hölder's inequality holds:

$$\int_\Omega |uv|\, dx \le \int_\Omega \left(\frac{u^p}{p} + \frac{v^q}{q}\right) dx.$$

In particular,

$$\int_\Omega |uv|\, dx \le \frac{1}{\inf_\Omega p} \int_\Omega u^p + \frac{1}{\inf_\Omega q} \int_\Omega v^q dx,$$

and

$$\int_\Omega |uv|\, dx \le \left(\frac{1}{\inf_\Omega p} + \frac{1}{\inf_\Omega q}\right) \|u\|_p \|v\|_q.$$

2.4.2 Dual of $L^\varphi(\Omega)$

Recall that if φ is proper the corresponding Musielak-Orlicz space $L^\varphi(\Omega)$ is a Banach function space. Accordingly, its associate space $(L^\varphi(\Omega))'$ is given by

$$(L^\varphi(\Omega))' = \left\{ g \in \mathcal{M}(\Omega) : \sup_{f \in L^\varphi(\Omega) \wedge \|f\|_\varphi \le 1} \int_\Omega |fg| d\mu < \infty \right\}.$$

Theorem 2.4.3. *Let $(\Omega, \mathcal{A}, \mu)$ be a σ-finite measure space and φ a proper Musielak-Orlicz function on Ω (i.e., $L^\varphi(\Omega)$ contains the class of simple functions). Then the set-theoretic equality*

$$L^{\varphi^*}(\Omega) = (L^\varphi(\Omega))'$$

holds and furthermore, if $u \in L^{\varphi^*}(\Omega)$ one has

$$\|u\|_{\varphi^*} \le \|u\|_{(L^\varphi(\Omega))'} \le 2\|u\|_{\varphi^*}. \tag{2.27}$$

Proof. If $u \in L^{\varphi^*}(\Omega)$ then Hölder's inequality yields, for any $w \in L^\varphi(\Omega)$:

$$\int_\Omega |u(x)w(x)| dx \le 2\|u\|_{\varphi^*}\|w\|_\varphi$$

and it is obvious from here that

$$L^{\varphi^*}(\Omega) \subseteq (L^{\varphi}(\Omega))'$$

and that the embedding

$$L^{\varphi^*}(\Omega) \hookrightarrow (L^{\varphi}(\Omega))'$$

is bounded with norm at most $2\|u\|_{\varphi^*}$.

To prove the remaining inclusion let $u \in (L^{\varphi}(\Omega))'$, set $q_1 = 0$ and let $(q_j) = (q_j)_{j\geq 1}$ be the sequence defined as

$$(q_j) = \mathbb{Q} \cap [0, \infty).$$

Let (Ω_k) be a nondecreasing sequence of sets with $\mu(\Omega_k) < \infty$ for each $k \in \mathbb{N}$ and such that

$$\Omega = \bigcup_1^{\infty} \Omega_k.$$

We define the sequence of functions (w_k) as

$$w_k(x) := \mathbf{1}_{\Omega_k}(x) \max_{1\leq j\leq k} \{q_j|u(x)| - \varphi(x, q_j)\}.$$

It is easy to verify that

(*i*) (w_k) is non-negative and non-decreasing pointwise in Ω $(q_1 = 0)$,

(*ii*) $w_k(x) \leq \varphi^*(x, |u(x)|)$ for any $x \in \Omega$.

In fact, pointwise in Ω, $w_k(x) \nearrow \varphi^*(x, |u(x)|)$ (trivially so if $|u(x)| = 0$). This is a direct consequence of the left-continuity of φ. To prove it, fix $\epsilon > 0$, $\nu > 0$ and $t_\nu > 0$ such that

$$\varphi^*(x, |u(x)|) - \frac{\epsilon}{3} < |u(x)|t_\nu - \varphi(x, t_\nu).$$

It follows immediately that for any $j \in \mathbb{N}$,

$$\begin{aligned}\varphi^*(x, |u(x)|) - \frac{\epsilon}{3} &< |u(x)|t_\nu - \varphi(x, t_\nu)\\ &= |u(x)|(t_\nu - q_j) - \varphi(x, t_\nu) + \varphi(x, q_j) + |u(x)|q_j - \varphi(x, q_j)\end{aligned}$$

and selecting M such that $q_M < t_\nu$ and

$$\max\{t_\nu - q_M, \varphi(x, q_M) - \varphi(x, t_\nu)\} < \frac{\epsilon}{3}$$

one concludes that if $k \geq M$

$$w_k(x) \geq w_M(x) \geq |u(x)|q_M - \varphi(x, q_M) > \varphi^*(x, |u(x)|) - \epsilon.$$

For each $k \in \mathbb{N}$ and $x \in \Omega_k$, let $j_k(x)$ be the smallest natural number such that

$$w_k(x) = q_{j_x(x)}|u(x)| - \varphi(x, q_{j_k(x)}).$$

With this definition, for each $k \in \mathbb{N}$ let

$$B_{j,k} = \{x \in \Omega_k : j = j_k(x)\}$$

and observe that for $x \in \Omega_k$

$$\begin{aligned} w_k(x) &= \sum_{j=1}^{k} (q_j|u(x)| - \varphi(x, q_j)) \mathbf{1}_{B_{j,k}} \\ &= s_k(x)|u(x)| - \varphi(x, s_k(x)), \end{aligned}$$

where

$$s_k = \sum_{j=1}^{k} q_j \mathbf{1}_{B_{j,k}}$$

is a simple function and hence belongs to $L^\varphi(\Omega)$ by assumption. Define $\Lambda_u \in (L^\varphi(\Omega))^*$ as

$$L^\varphi(\Omega) \ni u \longrightarrow \Lambda_u(v) = \int_\Omega u(x)v(x)d\mu$$

and recall the definition of the modular ρ_φ^* given in Theorem 1.3.12. We conclude from the above discussion that

$$\begin{aligned} \int_\Omega w_k(x)d\mu &= \int_\Omega (s_k(x)|u(x)| - \varphi(x, s_k(x)))\, d\mu \\ &= \int_\Omega s_k(x)|u(x)|d\mu - \rho_\varphi(s_k) \\ &\leq \sup_{v \in L^\varphi(\Omega)} \left\{ \int_\Omega u(x)v(x)d\mu - \rho_\varphi(v) \right\} \\ &= \rho_\varphi^*(\Lambda_u). \end{aligned}$$

A straightforward application of the monotone convergence theorem now yields

$$\rho_{\varphi^*}(u) = \int_\Omega \varphi^*(x, |u(x)|)d\mu \leq \rho_\varphi^*(\Lambda_u).$$

On the other hand, it is clear from Young's inequality (2.25) that

$$\begin{aligned} \rho_\varphi^*(\Lambda_u) &= \sup_{v \in L^\varphi(\Omega)} \left\{ \int_\Omega u(x)v(x)d\mu - \rho_\varphi(v) \right\} \\ &\leq \rho_{\varphi^*}(u). \end{aligned}$$

In all, it is concluded that

$$\rho_{\varphi^*}(u) = \rho_{\varphi}^*(\Lambda_u).$$

Consequently, if $\|u\|_{\varphi^*} \neq 0$,

$$\rho_{\varphi^*}\left(\frac{u}{\|\Lambda_u\|_{(L^\varphi(\Omega))^*}}\right) = \rho_{\varphi}^*\left(\frac{\Lambda_u}{\|\Lambda_u\|_{(L^\varphi(\Omega))^*}}\right) = 1,$$

whence

$$\|u\|_{\varphi^*} \leq \|\Lambda_u\|_{(L^\varphi(\Omega))^*} = \|u\|_{(L^\varphi(\Omega))'},$$

which completes the proof of the inequalities (2.27). □

The next theorem reveals the depth of the idea of conjugation of Musielak-Orlicz function and generalizes the well-known Riesz representation theorem for the classical L^p scale.

Theorem 2.4.4. *Let $(\Omega, \mathcal{A}, \mu)$ be a σ-finite measure space and let φ be a proper, locally integrable Musielak-Orlicz function on Ω. Denote the conjugate of φ by φ^* (see (2.24)). Then the map*

$$M : L^{\varphi^*}(\Omega) \longrightarrow (L^\varphi(\Omega))^*$$
$$\langle M(u), \phi\rangle = \int_\Omega u(x)\phi(x)d\mu$$

is an isomorphism if and only if for any $\lambda > 0$ and any $u \in L^\varphi(\Omega)$,

$$\rho_\varphi(\lambda u) < \infty.$$

Proof. The proof follows immediately from Theorem 1.2.18, Theorem 2.2.15 and Theorem 2.4.3. □

Corollary 2.4.5. *In the notation of Theorem 2.4.4 if $(\Omega, \mathcal{A}, \mu)$ is a σ-finite measure space and φ and its conjugate φ^* are proper, then $L^\varphi(\Omega)$ is reflexive.*

Proof. If M is as in Theorem 2.4.4, then

$$\left(M^{-1}\right)^* : \left(L^{\varphi^*}(\Omega)\right)^* \longrightarrow (L^\varphi(\Omega))^{**}$$

is an isomorphism. A further application of Theorem 2.4.4 in conjunction with the fact that $\varphi^{**} = \varphi$ reveals that, given $\Lambda \in (L^\varphi(\Omega))^{**}$ there exists $u_\lambda \in L^\varphi(\Omega)$ such that

$$\langle M^{-1}(\Lambda), \phi\rangle = \int_\Omega u_\lambda(x)\phi(x)d\mu,$$

for any $\phi \in L^{\varphi^*}(\Omega)$. Reflexivity follows immediately. □

Corollary 2.4.6. *If $\Omega \subseteq \mathbb{R}^n$ is open and $p : \Omega \longrightarrow (1, \infty)$ is Borel measurable with $p_+ < \infty$, then if $\frac{1}{p} + \frac{1}{q} = 1$ in Ω then the following isomorphism result holds:*

$$(L^p(\Omega))^* \approx (L^p(\Omega))' \approx (L^q(\Omega)).$$

If, in addition $p_- > 1$, then $L^p(\Omega)$ is reflexive.

2.5 Density of regular functions

In this section we consider the measure space $(\Omega, \mathcal{B}, \mu)$ where $\Omega \subseteq \mathbb{R}^n$ is an open set, $\mathcal{B}$ is the σ-algebra of the Borel subsets of Ω and μ is the Lebesgue's measure on Ω. Furthermore, we asume that φ is a locally integrable Musielak-Orlicz function on Ω satisfying the Δ_2 condition (Definition 2.2.2). It is further assumed that there exists $\gamma > 0$ such that

$$\gamma \leq \inf_{x \in \Omega} \varphi(x, 1). \tag{2.28}$$

Lemma 2.5.1. *Under the preceding conditions, the class $\mathcal{S}$ of simple functions supported on sets of finite measure on Ω is dense in $L^\varphi(\Omega)$.*

Proof. Consider first a non-negative function $u \in L^\varphi(\Omega)$ and a nondecreasing sequence (s_n) of nonnegative simple functions converging pointwise to u: for any $n \in \mathbb{N}$,

$$\varphi(x, |s_n(x) - u(x)|) \leq \varphi(x, 2|u(x)|);$$

equality (2.18) and the dominated convergence theorem imply that

$$\rho_\varphi(s_n - u) \nearrow 0 \text{ as } n \to \infty.$$

Thus $||s_n - u||_\varphi \longrightarrow 0$ as $n \to \infty$ by Corollary 1.3.9. This proves the claim in the case $u \geq 0$. Otherwise, we write as usual $u = u^+ - u^-$; by the first part, given $\epsilon > 0$ there are simple functions s and t such that $||u^+ - s||_\varphi < \frac{\epsilon}{2}$ and $||u^- - s||_\varphi < \frac{\epsilon}{2}$ respectively; this proves the claim. □

Lemma 2.5.2. *Let $u \in \mathcal{M}(\Omega)$ with $||u||_\infty \leq A$ and such that*

$$\mu(\{x : u(x) \neq 0\}) < \infty.$$

If $M = \{x : u(x) \neq 0\}$, then $u \in L^\varphi(M)$ and given $\epsilon > 0$ there exists $v \in C(\Omega)$, $||v||_\infty \leq A$ and $||u - v||_\varphi < \epsilon$.

Proof. Certainly $u \in L^\varphi(M)$. Indeed, due to the local integrability assumption on φ, it is obvious that

$$\int_M \varphi(x, |u(x)|)dx \leq \int_M \varphi(x, A)dx < \infty.$$

Next, fix $\eta > 0$ such that $\int_\Omega \varphi(x, \frac{2A}{\eta})dx \leq 1$. Lusin's Theorem yields a function $v \in C(\Omega)$, compactly supported such that if

$$\Omega_\eta = \{x \in \Omega : u(x) \neq v(x)\},$$

then $\mu(\Omega_\eta) < \eta$. Moreover, Lusin's Theorem asserts that

$$\|v\|_\infty \leq \|u\|_\infty \leq A.$$

Consequently, the local integrability of φ yields

$$\int_\Omega \varphi(x, |v(x)|)dx \leq \int_\Omega \varphi(x, A) < \infty.$$

The above inequality shows that $v \in L^\varphi(\Omega)$. Furthermore,

$$\begin{aligned}\int_\Omega \varphi\left(x, \frac{|v(x) - u(x)|}{\eta}\right) dx &= \int_{\Omega_\eta} \varphi\left(x, \frac{|v(x) - u(x)|}{\eta}\right) dx \\ &\leq \int_{\Omega_\eta} \varphi\left(x, \frac{2A}{\eta}\right) \leq 1;\end{aligned}$$

it is obvious then that

$$\|u - v\|_\varphi \leq \eta.$$

□

A direct consequence of the previous lemma is that simple functions on Ω can be approximated in the $L^\varphi(\Omega)$ norm, by continuous functions. More specifically,

Lemma 2.5.3. *Let Ω and φ be as specified above and $s \in \mathcal{S}$ be a simple function in $L^\varphi(\Omega)$. Then, given $\epsilon > 0$ there exists a function $v \in C(\Omega)$ such that $\|s - v\|_\varphi < \epsilon$.*

Proof. Select s as in the statement. By assumption, there are distinct real numbers $s_1, s_2, ..., s_k$ (which can be assumed to be non-zero) and disjoint measurable sets $\Omega_1, \Omega_2, ..., \Omega_k$ such that

$$s = \sum_{j=1}^{k} s_k \mathbf{1}_{\Omega_j}.$$

Because of the assumption $s \in L^\varphi(\Omega)$, there must exist $\lambda > 0$ for which $\int_\Omega \varphi(x, \lambda|s(x)|)\, dx < \infty$. Consequently, on account of the Δ_2 condition and the definition of Ω_j, one has:

$$\sum_{j=1}^{k} \int_{\Omega_j} \varphi(x, |s_j|)dx < \infty.$$

Since φ is assumed to satisfy condition (2.28) it is clear that for any $j : j = 1, 2, ..., k$ and for γ as in statement (2.28), one has

$$\mu(\Omega_j) \leq \gamma^{-1} \leq \int_{\Omega_j} \varphi(x, 1)dx < \infty.$$

Moreover, the support of s is contained in the union $\bigcup_{j=1}^{k} \Omega_j$, which the above discussion implies has finite measure and obviously

$$|s(x)| \leq \sup\{|s_j|, j = 1, 2, ..., k\}.$$

Lemma 2.5.2 implies now that s can be approximated at will, in norm $\|\cdot\|_\varphi$ by a continuous function. □

Lemmas 2.5.2 and 2.5.3 yield:

Lemma 2.5.4. *If φ is a Musielak-Orlicz function satisfying the Δ_2 condition and (2.28), the set $C(\Omega)$ is dense in $L^\varphi(\Omega)$.*

The next lemma follows as a consequence of the preceding discussion.

Lemma 2.5.5. *$C_0^\infty(\Omega)$ is dense in $C(\Omega) \cap L^\varphi(\Omega)$.*

Proof. Fix $u \in C(\Omega) \cap L^\varphi(\Omega)$ and let (K_j) be a non-decreasing sequence of compact subsets of Ω with $\bigcup K_j = \Omega$. Given $\epsilon > 0$ it follows from the Δ_2 condition that for any $u \in C(\Omega) \cap L^\varphi(\Omega)$

$$\varphi\left(x, \frac{2|u(x)|}{\epsilon}\right) \in L^1(\Omega).$$

The theorem of monotone convergence implies

$$\int_{K_j} \varphi\left(x, \frac{2|u(x)|}{\epsilon}\right) dx \nearrow \int_{\Omega} \varphi\left(x, \frac{2|u(x)|}{\epsilon}\right) dx \text{ as } j \to \infty,$$

and it is easy to derive from here that for some $j \in \mathbb{N}$,

$$\int_{\Omega \setminus K_j} \varphi\left(x, \frac{2|u(x)|}{\epsilon}\right) dx \leq 1.$$

Also, there exists a compact subset K_ϵ of Ω such that

$$\int_{\Omega \setminus K_\epsilon} \varphi\left(x, |u(x)|\frac{2}{\epsilon}\right) dx \leq 1. \tag{2.29}$$

Write $u_1 = u\mathbf{1}_{K_\epsilon}$. Then (2.29) implies

$$\|u - u_1\|_\varphi \leq \frac{\epsilon}{2}. \tag{2.30}$$

Next, select a mollifier function $\Psi \in C_0^\infty(\mathbb{R}^n)$ supported in the unit ball, that is $\Psi \geq 0$ in $\mathbb{R}^n$, $\int_{\mathbb{R}^n} \Psi(x)dx = 1$; it is well known that under these assumptions, for any $\delta > 0$ the mollified function

$$(u_1 * \Psi)_\delta(\cdot) = \delta^{-n} \int_\Omega \Psi\left(\frac{\cdot - y}{\delta}\right) u_1(y)dy$$

is in $C_0^\infty(\Omega)$ and converges uniformly to u_1 on every compact subset of Ω, as $\delta \to 0$. In fact, by definition u_1 is continuous and hence uniformly continuous on K_ϵ; for arbitrary δ_0, select δ so small that $|u_1(x-\delta) - u_1(x)| < \delta_0$ for any $x \in K_\epsilon$. Then it is easy to derive that

$$|u_1 * \Psi(x) - u_1(x)| \leq \delta_0.$$

Therefore, for $x \in K_\epsilon$,

$$\int_{K_\epsilon} \varphi\left(x, \frac{2|u_1 * \Psi(x) - u_1(x)|}{\epsilon}\right) dx \leq \int_{K_\epsilon} \varphi\left(x, \frac{2\delta_0}{\epsilon}\right) dx \tag{2.31}$$

and the latter tends to 0 as $\delta_0 \to 0$. It follows that

$$\|u_1 * \Psi - u_1\|_\varphi \leq \frac{\epsilon}{2}.$$

As an immediate consequence of (2.30) and (2.31), one concludes

$$\|u - u_1 * \Psi\|_\varphi \leq \|u - u_1\|_\varphi + \|u_1 * \Psi - u_1\|_\varphi \leq \frac{\epsilon}{2} + \frac{\epsilon}{2} = \epsilon.$$

□

The following theorem is an immediate consequence of Lemmas 2.5.1, 2.5.2, 2.5.3 and 2.5.5.

Theorem 2.5.6. *Let $\Omega \subseteq \mathbb{R}^n$ be open and φ be a locally integrable Musielak Orlicz function on Ω that satisfies the Δ_2 condition and (2.28). Then $C_0^\infty(\Omega)$ is dense in $L^\varphi(\Omega)$.*

Example 2.5.1. The Δ_2 condition in Theorem 2.5.6 is necessary. Indeed, $C_0^\infty(\Omega)$ need not have the density property. For example, let $n = 1, \Omega = (1, \infty)$

and $p(x) = x$ $(x \in \Omega)$; let f be the constant function on Ω that is everywhere equal to 1. Then $f \in L^{p}(\Omega)$ since

$$\int_{1}^{\infty} \lambda^{-x} dx = (\lambda \log \lambda)^{-1} \text{ if } \lambda > 1.$$

Let $g \in C_0^\infty(\Omega)$, so that supp $g \subset [1+\varepsilon, N]$ for some $\varepsilon > 0$ and $N > 1$. Hence

$$\|f - g\|_p \geq \inf \left\{ \lambda > 0 : \int_{N}^{\infty} \lambda^{-x} dx \leq 1 \right\},$$

and as the integral is finite only if $\lambda > 1$, it follows that $\|f - g\|_p \geq 1$. Thus $C_0^\infty(\Omega)$ is not dense in $L^{p}(\Omega)$.

Before closing this section we emphasize the following corollary to Theorem 2.5.6:

Corollary 2.5.7. *Let $\Omega \subseteq \mathbb{R}^n$ be open, $p \in \mathcal{P}(\Omega)$ and suppose that $p_+ < \infty$. Then the set of all bounded, measurable functions is dense in $L^p(\Omega)$, as is $C(\Omega) \cap L^p(\Omega)$; and $L^p(\Omega)$ is separable. Moreover, $C_0^\infty(\Omega)$ is dense in $L^p(\Omega)$.*

2.6 Uniform convexity of Musielak-Orlicz spaces

Recall from Definition 1.2.7 in Section 1.2.3 that a convex modular ρ on a vector space V is said to be uniformly convex if for every $\epsilon > 0$ there exists $\delta = \delta(\epsilon) > 0$ such that

$$(u \in V, v \in V, \rho(u) = \rho(v) = 1 \text{ and } \rho(u - v) > \epsilon) \Rightarrow \rho\left(\frac{u+v}{2}\right) < 1 - \delta(\epsilon).$$

It is only natural to expect a close connection between some form of convexity of the Musielak-Orlicz function φ and the uniform convexity of the modular ρ_φ. To that effect, the following definition is introduced.

Definition 2.6.1. *A φ be a Musielak-Orlicz function on Ω (Chapter 2) is said to be uniformly convex if there exists*

$$w : (0,1) \longrightarrow (0,1)$$

such that for any $r \in (0, \infty)$, $\alpha \in (0,1)$ and $\beta \in [0, \alpha]$ the inequality

$$\varphi\left(x, \frac{1+\beta}{2} r\right) \leq (1 - w(\alpha)) \frac{\varphi(x,r) + \varphi(x, \beta r)}{2}$$

holds a.e. in Ω.

In the interest of completeness we present the proof of the following theorem, following [24, 82].

Theorem 2.6.1. ([24, Theorem 2.4.11], [82, Theorem 11.6]) *Let $\Omega \subseteq \mathbb{R}^n$ be a domain. If the Musielak-Orlicz function*

$$\varphi : \Omega \times [0,\infty) \longrightarrow [0,\infty)$$

is uniformly convex and satisfies the Δ_2 condition, Then the Luxemburg norm $\|\cdot\|_{\rho_\varphi}$ is uniformly convex.

Proof. Fix $\epsilon : 0 < \epsilon < 1$ and functions u and v in $L^\varphi(\Omega)$ such that

$$\rho_\varphi(u) = \rho_\varphi(u) = 1$$

and $\rho_\varphi\left(\frac{u+v}{2}\right)$. Set

$$M = \left\{ x \in \Omega : |u(x) - v(x)| > \frac{\epsilon}{4} \max\{|u(x)|, |v(x)|\} \right\}.$$

Convexity yields

$$\int\limits_{\Omega\backslash M} \varphi\left(x, \frac{u+v}{2}\right) \leq \frac{1}{2}\left(\int\limits_{\Omega\backslash M} \varphi(x,u)dx + \int\limits_{\Omega\backslash M} \varphi(x,v)dx \right)$$

and it follows from here that

$$\gamma = \rho_\varphi\left(\mathbf{1}_{\Omega\backslash M}\frac{u}{2}\right) + \rho_\varphi\left(\mathbf{1}_{\Omega\backslash M}\frac{u}{2}\right) - \rho_\varphi\left(\mathbf{1}_{\Omega\backslash M}\frac{u+v}{2}\right) \geq 0.$$

Thus, invoking the uniform convexity of φ:

$$\begin{aligned} 1 - \rho_\varphi\left(\frac{u+v}{2}\right) &= \frac{1}{2}\left(\rho_\varphi(u) + \rho_\varphi(v)\right) - \rho_\varphi\left(\frac{u+v}{2}\right) \\ &= \frac{1}{2}\left(\rho_\varphi(\mathbf{1}_M u) + \rho_\varphi(\mathbf{1}_M v)\right) - \rho_\varphi\left(\mathbf{1}_M \frac{u+v}{2}\right) + \gamma \\ &\geq \frac{1}{2}\left(\rho_\varphi(\mathbf{1}_M u) + \rho_\varphi(\mathbf{1}_M v)\right) \\ &\quad - \frac{1 - w(1-\epsilon)}{2}\left(\rho_\varphi(\mathbf{1}_M u) + \rho_\varphi(\mathbf{1}_M v)\right) \\ &= w(1-\epsilon)\left(\rho_\varphi(\mathbf{1}_M u) + \rho_\varphi(\mathbf{1}_M v)\right). \end{aligned}$$

It follows from the very definition of M that

$$\begin{aligned} \rho_\varphi\left(\mathbf{1}_{\Omega\backslash M}(u-v)\right) &\leq \frac{\epsilon}{4}\left(\rho_\varphi\left(\mathbf{1}_{\Omega\backslash M} u\right) + \rho_\varphi\left(\mathbf{1}_{\Omega\backslash M} v\right)\right) \\ &\leq \frac{\epsilon}{4}\left(\rho_\varphi(u) + \rho_\varphi(v)\right) \leq \frac{\epsilon}{2}; \end{aligned}$$

consequently

$$\begin{aligned}\rho_\varphi\left((u-v)\mathbf{1}_M\right) &= \rho_\varphi(u-v) - \rho_\varphi\left((u-v)\mathbf{1}_{\Omega\setminus M}\right)\\ &\geq \rho_\varphi(u-v) - \frac{\epsilon}{2} > \frac{\epsilon}{2}.\end{aligned}$$

On account of condition (1.45) it follows that

$$\begin{aligned}\frac{\epsilon}{2} &< \rho_\varphi\left((u-v)\mathbf{1}_M\right)\\ &\leq \frac{1}{2}\left(\rho_\varphi\left(2u\mathbf{1}_M\right) + \rho_\varphi\left(2v\mathbf{1}_M\right)\right)\\ &\leq \frac{1}{2}K\left(\rho_\varphi\left(2u\mathbf{1}_M\right) + \rho_\varphi\left(2v\mathbf{1}_M\right)\right)\\ &\leq \frac{K}{2w(1-\epsilon)}\left(1-\rho_\varphi\left(\frac{u+v}{2}\right)\right).\end{aligned}$$

In conclusion

$$\rho_\varphi\left(\frac{u+v}{2}\right) < 1 - \frac{\epsilon}{k}w(1-\epsilon),$$

i.e., the modular δ is uniformly convex with $\delta(\epsilon) = \frac{\epsilon}{k}w(1-\epsilon)$. Next, fix $\epsilon : 0 < \epsilon < 1$ and take $\|u\|_{\rho_\varphi} = 1$ and $\|v\|_{\rho_\varphi} = 1$ with $\|u-v\|_{\rho_\varphi} > \epsilon$. By virtue of the Δ_2 condition necessarily $\rho_\varphi(u) = \rho_\varphi(v) = 1$ and

$$\rho_\varphi\left(\frac{u-v}{\|u-v\|_\varphi}\right) > 1.$$

Invoking the Δ_2 condition again it is readily seen that, for some constant $C = C(\epsilon) > 1$, $\rho_\varphi(u-v)$ is subject to the inequality:

$$1 < \rho_\varphi\left(\frac{u-v}{\|u-v\|_\varphi}\right) \leq \rho_\varphi\left(\frac{u-v}{\epsilon}\right) \leq C\rho_\varphi(u-v);$$

which, in turn, yields, on account of the uniform convexity of ρ_φ, the existence of $\delta = \delta(C(\epsilon)) > 0$ such that:

$$\rho_\varphi\left(\frac{u-v}{2}\right) < 1-\delta,$$

valid for some $\delta = \delta(\epsilon) > 0$. We claim that there exists $\eta : 0 < \eta(\epsilon) < 1$ such that

$$\left\|\frac{u+v}{2}\right\|_\varphi < 1-\eta.$$

For otherwise, as is easy to see, there would exist a sequence $(w_n) \in L^\varphi(\Omega)$ such that $\|w_n\|_\varphi \nearrow 1$ and

$$\rho_\varphi(w_n) < 1-\delta.$$

This quickly leads to a contradiction, as follows: Set $x_n = \frac{1}{\|w_n\|_\varphi} w_n$; then $\rho_\varphi(x_n) = 1$ and let C stand for the Δ_2 constant of ρ_φ. Then

$$\begin{aligned} 1 = \rho_\varphi(x_n) &= \rho_\varphi\left(2\left(\frac{1}{\|w_n\|_\varphi} - 1\right) w_n + \left(2 - \frac{1}{\|w_n\|_\varphi}\right) w_n\right) \\ &\le \left(\frac{1}{\|w_n\|_\varphi} - 1\right) C\rho_\varphi(w_n) + \left(2 - \frac{1}{\|w_n\|_\varphi}\right) \rho_\varphi(w_n). \end{aligned} \tag{2.32}$$

The right-hand side in (2.32) tends to $1 - \delta$ as $n \to \infty$, which concludes the proof. □

A direct consequence of the above discussion provides a criterion for the uniform convexity of $L^p(\Omega)$.

Corollary 2.6.2. *The variable exponent Lebesgue space $L^p(\Omega)$ (Section 1.5) is uniformly convex if*

$$1 < p_- = \operatorname*{ess\,inf}_\Omega p \le \operatorname*{ess\,inf}_\Omega p = p_+ < \infty. \tag{2.33}$$

Proof. It is easy to verify that for constant $q : q > 1$ the function

$$\begin{aligned} v &: (0,1) \longrightarrow (0,1) \\ v(s) &= \frac{(1+s)^q}{1+s^q} \end{aligned}$$

is increasing and that strictly less than 2^{q-1} ; consequently, for r, α and β as in Definition 2.6.1 one has

$$(1+\alpha^q)(1+\beta)^q \le (1+\alpha)^q(1+\beta^q) \tag{2.34}$$

and it follows easily from (2.34) that

$$2^q(1+\alpha^q)\left(\frac{1+\beta}{2}r\right)^q \le (1+\alpha)^q(r^q + (r\beta)^q).$$

Some elementary algebraic manipulations yield that as long as $q > 1$, for all $s : s \in (0,1)$ one has

$$w(s) = 2^{1-q}v(s) < 1. \tag{2.35}$$

Thus if (2.33) holds,

$$\begin{aligned} \varphi\left(x, \frac{1+\beta}{2}r\right) &= \left(\frac{1+\beta}{2}r\right)^{p(x)} \\ &\le \frac{(1+\alpha)^{p(x)}2^{1-p(x)}(1+\alpha^{p(x)})^{-1}}{2}(r^{p(x)} + (r\beta)^{p(x)}) \\ &= (1 - (1 - 2^{1-p(x)}v(\alpha))\frac{(\varphi(x,r) + \varphi(x,\beta r))}{2} \\ &\le (1 - (1 - 2^{1-p_-}v(\alpha))\frac{(\varphi(x,r) + \varphi(x,\beta r))}{2}. \end{aligned}$$

Thus the norm $\|\cdot\|_p$ is uniformly convex according to Definition 2.6.1 and on account of inequality (2.35). □

2.7 Carathéodory functions and Nemytskii operators on Musielak-Orlicz spaces

In the study of boundary value problems it is of paramount importance to understand the action of so-called Nemytskii-type operators, that is, operators of the form

$$X \ni u \longrightarrow f(x, u(x)),$$

where X is a given function space on a domain Ω and

$$f : \Omega \times \mathbb{R} \longrightarrow \mathbb{R}$$

is a measurable function.
In this section we analyze some fundamental questions related to the situation described above in the particular case when X is a Musielak-Orlicz space.
Let $\Omega \subseteq \mathbb{R}^n$ be an open set and consider a Musielak-Orlicz function and φ on Ω, as defined in Definition 2.1.1.

Definition 2.7.1. *Let $\Omega \subseteq \mathbb{R}^n$ be a domain. A function*

$$w : \Omega \times [0, \infty) \to [0, \infty)$$

such that $w(x, \cdot)$ is continuous a.e. $x \in \Omega$ and Lebesgue-measurable for every $t \in \mathbb{R}$ is said to be a Carathéodory function.

We remark at this point that a Carathéodory function w naturally defines an operator T_w on $\mathcal{M}(\Omega)$ by way of the assignment

$$u(\cdot) \longrightarrow T_w(u)(\cdot) = w(\cdot, |u(\cdot)|). \tag{2.36}$$

The following theorem can be found in [61]; we include its proof for the sake of completeness.

Theorem 2.7.1. *Let $\Omega \subseteq \mathbb{R}^n$ be a bounded domain and*

$$w : \Omega \to [0, \infty)$$

be a Carathéodory function. Let S be the subspace of $\mathcal{M}(\Omega)$ consisting of sequences of measurable functions, that are convergent in measure. Then S is invariant under the operator T_w defined in (2.36). That is, if $(u_j) \subseteq \mathcal{M}(\Omega)$ converges in measure, so does $(T_w(u_j))$.

Proof. Let $(u_n) \subseteq \mathcal{M}(\Omega)$ converge in measure to $u \in \mathcal{M}(\Omega)$. Fix $\epsilon > 0$. For any $j \in \mathbb{N}$ let

$$\Omega_j = \left\{ x \in \Omega : t - \frac{1}{j} < |u(x)| < t \Rightarrow |w(x, |u(x)|) - w(x,t)| < \epsilon \right\}.$$

It is plain that the sequence (Ω_j) is increasing and from the *a.e.* left-continuity assumption on $w(\cdot, s)$ it is readily concluded that

$$|\Omega| = \left| \bigcup_1^\infty \Omega_j \right|.$$

Thus,

$$|\Omega| = \lim_{j \to \infty} |\Omega_j|.$$

For any $\delta > 0$, then, one can select $j_0 \in \mathbb{N}$ such that $|\Omega_{j_0}| > |\Omega| - \frac{\delta}{2}$. Next, for each $j \in \mathbb{N}$ we set

$$E_j = \left\{ x \in \Omega : |u(x) - u_j(x)| < \frac{1}{j_0} \right\}.$$

Now, since (u_n) converges to u in measure one has, for large enough j,

$$\left| \left\{ x \in \Omega : |u(x) - u_j(x)| > \frac{1}{j_0} \right\} \right| < \frac{\delta}{2}$$

and it is therefore clear that for some $N \in \mathbb{N}$, $j > N$ implies

$$|E_j| > |\Omega| - \frac{\delta}{2}.$$

By definition, if $x \in \Omega_{j_0} \cap E_j$, then

$$|w(x, |u(x)|) - w(x, |u_j(x)|)| < \epsilon.$$

It follows then that

$$|\{x \in \Omega : |w(x, |u(x)|) - w(x, |u_j(x)|)| < \epsilon\}| \geq |\Omega_{j_0} \cap E_j|.$$

Since for $j > N$ it holds that

$$\begin{aligned} \left| E_j \bigcap \Omega_{j_0} \right| &= |E_j| + |\Omega_{j_0}| - \left| E_j \bigcup \Omega_{j_0} \right| \\ &\geq |\Omega| - \frac{\delta}{2}, \end{aligned}$$

it is clear that for $j > N$ one has:

$$|\{x \in \Omega : |w(x, |u(x)|) - w(x, |u_j(x)|)| \geq \epsilon\}| > \frac{\delta}{2}.$$

The theorem follows immediately from the last statement. □

For the next discussion we fix an Musielak-Orlicz function (Definition 2.1.1)

$$\varphi : \Omega \times [0,\infty) \longrightarrow [0,\infty)$$

on the domain $\Omega \subseteq \mathbb{R}^n$. We assume that both, φ and its conjugate φ^* (see (2.24) in Section 2.4.1) satisfy the Δ_2 condition and impose on φ the additional condition that it be an N-function, that is (see [2, 24]):

$$\lim_{t\to 0} \frac{\varphi(x,t)}{t} = 0 \ a.e. \quad x \in \Omega. \tag{2.37}$$

Theorem 2.7.2. *Let φ be an N-function (i.e., it satisfies (2.37)) and assume that*

$$\varphi_t(x,t) = \frac{\partial \varphi}{\partial t}(x,t)$$

exists a.e. $x \in \Omega$. Define the operator T_{φ_t} as

$$T_{\varphi_t} : \mathcal{M}(\Omega) \longrightarrow \mathcal{M}(\Omega)$$
$$T_{\varphi_t}(u) = \varphi_t(x,|u(x)|) = \frac{\partial \varphi}{\partial t}(x,|u(x)|).$$

Then, from the assumption

$$T_{\varphi_t}(L^{\varphi}(\Omega)) \subseteq L^{\varphi^*}(\Omega) \tag{2.38}$$

it follows that the operator

$$T_{\varphi_t} : L^{\varphi}(\Omega) \longrightarrow L^{\varphi^*}(\Omega)$$

is continuous and bounded.

Proof. Assume $(u_n) \subseteq L^{\varphi}(\Omega)$ converges to $u \in L^{\varphi}(\Omega)$. On $\Omega \times [0,\infty)$ define

$$w(x,t) = \varphi_t(x,|u(x)+t|) - \varphi_t(x,|u(x)|);$$

then on account of the assumption on φ_t, w is a Carathéodory function and $w(x,0) = 0$. If

$$T_w : L^{\varphi}(\Omega) \longrightarrow L^{\varphi^*}(\Omega)$$

is continuous at 0, then $T_w(u_n - u) \longrightarrow 0$ in $L^{\varphi^*}(\Omega)$ as $n \to \infty$. If φ^* satisfies the Δ_2 condition, the latter is equivalent to

$$\rho_{\varphi^*}(T_w(u_n - u)) = \int_{\Omega} \varphi^*(x, |\varphi_t(x,|u_n(x)|) - \varphi_t(x,|u(x)|)|) \to 0 \text{ as } n \to \infty,$$

that is

$$T_{\varphi_t}(u_n) \longrightarrow T_{\varphi_t}(u) \text{ in } L^{\varphi^*}(\Omega) \text{ as } n \to \infty.$$

Therefore, it is enough to show that T_{φ_t} is continuous at 0 under the assumption that $\varphi_t(x,0) = 0$ *a.e.* in Ω. Assume that T_{φ_t} is not continuous at 0; let $r > 0$ and let (u_n) be a sequence that converges to 0 in $L^{\varphi}(\Omega)$, for which

$$\|T_{\varphi_t}(u_n)\|_{\varphi^*} \geq r \text{ for any } n \in \mathbb{N}.$$

Since norm convergence implies modular convergence, one can, without loss of generality assume that

$$\max\{\rho_{\varphi}(u_n), \|u_n\|_{\varphi}\} < \frac{1}{2^n},$$

and hence that $\sum_{n=1}^{\infty} \int\limits_{\Omega} \varphi(x, |u_n(x)|)\, dx < \infty$. Due to the validity of the Δ_2 condition for φ^*, norm convergence and modular convergence are equivalent on $L^{\varphi^*}(\Omega)$. It follows that there exists $\epsilon(r) > 0$ such that $\rho_{\varphi^*}(T_{\varphi_t}(u_n)) \geq \epsilon(r)$ for any $n \in \mathbb{N}$. We next claim the existence of a sequence of real numbers (ϵ_k), a sequence (Ω_k) of subsets of Ω and a subsequence (u_{n_k}) of (u_n) satisfying the following conditions:

(*i*) $\epsilon_{k+1} < \frac{1}{2}\epsilon_k$,

(*ii*) $|\Omega_k| \leq \epsilon_k$,

(*iii*) $\int\limits_{\Omega_k} \varphi^*(x, |T_{\varphi_t}(u_{n_k})|)\, dx > \frac{2}{3}\epsilon(r)$.

(*iv*) If $E \subseteq \Omega$ is measurable and $|E| < 2\epsilon_{k+1}$, then

$$\int\limits_{E} \varphi^*(x, |T_{\varphi_t}(u_{n_k})|)\, dx < \frac{\epsilon(r)}{3}.$$

Set $\Omega_1 = \Omega$, $\epsilon_1 = |\Omega|$, $n_1 = 1$. We assume that ϵ_k, n_k and Ω_k are given, then, by assumption, $\varphi_t(\cdot, |u_{n_k}(\cdot)|) \in L^{\varphi^*}(\Omega)$ and on account of the Δ_2 condition one has

$$\int\limits_{\Omega} \varphi^*(x, |\varphi_t(x, |u_{n_k}(x)|)|)\, dx < \infty.$$

Since the measure

$$A \longrightarrow \mu(A) = \int\limits_{A} \varphi^*(x, |\varphi_t(x, |u_{n_k}(x)|)|)\, dx$$

defined on the Borel σ algebra $\mathcal{B}$ of subsets of Ω is absolutely continuous with respect to the Lebesgue measure, one can find ϵ_{k+1} such that any $X \in \mathcal{B}$ with $|X| < 2\epsilon_{k+1}$ satisfies $\int\limits_{X} \varphi^*(x, |\varphi_t(x, |u_{n_k}(x)|)|)\, dx < \frac{\epsilon(r)}{3}$. The assumption $\epsilon_k \leq 2\epsilon_{k+1}$ would contradict (*iii*). We now proceed to the construction of

Ω_{k+1} and n_{k+1}.
On account of Theorem 2.7.1, the strong convergence of (u_n) in $L^{\varphi}(\Omega)$ implies the convergence in measure of both $(T_{\varphi_t}(u_n))$ and $(T_{\varphi^*}(T_{\varphi_t}(u_n)))$. Consequently, there exists $n_{k+1} \in \mathbb{N}$ such that

$$\left|\left\{x \in \Omega : \left|T_{\varphi^*}(T_{\varphi_t}(u_{n_{k+1}}))\right| > \frac{\epsilon(r)}{3|\Omega|}\right\}\right| < \epsilon_{k+1} < \frac{\epsilon_k}{2} < \epsilon_k.$$

Define

$$\Omega_{k+1} = \left\{x \in \Omega : \left|T_{\varphi^*}(T_{\varphi_t}(u_{n_{k+1}}))\right| > \frac{\epsilon(r)}{3|\Omega|}\right\}.$$

Next,

$$\begin{aligned}\int\limits_{\Omega_{k+1}} \varphi^*(x, |T_{\varphi_t}(u_{n_{k+1}})|)\, dx &= \left(\int\limits_{\Omega} - \int\limits_{\Omega \setminus \Omega_{k+1}}\right) \varphi^*(x, |T_{\varphi_t}(u_{n_{k+1}})|)\, dx \\ &> \epsilon(r) - \frac{\epsilon(r)}{3} = \frac{2\epsilon(r)}{3}.\end{aligned}$$

By construction

$$\left|\bigcup_{j=k+1}^{\infty} \Omega_j\right| \leq \sum_{j=k+1}^{\infty} \epsilon_j < 2\epsilon_{k+1}. \tag{2.39}$$

Set

$$v(x) = \begin{cases} u_{n_k}(x) & \text{if } x \in \Omega_k \setminus \bigcup\limits_{j=k+1}^{\infty} \Omega_j \\ 0 & \text{otherwise.}\end{cases}$$

It is clear that $v \in L^{\varphi}(\Omega)$. Next, observe that:

$$\begin{aligned}\int\limits_{\Omega} \varphi^*(x, T_{\varphi_t}(v(x)))\, dx &\geq \sum_{k=1}^{\infty} \int\limits_{\Omega_k \setminus \cup_{j=k+1}^{\infty}\Omega_j} \varphi^*(x, T_{\varphi_t}(u_{n_k}(x)))\, dx \\ &= \sum_{k=1}^{\infty} \left(\int\limits_{\Omega_k} - \int\limits_{\cup_{j=k+1}^{\infty}\Omega_j}\right) \varphi^*(x, T_{\varphi_t}(u_{n_k}(x)))\, dx \\ &\geq \sum_{k=1}^{\infty} \left(\frac{2}{3}\epsilon(r) - \frac{1}{3}\epsilon(r)\right) \\ &= \infty,\end{aligned}$$

which follows by (2.39) and condition (iv). This contradicts assumption (2.38). Hence, T_{φ_t} is continuous at 0. □

2.8 Further properties of variable exponent spaces

Our next objective is to pause and exploit the general theory hitherto established to derive further properties in the particular case of the variable exponent Lebesgue spaces, such as uniform convexity. We follow the line of reasoning given in [76] and begin with a lemma that characterizes spaces with absolutely continuous norms. We fix a σ-finite measure space $(\Omega, \mathcal{A}, \mu)$. Recall that $\mathcal{P}(\Omega)$ stands for the vector space of all real-valued, Borel-measurable functions defined on Ω.

Lemma 2.8.1. *Let $p \in \mathcal{P}(\Omega)$. Then the following statements are equivalent:*
(i) $L^p(\Omega)$ has absolutely continuous norm (see Definition 1.2.3);
(ii) $p_+ < \infty$;
(iii) $L^p(\Omega)$ is separable.

Proof. First suppose that $p_+ = \infty$. If $|\Omega_\infty| = 0$, set:

$$\Omega^k = \{x \in \Omega : k \le p(x) < k+1\} \quad (k \in \mathbb{N})$$

and let $\{n_k\}_{k\in\mathbb{N}}$ be a sequence of natural numbers such that $|\Omega^{n_k}| > 0$ for each $k \in \mathbb{N}$. Let $c_k > 0$ be such that

$$\int_{\Omega^{n_k}} c_k^{p(x)} dx = 1 \quad (k \in \mathbb{N})$$

and put

$$u = \sum_{k=1}^{\infty} c_k \mathbf{1}_{\Omega^{n_k}}, \ E_j = \bigcup_{k=j}^{\infty} \Omega^{n_k};$$

thus $E_j \to \emptyset$. Since

$$\begin{aligned} \|u\|_p &= \inf\left\{\lambda > 0 : \sum_{k=1}^{\infty} \int_{\Omega^{n_k}} \left(\frac{c_k}{\lambda}\right)^{p(x)} dx \le 1\right\} \\ &\le \inf\left\{\lambda > 1 : \sum_{k=1}^{\infty} \left(\frac{1}{\lambda}\right)^{n_k} dx \le 1\right\} \le 2, \end{aligned}$$

it follows that $u \in L^p(\Omega)$. As

$$\left\|u\mathbf{1}_{E_j}\right\|_p \ge \inf\left\{\lambda > 1 : \int_{\Omega^{n_j}} \left(\frac{c_j}{\lambda}\right)^{p(x)} dx \le 1\right\} = 1,$$

u does not have absolutely continuous norm. If $|\Omega_\infty| > 0$, take any $A \subset \Omega_\infty$

with positive measure and observe that $\mathbf{1}_A$ evidently does not have absolutely continuous norm. Thus (i) implies (ii).

For the reverse implication, assume that $p_+ < \infty$, let $u \in L^p(\Omega)$ have norm 1, take $\varepsilon > 0$, let $\{E_k\}$ be a sequence of sets with $|E_k| \searrow 0$, let $j \in \mathbb{N}$ be such that $\|u\mathbf{1}_{E_j}\|_p \geq 1 - \varepsilon$, set $\phi = u\mathbf{1}_{\Omega\setminus E_j}$ and put $\psi = u\mathbf{1}_{E_j}$. By Proposition 1.5.2,

$$\int_\Omega \left|\frac{\phi}{\|\phi\|_p}\right|^p dx = \int_\Omega \left|\frac{\psi}{\|\psi\|_p}\right|^p dx = 1.$$

Thus

$$\|\phi\|_p^{p_+} \leq \int_\Omega |\phi|^p \, dx, \quad \|\psi\|_p^{p_+} \leq \int_\Omega |\psi|^p \, dx$$

and

$$\int_\Omega |\phi|^p \, dx + \int_\Omega |\psi|^p \, dx \leq 1.$$

Hence

$$\|\psi\|_p^{p_+} \leq \int_\Omega |\psi|^p \, dx \leq 1 - \int_\Omega |\phi|^p \, dx \leq 1 - \|\phi\|_p^{p_+} \leq 1 - (1-\varepsilon)^{p_+},$$

so that $\|\psi\|_p \leq \left(1 - (1-\varepsilon)^{p_+}\right)^{1/p_+}$ and the absolute continuity follows. Thus (i) and (ii) are equivalent. For the equivalence of (i) and (iii) we refer to [11, Corollary 1.5.6]. □

We are now ready to tackle the following theorem:

Theorem 2.8.2. *Let $p \in \mathcal{P}(\Omega)$. The following statements are equivalent:*
(i) $L^p(\Omega)$ is reflexive;
(ii) $L^p(\Omega)$ and $L^{p'(\cdot)}(\Omega)$ have absolutely continuous norms;
(iii) $L^p(\Omega)$ is uniformly convex;
(iv) $1 < p_- \leq p_+ < \infty$.

Proof. Since $L^p(\Omega)$ is a Banach function space, the equivalence of (i) and (ii) follows immediately from Theorem 1.2.19. In view of Lemma 2.8.1, (ii) implies (iv); and that (iii) implies (i) is obvious. All that is left is to show that (iv) implies (iii). Suppose that (iv) holds, let $\varepsilon \in (0,1)$ and let u, v have unit $L^p(\Omega)$ norm. Set $s = (u+v)/2, t = (u-v)/2$ and

$$\Gamma = \{x \in \Omega : p_- \leq p(x) \leq p_+\},$$

so that $|\Omega \setminus \Gamma| = 0$; put

$$S = \{x \in \Gamma : |t(x)| < \varepsilon |s(x)|\}, T = \{x \in \Gamma : |t(x)| \geq \varepsilon |s(x)|\}.$$

Then

$$\int_S |t|^p\,dx \le \int_S \varepsilon^p |s|^p\,dx \le \int_\Omega \varepsilon^p |s|^p\,dx \tag{2.40}$$

$$\le \varepsilon^{p_-} \int_\Omega |s|^p\,dx \le \varepsilon^{p_-}.$$

The function $\lambda \longmapsto \lambda^t$ is strictly convex on $\mathbb{R}$ whenever $1 < t < \infty$; hence

$$|\lambda|^t < \left(|\lambda+1|^t + |\lambda-1|^t\right)/2 \ (\lambda \in \mathbb{R}). \tag{2.41}$$

Thus the function $f : (q,\lambda) \longmapsto \left(|\lambda+1|^q + |\lambda-1|^q\right)/2 - |\lambda|^q$ $(q \in (1,\infty), \lambda \in \mathbb{R})$ is continuous and strictly positive on $(1,\infty)\times\mathbb{R}$. It follows that there exists $\alpha > 0$ such that $f(q,\lambda) \ge \alpha$ whenever $q \in [p_-, p_+]$ and $\lambda \in [-1/\varepsilon, 1/\varepsilon]$. We therefore see that

$$\frac{1}{2}\left(|\lambda+1|^{p(x)} + |\lambda-1|^{p(x)}\right) - |\lambda|^{p(x)} \ge \alpha$$

for all $x \in \Gamma$ and $\lambda \in [-1/\varepsilon, 1/\varepsilon]$, so that

$$\frac{1}{2}\left(|s(x)+t(x)|^{p(x)} + |s(x)-t(x)|^{p(x)}\right) \ge |s(x)|^{p(x)} + \alpha\,|t(x)|^{p(x)} \quad (x \in T),$$

while by (2.41),

$$\frac{1}{2}\left(|s(x)+t(x)|^{p(x)} + |s(x)-t(x)|^{p(x)}\right) \ge |s(x)|^{p(x)} \quad (x \in S).$$

Thus

$$1 = \int_\Omega \frac{1}{2}\left(|s+t|^p + |s-t|^p\right)dx$$

$$\ge \int_\Omega |s|^p\,dx + \alpha \int_T |t|^p\,dx,$$

and so

$$\int_T |t|^p\,dx < \varepsilon^{p_-} \text{ if } \int_\Omega |s|^p\,dx > 1 - \alpha\varepsilon^{p_-}.$$

Thus if $\delta := \alpha\varepsilon^{p_-}$ and $\|(u+v)/2\| > 1-\delta$, then $\int_\Omega |s|^p\,dx > 1-\delta$, from which with the aid of (2.40) it follows that

$$\int_\Omega |t|^p\,dx < 2\varepsilon^{p_-},$$

so that

$$\|u-v\|_p = \|2t\|_p \le 2(2\varepsilon^{p_-})^{1/p_+},$$

which establishes the uniform convexity of $L^p(\Omega)$. □

Recall from Section 1.2.2, if X be a Banach function space, three subspaces of X are to be distinguished, namely:
(i) the set of all elements of X with absolutely continuous norm (denote X_a);
(ii) the closure of the set of all bounded functions supported in sets of finite measure (denote X_b);
(iii) the space of all functions u with a continuous norm, by which we mean that for every $x \in \overline{\Omega}$ we have

$$\lim_{\varepsilon\to 0} \left\| u\mathbf{1}_{B(x,\varepsilon)\cap\Omega} \right\|_p = 0$$

(denote X_b).

From Lemma 1.2.17, we have that X_a is a closed subspace of X and $X_a \subset X_b$; it has been shown in [68] that X_c is a closed subspace of X and $X_a \subset X_c$. Now take $X = L^p(\Omega)$. Then in view of Lemma 2.8.1 it follows that

$$\text{if } p_+ < \infty, \quad \text{then } X_a = X_b = X_c = X. \tag{2.42}$$

The position when p_+ may be infinite is more complicated, as we now illustrate with the next two theorems from [35].

Theorem 2.8.3. *If Ω is bounded and $p \in \mathcal{P}(\Omega)$, then $X_a = X_c$.*

Proof. It suffices to show that $X_c \subset X_a$. First suppose that $|\Omega_\infty| = 0$. Assume that $u \notin X_a$. Then there exist $\alpha > 0$ and a decreasing sequence of sets $\{G_k\}_{k\in\mathbb{N}}$ such that $|G_k| \to 0$ and, for every $k \in \mathbb{N}$,

$$\|u\mathbf{1}_{G_k}\|_p = \inf\left\{ \lambda > 0 : \int_{G_k} (u/\lambda)^p \, dx \le 1 \right\} \ge \alpha.$$

With $\beta := \alpha/2$ this implies that

$$\int_{G_k} (f/\beta)^p \, dx > 1 \ (k \in \mathbb{N}).$$

If there existed m such that

$$\int_{G_m} (u/\beta)^p \, dx \le K < \infty,$$

then as the G_k are decreasing, with $|G_k| \to 0$, by dominated convergence we would have

$$\lim_{k\to\infty} \int_{G_k} (u/\beta)^p \, dx = 0,$$

a contradiction. Thus

$$\int_{G_k} (u/\beta)^p \, dx = \infty$$

for every $k \in \mathbb{N}$.

Since Ω is bounded, there is a closed cube Q with side length $\ell(Q)$ such that $\Omega \subset Q$. By subdivision of Q we obtain a decreasing sequence (Q_k) of closed cubes with $l(Q_k) = 2^{-k} l(Q)$ and

$$\int_{Q_k} (u/\beta)^p \, dx = \infty$$

for every k. Let $x \in \bigcap_{k=1}^{\infty} Q_k$ and $\varepsilon > 0$. Then $Q_k \subset B(x, \varepsilon)$ for large enough k and so

$$\int_{\Omega \cap B(x,\varepsilon)} (u/\beta)^p \, dx \geq \int_{\Omega \cap Q_k} (u/\beta)^p \, dx$$
$$= \int_{\Omega} (u \mathbf{1}_{Q_k}/\beta)^p \, dx = \infty.$$

Hence $\left\| u \mathbf{1}_{\Omega \cap B(x,\varepsilon)} \right\|_p > \beta$, and so $u \notin X_c$. This establishes the theorem when $|\Omega_\infty| = 0$.

Finally, we dispense with this assumption and take any $u \in X_c$. The norms $\|u\|_p$ and $\left\| u \mathbf{1}_{\Omega \backslash \Omega_\infty} \right\|_p +$ ess $\sup_{x \in \Omega_\infty} |u(x)|$ are equivalent. Since $u \in X_c$, $u(x) = 0$ on Ω_∞. Now apply the case of the theorem already proved to the set $\Omega \backslash \Omega_\infty$ and the function $p_1 := p \mathbf{1}_{\Omega_1}$ to obtain $u \in X_a$. □

Theorem 2.8.4. *Let $p \in \mathcal{P}(\Omega)$ and suppose that Ω is bounded; for each $k \in \mathbb{N}$ put $\Lambda_k = \{x \in \Omega : k \leq p(x) < k+1\}$.*
(i) If $p \in L^\infty(\Omega \backslash \Omega_\infty)$, then $X_b = X$;
(ii) if $|\Omega_\infty| = 0$ and $p_+ = \infty$, then $X_b \neq X$;
(iii) if $|\Omega_\infty| = 0$ and $p_+ = \infty$, then $X_a = X_b$ if and only if

$$\sum_{k=1}^{\infty} A^k |\Lambda_k| < \infty \text{ for all } A > 1. \tag{2.43}$$

Proof. (*i*) This is immediate from (2.42) and the fact that $L^\infty(\Omega_\infty) = X_b(\Omega_\infty)$.
(*ii*) For each $k \in \mathbb{N}$ let c_k be chosen so that $\int_{\Omega_k} c_k^p dx = 2^{-k}$ if $|\Lambda_k| > 0$, $c_k = 0$ otherwise; put $d_k = \max(c_k, 2k)$ if $c_k > 0$, $d_k = 0$ otherwise. There are sets $\Lambda_k' \subset \Lambda_k$ such that

$$\int_{\Lambda_k'} d_k^p dx = 2^{-k} \text{ if } |\Lambda_k| > 0.$$

Let $u = \sum\limits_{k=1}^{\infty} d_k \mathbf{1}_{\Lambda_k'}$. Since $u(x) = 0$ on Ω_∞,

$$\rho_p(u) = \int\limits_{\Omega\setminus\Omega_\infty} u^p dx = \sum_{k=1}^{\infty} \int\limits_{\Lambda_k'} d_k^p dx \leq \sum_{k=1}^{\infty} 2^{-k} = 1;$$

hence $\|f\|_p \leq 1$.

Now define

$$u_k(x) = \begin{cases} u(x), & 0 \leq u(x) \leq k, \\ k, & u(x) > k, \end{cases}$$

and let v be any measurable function on Ω with $|v| \leq k$. Since

$$|u(x) - v(x)| \geq |u - u_k(x)|,$$

we have $\|u - v\|_p \geq \|u - u_k\|_p$; moreover, $u - u_k = 0$ on Ω_∞. Thus

$$\begin{aligned}\int\limits_{\Omega\setminus\Omega_\infty} (4\,|u - u_k|)^p \, dx &\geq \sum_{j=k}^{\infty} 4^j \int\limits_{\Lambda_j'} (d_j - k)^p dx \\ &\geq \sum_{j=k}^{\infty} 4^j \int\limits_{\Lambda_j'} (d_j/2)^p dx \geq \sum_{j=k}^{\infty} \frac{4^j}{2^{j+1}} \int\limits_{\Lambda_j'} d_j{}^p dx.\end{aligned}$$

But $\int\limits_{\Lambda_j'} d_j{}^p dx = 2^{-j}$ for infinitely many j's. Hence the last sum is infinite and $\|u - v\|_p > 1/4$. Since v is an arbitrary bounded function, the proof of (ii) is complete.

(iii) Suppose that (2.43) holds and let u be a bounded function on Ω with $|u| \leq K$. Let $\{G_m\}_{m\in\mathbb{N}}$ be a decreasing sequence of measurable subsets of Ω with $|G_m| \to 0$, and let $\lambda \in (0, K)$. Then for every m,

$$\begin{aligned}\int\limits_{G_m} \left|\frac{f}{\lambda}\right|^p dx &\leq \int\limits_{G_m} \left(\frac{K}{\lambda}\right)^p dx = \sum_{k=1}^{\infty} \int\limits_{G_m\cap\Lambda_k} \left(\frac{K}{\lambda}\right)^p dx \\ &\leq \sum_{k=1}^{\infty} \left(\frac{K}{\lambda}\right)^{k+1} |G_m \cap \Lambda_k|.\end{aligned}$$

Since $K/\lambda > 1$, (2.43) implies that there exists k_0 such that

$$\sum_{k=k_0+1}^{\infty} \left(\frac{K}{\lambda}\right)^{k+1} |\Lambda_k| \leq \frac{1}{2};$$

as $|G_m| \to 0$, there exists m_0 such that if $m \geq m_0$, then

$$\sum_{k=1}^{\infty} \left(\frac{K}{\lambda}\right)^{k+1} |G_m \cap \Lambda_k| \leq \frac{1}{2}.$$

Thus,

$$\int_{G_m} \left|\frac{u}{\lambda}\right|^p dx \leq 1 \text{ if } m \geq n_0.$$

It follows that for any small $\lambda > 0$ we have $\|u\mathbf{1}_{G_m}\|_p \leq \lambda$ if $m \geq m_0$, so that $\lim_{m\to\infty} \|u\mathbf{1}_{G_m}\|_p = 0$. Hence $u \in X_a$.

Now assume that $u \in X_b$. Then there is a sequence of bounded functions u_m with $\|u - u_m\|_p \to 0$. By the last part of the proof, each $u_m \in X_a$; as X_a is closed, $u \in X_a$. Now suppose that there exists $A > 1$ such that $\sum_{k=1}^{\infty} A^k |\Lambda_k| = \infty$. Let v be the function that is identically 1 on Ω, and put $G_m = \bigcup_{k=m}^{\infty} \Lambda_k$. Then

$$\int_\Omega \left(\frac{u\mathbf{1}_{G_m}}{1/A}\right)^p dx = \sum_{k=1}^{\infty} \int_{\Lambda_k} A^p \mathbf{1}_{G_m} dx \geq \sum_{k=m}^{\infty} A^m |\Lambda_k| = \infty.$$

Hence $\|u\mathbf{1}_{G_m}\|_p > 1/A$ for any m. As the G_m are decreasing and $|G_m| \to 0$, it follows that $u \notin X_a$. Since plainly $u \in X_b$ the proof is complete. □

We underline that under the conditions (iii) of the preceding theorem another characterization of the equality of X_a and X_b can be given in terms of the non-increasing rearrangement p^* of the function p. In fact, since

$$\int_0^{|\Omega|} A^{p^*(t)} dt = \int_\Omega A^{p(x)} dx,$$

it is immediate that $X_a = X_b$ if and only if

$$\int_0^{|\Omega|} A^{p^*(t)} dt < \infty \quad \text{for all } A > 1.$$

By way of illustration, consider the case in which $n = 1$, $\Omega = (0, 1/e)$ and $p^*(x) = x^\alpha$ for some $\alpha < 0$. Then

$$\int_0^{1/e} A^{p^*(x)} dx = \int_0^{1/e} A^{x^\alpha} dx = \int_0^{1/e} \sum_{k=0}^{\infty} \frac{(x^\alpha \log A)^k}{k!} dx$$
$$= \infty$$

for all $A > 1$. Thus $X_a = X_b$.

On the other hand, if $n = 1$, $\Omega = (0, 1/e)$ and $p^*(x) = (\log x^{-1})^\alpha$ for some $\alpha \geq 0$, then

$$\int_0^{1/e} A^{p^*(x)} dx = \int_0^{1/e} \exp\left((\log x^{-1})^\alpha \log A\right) dx = \int_1^\infty \exp\left(y^\alpha \log A - y\right) dy$$
$$< \infty$$

for all $\alpha \in (0,1)$ and all $A > 1$: thus $X_a = X_b$ when $\alpha \in (0,1)$. However, if $\alpha \in [1,\infty)$, the choice $A = e$ gives

$$\int_0^{1/e} A^{p^*(x)} dx = \int_1^\infty \exp(y^\alpha \log A - y)\, dy = \infty,$$

and so $X_a \subsetneqq X_b$.

Our discussion so far has been focused on those aspects of the theory of spaces with variable exponent that emphasize the similarities between the properties of such spaces and those of their classical counterparts. Remarkable though these are, important differences remain, even when the exponent p is non-pathological. The most prominent of these relates to the matter of p-mean continuity. Recall that if p is a constant, $p \in [1,\infty)$, and Ω is a non-empty open bounded subset of $\mathbb{R}^n$, then every function $u \in L^p(\Omega)$, extended by 0 outside Ω, is p-mean continuous in the sense that given any $\varepsilon > 0$, there exists $\delta > 0$ such that

$$\left(\int_\Omega |u(x+h) - u(x)|^{p(x)}\, dx \right)^{1/p} < \varepsilon \text{ if } |h| < \delta.$$

By analogy with this, when p is any element of $\mathcal{P}(\Omega)$, we say that a function $u \in L^{p}(\Omega)$ is p-mean continuous if $\rho_p(u - \tau_h u) < \varepsilon$ whenever $|h| < \delta$: here $\tau_h f := u(\cdot + h)$. It turns out that elements of $L^{p}(\Omega)$ are not, in general, p-mean continuous: more precisely, in [60] the following theorem is given:

Theorem 2.8.5. *Let $p \in \mathcal{P}(\Omega)$ and suppose that Ω contains a ball $B(x_0,r)$ on which p is continuous and non-constant. Then there is a function $u \in L^{p}(\Omega)$ which is not p-mean continuous.*

Proof. Let $z \in B(x_0,r)$ be a point at which p does not have a local extremum. There are sequences (x_k), (y_k) of points of $B(x_0,r)$, each converging to z, with $p(x_k) < p(z) < p(y_k)$ for all $k \in \mathbb{N}$. Since p is continuous in $B(x_0,r)$, for each $k \in \mathbb{N}$ there exists $r_k > 0$ such that

$$p(x) < \frac{1}{2}(p(z) + p(x_k)) < p(z) \text{ if } x \in B(x_k, r_k) \tag{2.44}$$

and

$$p(x) > p(z) \text{ if if } x \in B(y_k, r_k). \tag{2.45}$$

For each $k \in \mathbb{N}$ put $q_k = \frac{1}{2}(p(z) + p(x_k))$, let u_k be a function on Ω such that $\operatorname{supp} u_k \subset B(x_k,r_k)$, $u_k \in L^{q_k}(B(x_k,r_k)) \setminus L^{p}(B(x_k,r_k))$ and $\|u\|_{q_k} = 1$, and define

$$u = \sum_{k=1}^\infty 2^{-k} u_k.$$

From (2.44) and Theorem 2.2.2 we have

$$\begin{aligned}\|u\|_p &\leq \sum_{k=1}^{\infty} 2^{-k} \|u_k\|_p \leq \sum_{k=1}^{\infty} 2^{-k} \|u_k\|_{q_k} (|B(x_k, r_k)| + 1) \\ &\leq 1 + \sup_k |B(x_k, r_k)| < \infty.\end{aligned}$$

Now put $h_k = y_k - x_k$ $(k \in \mathbb{N})$: from (2.45) and Theorem 2.2.2 we obtain

$$\begin{aligned}\|\tau_{h_k} u\|_p &\geq \left\|\mathbf{1}_{B(y_k, r_k)} \tau_{h_k} u\right\|_p \geq (1 + |B(y_k, r_k)|)^{-1} \left\|\mathbf{1}_{B(y_k, r_k)} \tau_{h_k} u\right\|_{p(z)} \\ &= (1 + |B(y_k, r_k)|)^{-1} \left\|\mathbf{1}_{B(y_k, r_k)} u\right\|_{p(z)} = \infty.\end{aligned}$$

It follows that $\tau_{h_k} u - u \notin L^p(\Omega)$. Thus u is not p-mean continuous. □

Despite this result, if $p_+ < \infty$ and $u \in L^p(\Omega)$ is bounded, with compact support, then there is a compact set K such that for all $h \in \mathbb{R}^n$ with $|h| < 1$, $\operatorname{supp} \tau_h u \subset K$. Thus by Theorem 2.2.2,

$$\lim_{|h| \to 0} \|\tau_h u - u\|_p \leq (1 + |K|) \lim_{|h| \to 0} \|\tau_h u - u\|_{p_+} = 0.$$

The lack of satisfactory behavior under translations is perhaps the biggest technical difficulty faced when dealing with spaces with variable exponent. Naturally it manifests itself in connection with convolutions. For example, if p is a constant, $p \in [1, \infty]$, then a special case of Young's inequality asserts that

$$\| u * v\|_{p,\mathbb{R}^n} \leq \|u\|_{p,\mathbb{R}^n} \|v\|_{1,\mathbb{R}^n}$$

for all $u \in L^p(\mathbb{R}^n)$ and $v \in L^1(\mathbb{R}^n)$; here $u * v$ denotes the convolution of u and v. However (see [24]) if $p \in \mathcal{P}(\Omega)$ and $1 < p_- \leq p_+ < \infty$, then there exists $c > 0$ such that

$$\| u * v\|_{p,\mathbb{R}^n} \leq c \|u\|_{p,\mathbb{R}^n} \|v\|_{1,\mathbb{R}^n}$$

for all $u \in L^p(\mathbb{R}^n)$ and all $v \in L^1(\mathbb{R}^n)$ if and only if p is constant. A more general form of Young's inequality for constant exponents asserts that

$$\| u * g\|_{r,\Omega} \leq c \|u\|_{p,\Omega} \|v\|_{q,\Omega},$$

when $\Omega \subset \mathbb{R}^n$ and p, q, r are constants lying in the interval $[1, \infty]$ and satisfying the equality

$$\frac{1}{r} + 1 = \frac{1}{p} + \frac{1}{q}.$$

Further insight into the severe restrictions that have to be imposed in order to obtain an analogue of the classical results is provided by the following result (see [23]), which also illustrates the usefulness of Theorem 2.8.5.

Theorem 2.8.6. *Let $p, r \in \mathcal{P}(\Omega) \cap C(\Omega)$ be such that $1 < p_- \leq p_+ < \infty$ and $1 < r_- \leq r_+ < \infty$; suppose that Ω is bounded and contains a ball on which r is not constant. Then*

$$\| u * v\|_{r,\Omega} \lesssim \|u\|_{p,\Omega} \|v\|_{1,\Omega} \quad (u \in L^p(\Omega), v \in L^1 1(\Omega))$$

if and only if $p_- \geq r_+$.

Proof. First suppose that $p_- \geq r_+$. Then from the classical form of Young's inequality and Theorem 2.2.2,

$$\| u * v\|_{r,\Omega} \lesssim \| u * v\|_{r_+,\Omega} \lesssim \|u\|_{r_+,\Omega} \|v\|_{1,\Omega} \lesssim \|u\|_{p,\Omega} \|v\|_{1,\Omega} .$$

For the converse, assume that $\| u * v\|_{r,\Omega} \lesssim \|u\|_{p,\Omega} \|v\|_{1,\Omega}$ and suppose that $p_- < r_+$. It is sufficient to deal with the case in which Ω is a ball on which r is not constant and $p(x) < r(x)$. Theorem 2.2.2 guarantees that

$$L^r(\Omega) \hookrightarrow L^p(\Omega);$$

it follows from Theorem 2.8.5 that there is a function $u \in L^p(\Omega)$ which is not r-mean continuous. Hence there exists $h \in \mathbb{R}^n$ such that, in the notation used in Theorem 2.8.5, $u_h \notin L^r(\Omega)$. Now let ϕ be a Friedrichs mollifier and for each $\varepsilon > 0$ define ϕ_ε as usual by $\phi_\varepsilon(x) = \varepsilon^{-n}\phi((x-h)/\varepsilon)$, $x \in \mathbb{R}^n$. Then as $\varepsilon \to 0$, $u * \phi_\varepsilon \to u_h$ in $L^1(\Omega)$; and by the continuity of the convolution operator,

$$\| u * \phi_\varepsilon\|_{r,\Omega} \lesssim \|u\|_{p,\Omega} \|\phi_\varepsilon\|_{1,\Omega} = \|u\|_{p,\Omega} .$$

As $L^r(\Omega)$ is reflexive, there is a sequence (ε_k), with $\varepsilon_k \to 0$, such that $(u * \phi_{\varepsilon_k})$ converges weakly in $L^r(\Omega)$; evidently the limit is u_h. But $u_h \notin L^r(\Omega)$ and we have a contradiction. □

Yet another difficulty arises in connection with the Hardy-Littlewood maximal operator M, defined for all functions u that are locally integrable on Ω, by

$$(Mu)(x) = \sup |B|^{-1} \int_{B\cap\Omega} |u(y)|\, dy \quad (x \in \Omega),$$

where the supremum is taken over all balls B that contain x and for which $|B \cap \Omega| > 0$. It is well known that if p is a constant lying in the interval $(1, \infty)$, then M maps $L^p(\Omega)$ boundedly into itself: there is a positive constant $c = c(p, \Omega)$ such that for all $u \in L^p(\Omega)$,

$$\|Mu\|_p \leq c \|u\|_p . \tag{2.46}$$

However, (2.46) does not hold for all $p \in \mathcal{P}(\Omega)$. A sufficient condition for its validity, when Ω is open and bounded, is that for some constant $C > 0$,

$$|p(x) - p(y)| \leq -\frac{C}{\log|x-y|} \quad \text{for all } x, y \in \Omega \text{ with } 0 < |x-y| < 1/2.$$

A detailed discussion of this is given in [24]. As Theorem 2.8.2 settles the question of uniform convexity for the variable exponent Lebesgue's spaces, we now turn to the modular uniform convexity. We recall the Definition of $UCC2$ (see Definition 1.3.9).

Theorem 2.8.7. *Let $\Omega \subseteq \mathbb{R}^n$ be open and $\in \mathcal{P}(\Omega)$. If $p_- > 1$ then the modular*

$$\rho_p : L^p(\Omega) \longrightarrow [0, \infty)$$

$$\rho_p(u) = \int_\Omega |u(x)|^{p(x)} dx$$

satisfies the $UCC2$ condition.

Proof. The proof follows along the same lines as that of Theorem 1.4.32: We include the details for the sake of completeness. Let $\Omega \subseteq \mathbb{R}^n$ be a domain and $p \in \mathcal{P}(\Omega)$. We retain the terminology from Section 1.5; in addition, for a subset $A \subseteq \Omega$ and $u \in \mathcal{M}(\Omega)$, we write

$$\rho_p^A(u) = \int\limits_A |u|^{p(x)} dx.$$

Let $r > 0$, $\epsilon > 0$ and consider $u, v \in L^p(\Omega)$ such that

$$\rho_p(u) \leq r\,,\ \rho_p(v) \leq r\,,\ \rho_p\left(\frac{u-v}{2}\right) \geq \epsilon r.$$

On account of the convexity of ρ_p we have $\epsilon \leq 1$: indeed,

$$r\epsilon \leq \rho_p\left(\frac{u-v}{2}\right) \leq r.$$

Let $\Omega_1 := \{x \in \Omega : p(x) \geq 2\}$. Then either

$$\int\limits_{\Omega_1} \left|\frac{u(x)-v(x)}{2}\right|^{p(x)} dx \geq \frac{r\epsilon}{2} \tag{2.47}$$

or

$$\int\limits_{\Omega\setminus\Omega_1} \left|\frac{u(x)-v(x)}{2}\right|^{p(x)} dx \geq \frac{r\epsilon}{2}. \tag{2.48}$$

In case (2.47) holds, one has, by virtue of inequality (1.52):

$$\int\limits_{\Omega_1} \left|\frac{u(x)-v(x)}{2}\right|^{p(x)} dx + \int\limits_{\Omega_1} \left|\frac{u(x)+v(x)}{2}\right|^{p(x)} dx$$

$$\leq \frac{1}{2}\left(\int\limits_{\Omega_1} |u(x)|^{p(x)} dx + \int\limits_{\Omega_1} |v(x)|^{p(x)} dx\right).$$

It is thus concluded that

$$\int_{\Omega_1} \left| \frac{u(x)+v(x)}{2} \right|^{p(x)} dx \le \frac{1}{2} \left(\int_{\Omega_1} |u(x)|^{p(x)} dx + \int_{\Omega_1} |v(x)|^{p(x)} dx \right) - \frac{r\epsilon}{2}.$$

Thus,

$$\begin{aligned}
\rho_p\left(\frac{u+v}{2}\right) &= \int_{\Omega_1} \left| \frac{u(x)+v(x)}{2} \right|^{p(x)} dx + \int_{\Omega\setminus\Omega_1} \left| \frac{u(x)+v(x)}{2} \right|^{p(x)} dx \\
&\le \frac{1}{2} \left(\int_{\Omega_1} |u(x)|^{p(x)} dx + \int_{\Omega_1} |v(x)|^{p(x)} dx \right) - \frac{r\epsilon}{2} \\
&\quad + \frac{1}{2} \left(\int_{\Omega\setminus\Omega_1} |u(x)|^{p(x)} dx + \int_{\Omega\setminus\Omega_1} |v(x)|^{p(x)} dx \right) \\
&= \frac{1}{2} \left(\rho_p(u) + \rho_p(v)\right) - \frac{r\epsilon}{2} \\
&\le r\left(1 - \frac{\epsilon}{2}\right).
\end{aligned}$$

In case (2.48) holds, we define

$$\Omega_2 := \left\{ x \in \Omega \setminus \Omega_1 : |u(x) - v(x)| \le \frac{\epsilon}{4}(|u(x)| + |v(x)|) \right\}.$$

It follows that

$$\begin{aligned}
\int_{\Omega_2} \left| \frac{u(x)-v(x)}{2} \right|^{p(x)} dx &\le \frac{\epsilon}{8} \left(\int_{\Omega_2} |u(x)|^{p(x)} dx + \int_{\Omega_2} |v(x)|^{p(x)} \right) \\
&\le \frac{\epsilon}{8} \left(\rho_p(u) + \rho_p(u)\right) \le \frac{r\epsilon}{8}.
\end{aligned}$$

The validity of (2.48) implies in particular that

$$\begin{aligned}
\int_{\Omega\setminus(\Omega_1\cup\Omega_2)} \left| \frac{u(x)-v(x)}{2} \right|^{p(x)} &= \int_{\Omega\setminus\Omega_1} \left| \frac{u(x)-v(x)}{2} \right|^{p(x)} - \int_{\Omega_2} \left| \frac{u(x)-v(x)}{2} \right|^{p(x)} \\
&\ge \frac{r\epsilon}{2} - \frac{r\epsilon}{4} = \frac{r\epsilon}{4}.
\end{aligned}$$

It follows from (1.53) that if $x \in \Omega \setminus (\Omega_1 \cup \Omega_2)$, one has

$$\begin{aligned}
&\left|\frac{u(x)+v(x)}{2}\right|^{p(x)} + (p_- - 1)\frac{\epsilon}{8}\left|\frac{u(x)-v(x)}{2}\right|^{p(x)} \\
&\le \left|\frac{u(x)+v(x)}{2}\right|^{p(x)} + \frac{p(x)(p(x)-1)}{2}\left(\frac{\epsilon}{4}\right)^{2-p(x)}\left|\frac{u(x)-v(x)}{2}\right|^{p(x)} \\
&\le \left|\frac{u(x)+v(x)}{2}\right|^{p(x)} + \frac{p(x)(p(x)-1)}{2}\left|\frac{u(x)-v(x)}{|u(x)|+|v(x)|}\right|^{2-p(x)}\left|\frac{u(x)-v(x)}{2}\right|^{p(x)} \\
&\le \frac{1}{2}(|u(x)|^{p(x)} + |v(x)|^{p(x)}).
\end{aligned}$$

Integrating the last inequality over $\Omega \setminus (\Omega_1 \cup \Omega_2)$ it follows that

$$\begin{aligned}
&\int_{\Omega\setminus(\Omega_1\cup\Omega_2)} \left|\frac{u(x)+v(x)}{2}\right|^{p(x)} dx \\
&\quad + \left(\inf_{\Omega} p(x) - 1\right)\frac{\epsilon}{8}\int_{\Omega\setminus(\Omega_1\cup\Omega_2)} \left|\frac{u(x)-v(x)}{2}\right|^{p(x)} dx \\
&\quad \le \frac{1}{2}\left(\int_{\Omega\setminus(\Omega_1\cup\Omega_2)} |u(x)|^{p(x)} dx + \int_{\Omega\setminus(\Omega_1\cup\Omega_2)} |v(x)|^{p(x)} dx\right).
\end{aligned}$$

We arrive thus at

$$\begin{aligned}
&\int_{\Omega\setminus(\Omega_1\cup\Omega_2)} \left|\frac{u(x)+v(x)}{2}\right|^{p(x)} dx \\
&\le \frac{1}{2}\left(\int_{\Omega\setminus(\Omega_1\cup\Omega_2)} |u(x)|^{p(x)} dx + \int_{\Omega\setminus(\Omega_1\cup\Omega_2)} |v(x)|^{p(x)} dx\right) - (p_- - 1)\frac{\epsilon^2}{32}r.
\end{aligned}$$

In all

$$\begin{aligned}
\rho_p\left(\frac{u+v}{2}\right) &= \int_{\Omega_1\cup\Omega_2} \left|\frac{u(x)+v(x)}{2}\right|^{p(x)} dx + \int_{\Omega\setminus(\Omega_1\cup\Omega_2)} \left|\frac{u(x)+v(x)}{2}\right|^{p(x)} dx \\
&\leq \frac{1}{2}\left[\int_{\Omega_1\cup\Omega_2} |u(x)|^{p(x)} dx + \int_{\Omega_1\cup\Omega_2} |v(x)|^{p(x)} dx\right] \\
&+ \frac{1}{2}\left[\int_{\Omega\setminus(\Omega_1\cup\Omega_2)} |u(x)|^{p(x)} dx + \int_{\Omega\setminus\Omega_1\cup\Omega_2} |v(x)|^{p(x)} dx\right] \\
&- (p_- - 1)\frac{\epsilon^2}{32} r \\
&\leq r - (p_- - 1)\frac{\epsilon^2}{32} r = r\left(1 - (p_- - 1)\frac{\epsilon^2}{32}\right).
\end{aligned}$$

We conclude that for any $r > 0$, $\epsilon > 0$ and arbitrary $u, v \in D(r,\epsilon)$ as specified in Definition 1.21, it holds that

$$1 - \frac{1}{r}\rho_p\left(\frac{u+v}{2}\right) \geq \min\left\{\frac{\epsilon}{2}, (p_- - 1)\frac{\epsilon^2}{32}\right\} > 0,$$

and $L^p(\Omega)$ is $UUC2$. □

2.8.1 Duality maps on spaces of variable integrability

Finally we turn to the determination of duality maps on $L^p(\Omega)$. When p is a constant in the interval $(1,\infty)$, it is a familiar fact that the Gâteaux derivative of $\|\cdot\|_p$ is given by

$$\left(\text{grad}\,\|u\|_p\right)(x) = \|u\|_p^{1-p}\,|u(x)|^{p-1}\ \text{sgn}\ u(x) \quad (u \neq 0, x \in \Omega),$$

from which the duality map J with gauge function μ may be calculated by means of the formula

$$Jf = \mu\left(\|u\|_p\right)\text{grad}\,\|u\|_p\,.$$

Now suppose that $p \in \mathcal{P}(\Omega)$ and $1 < p_- \leq p_+ < \infty$. By Theorems 1.2.25 and 2.8.2, $L^p(\Omega)$ is uniformly convex and uniformly smooth, and so its norm is even Fréchet-differentiable.

Theorem 2.8.8. *Let Ω be a bounded open subset of $\mathbb{R}^n$ and let $p \in \mathcal{P}(\Omega)$ be such that $1 < p_- \leq p_+ < \infty$. Then for every $u \in L^p(\Omega)\setminus\{0\}$,*

$$(\text{grad}\,\|u\|_p)(z) = \frac{p(z)\,\|u\|_p^{-p(z)}\,|u(z)|^{p(z)-1}\,\text{sgn}\,u(z)}{\int_\Omega p\,\|u\|_p^{-p-1}\,|u|^p\,dx}. \tag{2.49}$$

Proof. Put

$$A(x) = \left(\frac{|u(x)|}{\|u\|_p} \right)^{p(x)-1} \operatorname{sgn} u(x), \; B = \int_\Omega p \, \|u\|_p^{-p-1} \, |u|^p \, dx.$$

Thus the right-hand side of (2.49) equals $p(x) \, \|u\|_p^{-1} \, A(x)/B$. Note that

$$\frac{p_-}{\|u\|_p} = \frac{p_-}{\|u\|_p} \phi_p \left(u / \, \|u\|_p \right) \leq B \leq \frac{p_+}{\|u\|_p}.$$

Moreover,

$$\phi_{p'}(A) = \int_\Omega \left(\frac{|u|}{\|u\|_p} \right)^p dx = 1.$$

Hence the right-hand side of (2.49) represents an element of $L^{p'}(\Omega)$ and so can be identified with an element of the dual of $L^p(\Omega)$, the value of which at u is $\|u\|_p$. The result follows from this observation. □

2.9 The Matuszewska-Orlicz index of a Musielak-Orlicz space

We begin this section with a few elementary facts concerning subadditive functions. For an exhaustive treatment of the topic we refer the reader to [78]. See also [11].

Lemma 2.9.1. *Let*

$$g : \mathbb{R} \longrightarrow (0, \infty)$$

be measurable, increasing and subadditive (i.e., $g(s + t) \leq g(t) + g(s)$ for all $(s, t) \in \mathbb{R}^2$) and assume $g(0) = 0$. Then

$$0 < \lim_{t \to \infty} \frac{g(t)}{t} = \inf_{t > 0} \frac{g(t)}{t} < \infty. \tag{2.50}$$

Proof. For one thing it is clear that

$$L = \inf_{t > 0} \frac{g(t)}{t} \leq 1.$$

Given a fixed $\delta > 0$, let t be chosen so that

$$L \leq \frac{g(t)}{t} < L + \delta,$$

then for some natural number k it holds

$$(1+k^{-1})\frac{g(t)}{t} < L+\delta.$$

For any real number s with $s \geq kt$, let n be chosen such that

$$\frac{s}{t} \in [n, n+1)$$

(obviously, $n \geq k$). From the conditions on g it follows that

$$g(s) \leq (n+1)g(t).$$

In all

$$\frac{g(s)}{s} \leq \frac{(n+1)g(t)}{nt} \leq \left(1+\frac{1}{n}\right)\frac{g(t)}{t} \leq \left(1+\frac{1}{k}\right)\frac{g(t)}{t} < L+\delta.$$

The latter shows both the existence of the limit and the equality in (2.50). □

Definition 2.9.1. *For a Musielak-Orlicz function* φ *and each* $x \in \Omega$, *set*

$$M_\varphi(x,t) = \limsup_{u\to\infty} \frac{\varphi(x,tu)}{\varphi(x,u)}. \tag{2.51}$$

The Matuszewska index of φ *(or Matuszewska-Orlicz index) is defined to be*

$$m_\varphi(x) = \lim_{t\to\infty} \frac{\ln M_\varphi(x,t)}{\ln t} = \inf_{t>1} \frac{\ln M_\varphi(x,t)}{\ln t}. \tag{2.52}$$

In particular, for $t, s > 1$,

$$\limsup_{u\to\infty} \frac{\varphi(x,tsu)}{\varphi(x,u)} \leq \limsup_{u\to\infty} \frac{\varphi(x,tsu)}{\varphi(x,tu)} \limsup_{u\to\infty} \frac{\varphi(x,tu)}{\varphi(x,u)};$$

this yields

$$M_\varphi(x,ts) \leq M_\varphi(x,ts)M_\varphi(x,ts).$$

We remark that for fixed $x \in \Omega$,

$$M^{(\varphi)} : [0,\infty) \longrightarrow [0,\infty)$$
$$M^{(\varphi)}(x,s) = M_\varphi(x,e^s)$$

is subadditive and that

$$\lim_{t\to\infty} \frac{M^{(\varphi)}(x,t)}{\ln t} = \lim_{s\to\infty} \frac{M_\varphi(x,e^s)}{s}$$
$$= \inf_{s>0} \frac{M_\varphi(x,e^s)}{s} = \inf_{t>1} \frac{M_\varphi(x,t)}{\ln t}.$$

Example 2.9.1. It is easy to verify that if $\varphi(x,t) = t^{p(x)}$, then $m_\varphi(x) = p(x)$.

Example 2.9.2. If $\varphi(x,t) = t^{p(x)}(\log(1+t))^{q(x)}$, then $m_\varphi(x) = p(x)$.

The Matuszewska-Orlicz index of an Orlicz function was first introduced in [79]. We also refer the interested reader to [53, 78, 91, 92, 102] for a more detailed account and further ramifications of this topic.

Definition 2.9.2. *Let φ_1 and φ_2 be Musielak-Orlicz functions defined on $\Omega \subset \mathbb{R}^n$. The symbol*

$$\varphi_1 \sim \varphi_2$$

will be denoted to indicate that there exist positive constants A, B, C, D, T such that for any $t \geq T$ the inequalities

$$A\varphi_1(x, Bt) \leq \varphi_2(x,t) \leq C\varphi_1(x, Dt)$$

hold uniformly in Ω.

Proposition 2.9.2. *Let φ_i, $i = 1, 2$ be Musielak-Orlicz functions on $\Omega \subset \mathbb{R}^n$ with $\varphi_1 \sim \varphi_2$. Then*

$$m_{\varphi_1} = m_{\varphi_2}.$$

Proof. The proof is elementary: For any $t > 0$, $u > 0$ it follows from Definition 2.9.2 that uniformly in Ω one has:

$$\frac{\varphi_2(x,tu)}{\varphi_2(x,u)} \leq CA^{-1}\frac{\varphi_1(x,Dtu)}{\varphi_1(x,Bu)} \leq CA^{-1}\frac{\varphi_2(x,B^{-1}tu)}{\varphi_2(x,BD^{-1}u)}.$$

The claim follows from the last string of inequalities and Definition 2.9.1. □

It is well known (and the proof is elementary, see [78]) that the finiteness of the Matuszewska-Orlicz index of an *Orlicz* function φ is equivalent to the validity of the Δ_2 condition for φ. The following example shows that this is not so in the general case of a Musielak-Orlicz function φ on a domain Ω.

Example 2.9.3. Let Ω stand for the Euclidean unit ball in $\mathbb{R}^6$ and for $n \in \mathbb{N}$, let B_n be the ball of radius 2^{-n-2} centered at $x_n = (2^{-n}, 0, 0, 0, 0, 0)$, denote the ball concentric with B_n of radius 2^{-n-3} with B_n^- and set

$$B_n^+ = B_n \setminus B_n^-.$$

For each $n \in \mathbb{N}$ choose $z_n = 2^{3(n+3)}$ and fix $s_n > z_n$ such that

$$s_n^3 - z_n^2 = 3s_n^2(s_n - z_n)$$

and consider the function:

$$\phi_{z_n}(t) = \begin{cases} t^2 & \text{if } t \in (0, z_n) \\ 3s_n^2(t - z_n) + z_n^2 & \text{if } t \in (z_n, s_n) \\ t^3 & \text{if } t \in (s_n, \infty). \end{cases}$$

Next, we define

$$\varphi : \Omega \times [0, \infty) \to [0, \infty)$$

as

$$\varphi(x, t) = \begin{cases} t^3 & \text{if } x \in \Omega \setminus \bigcup_{j \in \mathbb{N}} B_j^+ \\ \phi_{z_n}(t) & \text{if } x \in B_n^+. \end{cases}$$

It is a matter of course to verify that φ is an Musielak-Orlicz function on Ω; in addition, we claim that φ does not fulfill the Δ_2 condition. Indeed, for any $x \in B_n^+$

$$\varphi(x, z_n) = \varphi(x, 2^{3(n+3)}) = \phi_{2^{3(n+3)}}\left(2^{3(n+3)}\right) = 2^{6(n+3)},$$

whereas

$$\begin{aligned} \varphi(x, 2z_n) &= \varphi(x, 2^{3n+10}) = \phi_{2^{3(n+3)}}\left(2^{3n+10}\right) \\ &\geq 3s_n^2(2^{3n+10} - 2^{3(n+3)}) + 2^{6(n+3)} \\ &\geq 2^{3(n+3)}(3s_n^2 + 2^{3(n+3)}) \\ &= \varphi(x, z_n)\frac{3s_n^2 + 2^{3(n+3)}}{2^{3n+9}}. \end{aligned} \tag{2.53}$$

Since $s_n > 2^{3n+9}$ and, by definition $s_n \to \infty$ as $n \to \infty$, inequality (2.53) shows that φ fails the Δ_2 condition, as claimed.
Yet, it is easy to show that the Matuszewska index of φ is equal to 3 in Ω and that

$$\begin{aligned} &\varphi(x, t) \longrightarrow t^2 \text{ as } x \longrightarrow (0, 0, 0, 0, 0, 0), \text{ on } B_n^+ \\ &\varphi(x, t) \longrightarrow t^3 \text{ as } x \longrightarrow (0, 0, 0, 0, 0, 0) \text{ on } \Omega \setminus \bigcup_{j \in \mathbb{N}} B_j^+. \end{aligned}$$

A simple calculation reveals that in the preceding example the limit in Definition 2.9.1 is not uniform in Ω.

Lemma 2.9.3. *Let $\Omega \subset \mathbb{R}^n$ be a bounded domain and φ an MO function as described above. If the convergence to the limits (2.51) and (2.52) is uniform in Ω, then φ satisfies the Δ_2 condition.*

Proof. Fix $\delta > 0$, then, owing to the uniformity of (2.52), one has for some $T_0 > 1$ and any $t \geq T_0$:

$$t^{m(x)-\delta} < M(x, t) < t^{m(x)+\delta}$$

uniformly in Ω. By definition of $M(x,t)$ and by virtue of the uniformity assumption of the infimum (2.51), there exists a positive number N for which, uniformly for $t \geq T_0$ and $x \in \Omega$, it holds that

$$\sup_{s \geq N} \frac{\varphi(x, st)}{\varphi(x, s)} < t^{m(x)+\delta}.$$

In particular, for all $s \geq N$:

$$\varphi(x, sT_0) \leq T_0^{\sup_\Omega m(x)+\delta} \varphi(x, s);$$

by virtue of Lemma 2.2.11 it is clear that φ satisfied the Δ_2 condition.

□

2.10 Historical notes

"The theorem commonly known as Parseval's Theorem, which in its latest form as extended by Fatou, asserts that if $f(x)$ and $g(x)$ are two functions whose squares are summable and whose Fourier constants are a_n, b_n and α_n, β_n, then the series

$$\frac{1}{2} a_0 \alpha_0 + \sum_{n=1}^{\infty} (a_n \alpha_n + b_n \beta_n)$$

converges absolutely and has for its sum

$$\frac{1}{\pi} \int_{-\pi}^{\pi} f(x) g(x) \, dx,$$

must be regarded as one of the most important results in the whole of the theory of Fourier series."

Such is the opening paragraph in Young's work "On a class of parametric integrals and their application to the theory of Fourier Series", [104].

The author then proceeds to generalize the above result to the case when $f \in L^p((-\pi, \pi))$, $g \in L^q((-\pi, \pi))$ with p and q Hölder conjugate exponents and "the series in question is summed in the Césaro way." Far from being our intention to discuss the details of this work, we call the reader's attention to [104, Paragraph 7]. The following observation (whose proof is elementary) is found there: If $p > 0$, $sp = r$, then for $U > 0$, $V > 0$

$$(1 + p)UV \leq pU^{1+\frac{1}{p}} + V^{1+p}.$$

About a year later, in [105] the author recalls this result by observing that it can be easily generalized in te following fashion: If

$$U : \mathbb{R} \longrightarrow \mathbb{R}$$

is increasing and positive with inverse V, then

$$uv - aU(a) \leq \int_a^u U(t)\,dt + \int_{U(a)}^v V(t)\,dt.$$

As an obvious application, the author points out almost in passing what constitutes perhaps the first public appearence of what decades later came to be known as Hölder's inequality for Orlicz spaces: If u, v are functions of x and $u \ln u$ and e^v are integrable, then so is uv.
In this latter work, Young explicitly directs his attention to the class of functions u for which $U(u)$ is integrable, of which class the set of functions in L^p is a particular case. Moreover, he suggests that generalizations of the known integration results are possible for functions in this wider class.
A further line of research in the second decade of the twentieth century inevitably led back to the idea introduced by Young. In 1915, de la Vallée Poussin [25] arrived to the class of "functions a function of which is integrable" while dealing with interchanging limits and integrals.
More specifically (see [25, Theorem VI]) let $E \subseteq \mathbb{R}$ be a Borel-measurable set and a sequence of non-negative functions $(f_j) \subset L^1(E)$ such that

$$\lim_{n\to\infty} f_n(x) = f(x)$$

for every $x \in E$. Then $f \in L^1(E)$ and

$$\lim_{n\to\infty} \int_E f_n dx = \int_E f dx$$

provided there exists a non-decreasing function

$$\psi : [0,\infty) \to [0,\infty) \ , \lim_{\xi\to\infty} \psi(x) = \infty$$

for which

$$\sup_n \left\{ \int_E f_n \psi(f_n) dx \right\} < \infty.$$

The latter leads to what came to be known as the de la Vallée Poussin-Lemma.
A family of measurable random variables $\mathcal{F} = (X_\alpha)_\alpha \subseteq L^1(\Omega)$ on a probability space $(\Omega, \mathcal{A}, \mu)$ is said to be **uniformly integrable** iff

$$\lim_{a\to\infty} \sup_{g\in\mathcal{F}} \left\{ \int_{|g|>a} |g|\,d\mu \right\} = 0.$$

Theorem 2.10.1. *In the terminology of the preceding paragraph, $\mathcal{F}$ is uniformly integrable if and only if there exists a convex, even function*

$$\varphi : \mathbb{R} \longrightarrow \mathbb{R}$$

with $\varphi(0) = 0$, $\lim_{x\to\infty} \frac{\varphi(x)}{x} = \infty$ *and*

$$\sup_{g\in\mathcal{F}} \int_{\Omega} \varphi(|g|)\, d\mu < \infty.$$

The interested reader might refer to [57] for the proof of this theorem and further details.

These early versions of what would later be called Orlicz spaces reappear in 1926, in an article by W. Young [106]. In this work, Young specifically mentions that this enhanced class of functions includes as particular cases the family of measurable functions u for which u^p is integrable for some exponent $p : 1 \leq p < \infty$, as well as the class of those functions u for which $u \ln \ln \ldots \ln u$ is integrable.

In the second half of 1930 the field received new impetus with the publication of the work by Birnbaum and Orlicz [13]. By writing

$$M(u) = |u|^p \ , \ N(v) = |v|^{\frac{p}{p-1}},$$

the authors point out that the classical Hölder's inequality can be viewed as the assertion that if f and g are measurable on the interval $(0, 1)$ and both

$$\int_0^1 M(f)dx < \infty \tag{2.54}$$

and

$$\int_0^1 N(g)dx < \infty \tag{2.55}$$

then

$$\int_0^1 fgdx < \infty \tag{2.56}$$

and that conversely, if (2.56) holds for all f satisfying (2.54), then g must a fortiori fulfill (2.55). Next, the authors go on to challenge the necessity of confining themselves to the specific forms (2.54) and (2.55) for M and N, and open the question of what other functions M and N would be admissible to guarantee the validity of the above conditions. Birnbaum and Orlicz devote the rest of this 67-page long article to settle the issue. In Theorem 6 in Chapter I and Theorem 3 in Chapter II, necessary and sufficient conditions are given for the existence of a conjugate function to a suitable function M (we refer the reader to the actual article for the specifics) and Chapter 3 is devoted to the carrying over of further properties of conjugate functions, properties

known at the time to hold for the L^p setting. Though much of the modern terminology is still not included in their work, in the process, they unleashed a number of ideas that would impregnate the field to the very day the present book is being written.
At this juncture the field was already mature for further development. In 1932, almost in tandem with [13], W. Orlicz published a research article on "sequences with variable exponents," as he himself calls them (Über konjugierte Exponentenfolgen, [85]). To begin his analysis he adduces the fact that "in the Theory of Orthogonal expansion one encounters series which satisfy

$$\sum_{\nu=1}^{\infty} |b_\nu|^\beta < \infty$$

but in which the exponents are not all equal." Following this line of thought he defines the ℓ^p spaces (for $p > 1$) and proves the corresponding version of Hölder's inequality for conjugate exponents. As an application of other theorems in this work, he corrects the following statement made by S. Banach in [8]:

Theorem 2.10.2. *Let $(\epsilon_n)_n$ be a sequence of positive real numbers converging to 0. Then, there exists a continuous function $u \in C([-\pi, \pi])$ with Fourier expansion*

$$\frac{a_0}{2} + \sum_{1}^{\infty} (a_n \cos nx + b_n \sin nx)$$

such that

$$\sum_{1}^{\infty} \left(|a_n|^{2-\epsilon_n} + |b_n|^{2-\epsilon_n}\right)$$

diverges.

Theorem 3′ (Satz 3′) in [85] shows that this is in general not the case. In fact, Theorem **b** in [8] sates that if

$$0 < \lambda_n \longrightarrow \infty \text{ as } n \longrightarrow \infty,$$

then one can find an integrable function u with Fourier coefficients $(a_n)_n$ and $(b_n)_n$ for which the series

$$\sum_{n=1}^{\infty} \left(|a_n|^{\lambda_n} + |b_n|^{\lambda_n}\right)$$

diverges. Based on a number of results on series with variable exponent, Orlicz goes further and generalizes this theorem by observing that the particular choice of the trigonometric system is immaterial: any orthonormal system of uniformly bounded functions would do and that the condition on (λ_n) can be weakened to require that

$$\sum_{n=1}^{\infty} k^{\lambda_n}$$

converges for every real number $k : 0 < k < 1$ (Theorem 8).
The second section in [85] should inevitably attract the attention of the specialist interested in the history of Musielak-Orlicz spaces. In what seems to be the first reference to variable exponent Lebesgue spaces, Orlicz defines "$f(x)$ being integrable with $\alpha(x) > 1$" (straight translation from the German original) for two measurable functions f and α by requiring

$$\int_0^1 |f(x)|^{\alpha(x)}\, dx < \infty.$$

"A variety of results about integrable functions with constant α," in Orlicz' own words, "can be generalized to certain classes of functions that are integrable with a function α." Putting his words into action he sets out to prove several theorems that are now part of the standard theory of Lebesgue spaces with variable exponent. [85, Theorem 4] is essentially the completeness of $L^{\alpha(\cdot)}$, [85, Theorem 5] relates to the weak-compactness of the unit ball in $L^{\alpha(\cdot)}$ and [85, Theorem 6] is the generalized Hölder's inequality.
Finally, in 1932, Orlicz formally defines the spaces L^M in the splendorous elegance of its modern form [86]. The opening paragraph of this work might add to the mathematical understanding of the history of the emergence of Orlicz spaces in Analysis. As Orlicz puts it:
The problem concerning the generalization of the concept of conjugate functions poses the question of the extent to which theorems whose validity is known for the class S^α (functions whose α^{th} power is integrable) can be carried over to other classes of functions. Thus, the question of whether a given functional class can be regarded as a B-space[1] *of similar character to that of S^α plays an important role in Functional Analysis.* A series of Orlicz' works throughout the 1930s [85, 87, 88, 89, 90] reflects the attention given to problems arising in the Theory of Fourier series at the time. The last article in the sequence is of particular interest from the point of view of the history of Orlicz spaces. In [90] the author sets out to characterize the sequences of real numbers that are actually Fourier coefficients of functions in some class. As Orlicz himself points out [90], the consideration of functions in the class L^2 made this characterization extremely difficult.
On the spot he invokes the spaces in [85] and proceeds to prove several theorems aiming at characterizing the collection of sequences of real numbers that are Fourier coefficients of functions in (among other spaces) L^M.
During the 1950 the mathematical importance of these "L^M" spaces as well as their natural role as generalizations of the Lebesgue spaces that had been introduced by Riesz [93, 94], seem to have been realized and accordingly, considerable amount of research was devoted to them: gradually these spaces start to emerge not only as examples or auxiliary elements, as had generally been hitherto the case. Mazur, Musielak and Orlicz, specifically were at this point interested in these spaces per se. Are the L^M-spaces linear? What

[1]Orlicz is using the expression "B-space" for what is known today as a Banach space.

conditions on M would make L^M an F-space? It is only natural that these questions were first posed and answered in the discrete situation [80] and almost immediately thereafter in the general measure-theoretic realm [81]. It is worthwhile noticing, however that the classes of functions introduced in the research mentioned in the preceding paragraph did not include the variable-exponent Lebesgue spaces that had been mentioned in [85].

There seems to be general agreement in the field about the fact that the first abstract generalization of the variable-exponent Lebesgue spaces introduced in [85] appeared in 1950 and is due to Nakano [83], who included such generalization in a rather arcane work on part of which he claims to have been working during the Second World War. It is plausible that Nakano had introduced these spaces in an unpublished work he mentions among the bibliographic references in [83].

It looks in retrospective that the introduction of some Sobolev version of Orlicz spaces was imminent. Inevitably, this finally happened in the mid sixties. According to Adams [2], Dankert [22] and Donaldson obtained the first Sobolev-type embedding theorems for Sobolev-Orlicz spaces. In 1971 this line of work was continued by Donaldson and Trudinger [30].

The systematic study of Sobolev spaces based on the Nakano class had to wait a little longer and seems to have been initiated by Hudzik in the mid seventies [45, 44, 46, 47, 48, 50, 51]. In 1983 Musielak published his celebrated work [82] that was bound to become the standard reference for Musielak-Orlicz spaces (the name with which the Nakano class came to be known ever since) to this day.

Chapter 3

Sobolev spaces of Musielak-Orlicz type

3.1 Sobolev spaces: definition and basic properties 161
3.1.1 Examples 164
3.2 Separability 165
3.3 Duality of Sobolev spaces of Musielak-Orlicz type 166
3.4 Embeddings, compactness, Poincaré-type inequalities 166

3.1 Sobolev spaces: definition and basic properties

In this chapter we retain the notation introduced in Chapter 1 but unless otherwise indicated, $\Omega \subset \mathbb{R}^n$ will stand for an open set and we will consider the measure space $(\Omega, \mathcal{B}, \mu)$ where $\mathcal{B}$ stands for the Borel σ-algebra of subsets of Ω, μ denotes the restriction of the Lebesgue measure to the Borel σ-algebra of Ω and

$$\varphi : \Omega \times [0, \infty) \longrightarrow [0, \infty)$$

denotes an MO-function on Ω. Recall (Definition 2.1.2) that the MO-function φ is said to be locally integrable if for any $t > 0$ and any $\mathcal{W} \in \mathcal{A}$ with $\mu(\mathcal{W}) < \infty$,

$$\int_{\mathcal{W}} \varphi(x, t)\, dx < \infty.$$

Lemma 3.1.1. *Let φ be an MO function on the domain $\Omega \subseteq \mathbf{R}^n$. If*

$$\gamma = \operatorname*{ess\,inf}_{x \in \Omega} \varphi(x, 1) > 0, \tag{3.1}$$

then every function in the MO space $L^{\varphi}(\Omega)$ is locally integrable in Ω.

Proof. It follows from the uniformity condition (3.1) and convexity that for any $s \geq 1$ and *a.e.* $x \in \Omega$,

$$0 < \gamma \leq \varphi(x, s\frac{1}{s}) \leq \frac{1}{s}\varphi(x, s).$$

We underline the fact, due to convexity, that for a fixed $x \in \Omega$ and any measurable $v \in L^{\varphi}(\Omega)$ for which $|v(x)| < 1$ it holds that:

$$\varphi\left(x, |v(x)|\right) \leq |v(x)|\varphi\left(x, |v(x)|\right).$$

Therefore, for any compact $\Omega_0 \subset \Omega$ and any $u \in L^{\varphi}(\Omega)$:

$$\begin{aligned}\int_{\Omega_0} \frac{|u(x)|}{\|u\|_{\varphi}}\, dx &= \left(\int_{\{x:|u(x)|/\|u\|_{\varphi}\geq 1\}} + \int_{\{x:|u(x)|/\|u\|_{\varphi}<1\}} \right) \frac{|u(x)|}{\|u\|_{\varphi}}\, dx \\ &\leq \int_{\Omega_0} \gamma^{-1}\varphi\left(x, \frac{|u(x)|}{\|u\|_{\varphi}}\right) dx + \mu(\Omega_0), \\ &<\infty.\end{aligned}$$

The right-hand side is positive and depends only on u, φ and Ω_0; denote it by C to conclude that

$$\int_{\Omega_0} |u(x)|\, dx \leq C\|u\|_{\varphi} < \infty.$$

The last statement yields the local integrability of u. □

Lemma 3.1.1 immediately yields the following corollary:

Corollary 3.1.2. *If φ is a locally integrable MO-function satisfying the condition of Lemma 3.1.1, then the set-theoretic inclusion*

$$L^{\varphi}(\Omega) \subset \mathcal{D}'(\Omega)$$

holds, where the right-hand symbol stands for the space of distributions in Ω.

As is customary, for any multiindex $\alpha = (\alpha_1, ...\alpha_n)$ of nonnegative integers α_j and a distribution T in Ω, we let $D^{\alpha}T$ denote the corresponding derivative of T. It is a matter of routine to verify that given a fixed multiindex α, the class of Borel-measurable functions on Ω whose distributional derivative $D^{\alpha}(u)$ are locally integrable is a real vector space with the usual pointwise operations. The next definition is justified by Lemma 3.1.1.

Definition 3.1.1. *Let φ be a Musielak-Orlicz function and $k \in \mathbb{N}$. We define the Musielak-Orlicz-Sobolev space $W^{k,\varphi}(\Omega)$ as*

$$W^{k,\varphi}(\Omega) = \{u \in L^{\varphi}(\Omega) : D^{\alpha}u \in L^{\varphi}(\Omega) \text{ for any multiindex } \alpha : |\alpha| \leq k\}$$

endowed with the norm

$$\|u\|_{k,\varphi} = \|u\|_{\varphi} + \sum_{|\alpha|=1}^{k} \|D^{\alpha}u\|_{\varphi}.$$

It will be understood from now on that $W^{0,\varphi}(\Omega) = L^{\varphi}(\Omega)$.

Definition 3.1.2. *$W_0^{k,\varphi}(\Omega)$ is the closure of $C_0^{\infty}(\Omega)$ in $W_0^{k,\varphi}(\Omega)$.*

Remark 3.1.3. It is evident from the preceding Definition 3.1.1 that the Sobolev space $W^{k,\varphi}(\Omega)$ could have been obtained by considering the modular

$$\rho_{\varphi}^{k} : \mathcal{M} \longrightarrow [0, \infty]$$
$$\rho_{\varphi}^{k}(u) = \sum_{|\alpha| \leq k} \rho_{\varphi}(D^{\alpha}u)$$

and the vector space

$$V_{\rho_{\alpha}^{k}} = \{u \in \mathcal{M} : \rho_{\alpha}^{k}(\lambda u) < \infty \text{ for some } \lambda > 0\}$$

and defining $W^{k,\varphi}(\Omega)$ as $V_{\rho_{\alpha}^{k}}$ endowed with the Luxemburg norm corresponding to ρ_{α}^{k}. The verification of the equivalence of the two definitions is a straightforward matter.

Theorem 3.1.4. *If φ is locally integrable and satisfies the condition (3.1) then for any $k \in \mathbb{N}$ the Sobolev space $W^{k,\varphi}(\Omega)$ is a Banach space.*

Proof. Let (u_j) be a Cauchy sequence in $W^{k,\varphi}(\Omega)$. Since $L^{\varphi}(\Omega)$ is complete, it is easily concluded that for each multiindex α, $0 \leq |\alpha| \leq k$ there exists $u_{\alpha} \in L^{\varphi}(\Omega)$ with

$$D^{\alpha}u_j \longrightarrow u_{\alpha} \text{ as } j \longrightarrow \infty.$$

Let $w \in C_0^{\infty}(\Omega)$ supported in Ω_0. As was shown in Lemma 3.1.1, for any multiindex α as above:

$$\left| \int_{\Omega} (D^{\alpha}u_0(x) - D^{\alpha}u_j(x))\, dx w(x) \right| \leq$$

$$C(\alpha)\|u - u_0\|_{\varphi} \left(\int_{\Omega_0} \gamma^{-1}\varphi\left(x, \frac{|u_j(x) - u_0(x)|}{\|u_j - u_0\|_{\varphi}}\right) dx + \int_{\Omega_0} \gamma^{-1}\varphi(x,1)\, dx \right)$$

for a certain positive constant $C(\alpha)$ depending only on α and w. It is apparent that the right-hand side of the last inequality tends to 0 as $j \longrightarrow \infty$. Therefore:

$$\left| \int_{\Omega} (D^{\alpha}u_0(x) - u_{\alpha})\, dx w(x)\, dx \right| = 0$$

and from the arbitrariness of w one rapidly derives the equality

$$D^{\alpha}u_0(x) = u_{\alpha}.$$

In particular, each weak derivative of u_0 up to the kth order is in $L^{\varphi}(\Omega)$, i.e.,

$$u_0 \in W^{k,\varphi}(\Omega)$$

and the above reasoning shows that by definition

$$u_j \longrightarrow u_0 \text{ in } W^{k,\varphi}(\Omega).$$

Hence, the sequence (u_j) converges and it follows that $W^{k,\varphi}(\Omega)$ is a Banach space. □

3.1.1 Examples

As it transpires from the next Example, the conditions of Lemma 3.1.1 cannot be relaxed.

Example 3.1.1.

$$\begin{aligned} \varphi :&(0,1) \longrightarrow [0,\infty) \\ \varphi(x) &= 1 - \mathbf{1}_{(\frac{1}{3},\frac{2}{3})}. \end{aligned}$$

Then the sequence

$$u_n(x) = \begin{cases} 0 & \text{if } x \in (0, \frac{1}{2} - \frac{1}{n}) \\ \frac{n}{2}(x - \frac{1}{2}) + \frac{1}{2} & \text{if } x \in (\frac{1}{2} - \frac{1}{n}, \frac{1}{2} + \frac{1}{n}) \\ 1 & \text{if } x \in (\frac{1}{2} + \frac{1}{n}, 1). \end{cases}$$

is Cauchy, yet it is not convergent $W^{1,\varphi}((0,1))$

Example 3.1.2. It is manifest from the preceding discussion that if $p : \Omega \longrightarrow (1,\infty)$ is Borel measurable $k \in \mathbb{N}$ and $\varphi_p(x,t) = t^{p(x)}$, then

$$W^{1,p}(\Omega) := W^{k,\varphi_p}(\Omega)$$

is a Banach space of distributions with respect to the norm

$$\|u\|_{k,p} := \|u\|_{1,\varphi_p} = \|u\|_p + \sum_{|\alpha|=1}^{n} \|D^\alpha u\|_p.$$

Lemma 3.1.5. *Let $\Omega \subseteq \mathbb{R}^n$ be an open set and φ be a locally integrable Musielak-Orlicz function on Ω and $k \in \mathbb{N}$. Then $C_0^\infty(\Omega) \subset W^{k,\varphi}(\Omega)$.*

Proof. If $u \in C_0^\infty(\Omega)$ then for any multiindex α, the derivative $D^\alpha u$ is bounded in Ω and the locally integrability assumption yields

$$\int_\Omega \varphi\left(x, |D^\alpha u(x)|\right) dx < \infty,$$

from which the claimed set-theoretical inclusion follows at once. □

Theorem 2.5.6 in conjunction with Lemma 3.1.5 yields:

Theorem 3.1.6. *The set $W^{1,\varphi}(\Omega)$ is dense in $L^{\varphi}(\Omega)$.*

3.2 Separability

In this section we address the problem of separability of the Sobolev spaces of Musielak-Orlicz type. We refer the reader to [48] for further discussions on this topic.

Theorem 3.2.1. *If $k \in \mathbb{N}$ and φ is a locally integrable Musielak-Orlicz function satisfying condition (3.1) and the Δ_2 condition, then $W^{k,\varphi}(\Omega)$ is separable.*

Proof. Consider the collection of all multiindexes $\alpha = (\alpha_1, \alpha_2, ...\alpha_n)$ with $\sum_{j=1}^{n} \alpha_j \leq k$. Let there be m such multiindexes and consider the cartesian product

$$\Pi_{j=1}^{m} L^{\varphi}(\Omega).$$

The modular

$$\rho_{k,\varphi}(u_1, u_2, ..., u_m) = \sum_{j=0}^{m} \rho_{\varphi}(u_j)$$

yields the Luxemburg norm

$$\|(f_j)_{1\leq j\leq m}\|_{k,\varphi}$$

and it is obvious that $(\Pi_{j=1}^{m} L^{\varphi}(\Omega), \|\cdot\|_{k,\varphi})$ is a Banach space, which, by virtue of Theorem 2.3.2 is separable.

Select and fix a one-to-one correspondence ω between the set of all subindexes $(\alpha_1, ...\alpha_n)$ with $\sum_{j=1}^{n} \alpha_j \leq k$ and the set of natural numbers $\{1, 2, ...m\}$. Define

$$P : W^{k,\varphi}(\Omega) \longrightarrow \Pi_{j=1}^{m} L^{\varphi}(\Omega)$$
$$P(u) = \left(D^{\omega^{-1}(1)}u, D^{\omega^{-1}(2)}u, ..., D^{\omega^{-1}(m)}u\right),$$

then P is an isometry and $P(W^{k,\varphi}(\Omega)) \subset \Pi_{j=1}^{m} L^{\varphi}(\Omega)$ is separable, being a subset of a separable metric space. Let (y_k) be a countable subset of $P(W^{k,\varphi}(\Omega))$ which is dense in the topology of $\Pi_{j=1}^{m} L^{\varphi}(\Omega)$. Then it follows at once that $\left(P^{-1}(y_k)\right)_k$ is dense in $W^{k,\varphi}(\Omega)$ which completes the proof of separability.
The completeness of $W^{k,\varphi}(\Omega)$ implies that $P\left(W^{k,\varphi}(\Omega)\right)$ is a closed subspace of $\Pi_{j=1}^{m} L^{\varphi}(\Omega)$; it follows then that $W^{k,\varphi}(\Omega) = P^{-1}\left(P\left(W^{k,\varphi}(\Omega)\right)\right)$ is separable. □

3.3 Duality of Sobolev spaces of Musielak-Orlicz type

Recall from Section 2.4.1 that φ^* stands for the conjugate function of the Musielak-Orlicz function φ. On account of Theorem 2.4.4 in concert with the Hahn Banach Theorem, it is easy to see that given any bounded linear functional $\Lambda \in \left(W^{k,\varphi}(\Omega)\right)^*$ there exist $v_j \in L^{\varphi^*}(\Omega)$, $0 \leq j \leq m$ such that for any $\phi \in W^{k,\varphi}(\Omega)$,

$$\langle \Lambda, \phi \rangle = \sum_{j=0}^{m} \int_{\Omega} v_j(x) D^{w^{-1}(j)} \phi(x)\, dx$$

and

$$\|(v_j)_{1 \leq j \leq m}\|_{\left(L^{\varphi^*}(\Omega)\right)^m} \leq \|\Lambda\|_{(W^{k,\varphi}(\Omega))^*}.$$

Conversely, it is immediately seen that for any sequence $(v_j)_{1 \leq j \leq m} \in \left(L^{\varphi^*}(\Omega)\right)^m$, the assignment

$$\phi \longrightarrow \langle \Lambda, \phi \rangle = \sum_{j=0}^{m} \int_{\Omega} v_j(x) D^{w^{-1}(j)} \phi(x)\, dx$$

defines a bounded, linear functional on $W^{k,\varphi}(\Omega)$ with

$$\|\Lambda\|_{(W^{k,\varphi}(\Omega))^*} \leq \|(v_j)_{1 \leq j \leq m}\|_{\left(L^{\varphi^*}(\Omega)\right)^m}.$$

Theorem 3.3.1. *Let $(\Omega, \mathcal{A}, \mu)$ be a measure space; assume that μ is σ-finite. Let φ be an MO function on $(\Omega, \mathcal{A}, \mu)$. Assume that both φ and its conjugate φ^* are proper (Definition 2.1.5). Then $W^{1,\varphi}(\Omega)$ and $W_0^{1,\varphi}(\Omega)$ are reflexive.*

Proof. It was shown in the course of the proof of Theorem 3.2.1 that $W^{1,\varphi}(\Omega)$ is isometrically isomorphic to a closed subspace of a reflexive Banach space, i.e., it is reflexive. By definition, $W_0^{1,\varphi}(\Omega)$ is a closed subspace of $W^{1,\varphi}(\Omega)$, *i.e.*, it is also reflexive. □

3.4 Embeddings, compactness, Poincaré-type inequalities

The following theorem settles the question of embeddings among Sobolev spaces of corresponding to different MO-functions ϕ and φ.

Theorem 3.4.1. *Let $\Omega \subseteq \mathbb{R}^n$ be open, consider the measure space $(\Omega, \mathcal{B}, \mu)$ where $\mathcal{B}$ stands for the Borel σ-algebra of subsets of Ω and μ is the Lebesgue measure. Let φ and ϕ be locally integrable Musielak-Orlicz functions satisfying (2.28). Then the embedding*

$$W^{k,\phi}(\Omega) \hookrightarrow W^{k,\varphi}(\Omega)$$

is continuous if and only if $\varphi \precsim \phi$, that is if there exists a positive constant B and $0 \le h \in L^1(\Omega)$ such that

$$\varphi(x,t) \le \phi(x, Bt) + h(t)$$

for all $t \in [0, \infty)$ and a.e. $x \in \Omega$.

Proof. The proof follows at once from Theorem 2.2.1 and the definition of $W^{k,\varphi}(\Omega)$. □

Let us recall the following classical embedding theorems (see [2, 15, 42, 108]):

Theorem 3.4.2. *Let $\Omega \subset \mathbf{R}^n$ be a bounded domain, $p \in (1, n)$. Then the space $W_0^{1,p}(\Omega)$ is compactly embedded in $L^q(\Omega)$ for any $q \in (1, \frac{np}{n-p})$. If $p > n$, $W_0^{1,p}(\Omega)$ embeds compactly in $C(\overline{\Omega})$ (and hence in $L^q(\Omega)$ for any $q \in (1, \infty)$). Finally, $W_0^{1,n}(\Omega)$ is compactly embedded in $L^q(\Omega)$ for any $q \in [1, \infty)$. Moreover, for any $p \in (1, \infty)$, the embedding*

$$W_0^{1,p}(\Omega) \hookrightarrow L^p(\Omega) \tag{3.2}$$

is compact.

In what follows, we aim at a Musielak-Orlicz version of the embedding (3.2). Specifically, we undertake the task of finding conditions on a Musielak-Orlicz function φ that guarantee the compactness of the embedding

$$W_0^{1,\varphi}(\Omega) \hookrightarrow L^\varphi(\Omega). \tag{3.3}$$

In the particular case of Orlicz spaces, that is when φ is independent of x related results were obtained by [22, 30]. The variable exponent Sobolev embedding theorem, $\varphi(x,t) = t^{p(x)}$ was anounced in [60, Theorem 3.10] (see also [66] for a detailed proof.)

Example 3.4.1. We define the Musielak-Orlicz function

$$\varphi : (0,1) \longrightarrow [0, \infty)$$

as

$$\varphi(x,t) = \begin{cases} t \text{ if } x \in [\frac{1}{n+1}, \frac{10n+1}{10n(n+1)}] \cup [\frac{10n+9}{10n(n+1)}, \frac{1}{n}] \\ t^n \text{ if } x \in [\frac{10n+1}{10n(n+1)}, \frac{10n+9}{10n(n+1)}] \end{cases}$$

and consider the sequence (u_n) defined by

$$u_n(x) = \begin{cases} 1 \text{ for } x \in [\frac{10n+1}{10n(n+1)}, \frac{10n+9}{10n(n+1)}] \\ 10n(n+1)\left(x - \frac{1}{n+1}\right) \text{ if } x \in [\frac{1}{n+1}, \frac{10n+1}{10n(n+1)}] \\ -10n(n+1)\left(x - \frac{1}{n}\right) \text{ if } x \in [\frac{10n+9}{10n(n+1)}, \frac{1}{n}] \end{cases}$$

Example 3.4.2. Here we set

$$\varphi : \Omega = (0,1) \longrightarrow [0, \infty)$$

defined by

$$\varphi(x,t) = \begin{cases} t^2 \text{ if } x \in [\frac{1}{n+1} + \frac{1}{10n(n+1)}, \frac{1}{n} - \frac{1}{10n(n+1)}] \\ \frac{1}{n^5}t^2 \text{ if } x \in [\frac{1}{n+1}, \frac{10n+1}{10n(n+1)}] \cup [\frac{10n+9}{10n(n+1)}, \frac{1}{n}]. \end{cases}$$

In this case we define the sequence

$$u_n(x) = \begin{cases} n \text{ for } x \in [\frac{10n+1}{10n(n+1)}, \frac{10n+9}{10n(n+1)}] \\ 10n^2(n+1)\left(x - \frac{1}{n+1}\right) \text{ if } x \in [\frac{1}{n+1}, \frac{10n+1}{10n(n+1)}] \\ -10n^2(n+1)\left(x - \frac{1}{n}\right) \text{ if } x \in [\frac{10n+9}{10n(n+1)}, \frac{1}{n}] \end{cases}$$

In both examples above, no subsequence of (u_n) is a Cauchy sequence in $L^{\varphi}((0,1))$, which implies that no subsequence can converge in the latter space. Yet, it is readily verified that both sequences are bounded in $W_0^{1,\varphi}((0,1))$. We highlight the fact that in Example 3.4.1, the Matuszewska-Orlicz index (see Section 2.9) of φ is unbounded on Ω, whereas in Example 3.4.2, the Matuszewska-Orlicz index is identically equal to 2 on Ω. As we shall see, these observations are by no means incidental, for the behavior of the Matuszewska-Orlicz index in fact determines the validity of the Sobolev Embedding (3.3).

In preparation for the main result in this section we prove:

Lemma 3.4.3. *If φ is an MO function for which the limits*

$$M_{\varphi}(x,t) = \limsup_{u \to \infty} \frac{\varphi(x,tu)}{\varphi(x,u)} \tag{3.4}$$

and

$$m_{\varphi}(x) = \lim_{t \to \infty} \frac{\ln M_{\varphi}(x,t)}{\ln t} = \inf_{t > 1} \frac{\ln M_{\varphi}(x,t)}{\ln t} \tag{3.5}$$

are uniform on Ω, then, there are constants $C_1 > 1$, $C_2 > 1$ and $S_0 > 1$ for which

$$\varphi(x, C_1 s) \leq C_2 \varphi(x,s) \ \textit{for} \ s \geq S_0.$$

Proof. A straightforward calculation shows that if $\delta > 0$ then there exists a constant $C_1 > 1$ for which $t \geq C_1$ implies

$$M(x,t) < t^{m(x)+\delta};$$

the assumed uniformity of the limit yields the existence of $S_0 > 1$ for which

$$\sup_{s \geq S_0} \frac{\varphi(x,ts)}{\varphi(x,s)} < t^{m(x)+\delta} + \frac{1}{2} C_1^{[\sup_{x\in\Omega} m(x)+\delta]} \tag{3.6}$$

whenever $s \geq S_0$, $t \geq C_1$; in particular, setting $t = C_1$ in (3.6) one easily sees that for $s \geq S_0$ it holds that

$$\varphi(x, C_1 s) \leq \frac{3}{2} C_1^{[\sup_{x\in\Omega} m(x)+\delta]} \varphi(x,s),$$

whence the lemma follows immediately. □

Theorem 3.4.4. *Let $\Omega \subset \mathbb{R}^n$ be a bounded domain and*

$$\varphi : \Omega \times [0,\infty) \longrightarrow \mathbb{R}$$

a locally integrable Musielak-Orlicz function. Assume

(i)

$$\operatorname*{ess\,inf}_{x\in\Omega} \varphi(x,1) > 0.$$

(ii) The limits (3.4) and (3.5) are uniform on Ω.

(iii) The Matuszewska-Orlicz index m_φ is the restriction to Ω of a continuous function $\tilde{m}$ defined on the closure of Ω.

(iv)

$$1 < m_- := \inf_{\Omega} m_\varphi.$$

(v) There exists a function

$$\beta : (0,\infty) \longrightarrow (0,\infty)$$

such that the inequality

$$\varphi(x,t) \leq \beta(t) \tag{3.7}$$

holds uniformly in Ω.

Then the embedding

$$W_0^{1,\varphi}(\Omega) \hookrightarrow L^{\varphi}(\Omega) \tag{3.8}$$

is compact.

Proof. Assume first $m_- < n$. We remark the technical point that if γ satisfies the inequalities

$$0 < \gamma < \frac{m_-(n - m_-)}{2n - m_-},$$

then it must hold:

$$\frac{m_- - \gamma}{n - m_- + \gamma} > \frac{1}{2} \frac{m_-}{n - m_-}.$$

Set

$$\mu := \frac{1}{2} \frac{m_-}{n - m_-}.$$

Let $r > 0$ be small enough to guarantee the inequality

$$\frac{n(n-r)}{r} > n + 2r \tag{3.9}$$

and set

$$\epsilon < \min\{\mu, 5r\} \ , \ 0 < \gamma < \min\left\{\frac{\epsilon}{30}, \frac{m_-(n - m_-)}{2n - m_-}, \frac{m_- - 1}{2}\right\}. \tag{3.10}$$

Then, obviously

$$w(x) := m(x) - \gamma < m(x) < m(x) - \gamma + \frac{\epsilon}{20} = w(x) + \frac{\epsilon}{20}.$$

The uniformity conditions (3.5) thus yield the existence of γ satisfying the second inequality in (3.10) and of a constant $T_0 > 1$ such that $t \geq T_0$ implies that the following inequality holds uniformly for all $t \geq T_0$, $x \in \Omega$:

$$t^{m(x)-\gamma} = t^{w(x)} < M(x,t) < t^{m(x)-\gamma+\epsilon/20} = t^{w(x)+\epsilon/20}.$$

The uniformity of the infimum (3.5) with respect to t and x yields a positive number S_0 (without loss of generality it can be assumed $S_0 > 1$) such that, for all $(x,t) \in \Omega \times [0,\infty)$, one has, for $t \geq T_0 > 1$ and any δ such that

$$\gamma < \delta < \frac{\epsilon}{20}:$$

$$M(x,t) - \frac{1}{2}T_0^{m_- - \delta} < \frac{\varphi(x, tS_0)}{\varphi(x, S_0)} < M(x,t) + \frac{1}{2}T_0^{m_- - \delta}.$$

In all, uniformly in Ω and for $t \geq T_0$:

$$\frac{1}{2}t^{w(x)} < \frac{\varphi(x, tS_0)}{\varphi(x, S_0)} < \frac{3}{2}t^{w(x)+\epsilon/20}$$

holds uniformly in Ω for any $t \geq T_0$. Setting $tS_0 = s$, it is readily seen that for $s \geq T_0 S_0$, one has

$$\frac{1}{2}\varphi(x, S_0)\left(\frac{s}{S_0}\right)^{w(x)} \leq \varphi(x, s) \leq \frac{3}{2}\varphi(x, S_0)\left(\frac{s}{S_0}\right)^{w(x)+\frac{\epsilon}{20}}. \tag{3.11}$$

Furthermore, assumptions (i) and (v) yield positive constants A, B such that for all $x \in \Omega$ one has

$$(\varphi(x, S_0))^{\frac{1}{w(x)}} \leq \left(\sup_{\Omega} \varphi(x, S_0)\right)^{\frac{1}{w(x)}} \leq B$$

and

$$A \leq \left(\inf_{\Omega} \varphi(x, S_0)\right)^{\frac{1}{w(x)}} \leq (\varphi(x, S_0))^{\frac{1}{w(x)}}.$$

Consequently, (3.11) implies the existence of positive constants c_1, c_2 for which

$$c_1 s^{w(x)} < \varphi(x, s) < c_2 s^{w(x)+\frac{\epsilon}{20}}, \tag{3.12}$$

valid for all $x \in \Omega$ and $s \geq S_0 > 1$. By assumption and by virtue of Tietze's extension theorem, w is the restriction to Ω of a continuous function

$$p : \mathbb{R}^n \longrightarrow [w_-, w_+].$$

Let $p_1 = w_-$. For $k > 1$ if $p_{k-1} < n$, set

$$p_k = \frac{np_{k-1}}{n - p_{k-1}} - \frac{\epsilon}{5}.$$

Clearly, if $p_{j-1} < n$, $p_j > p_1 + (j-1)\frac{4}{5}\epsilon$ and $p_{j-1} < p_j$. Let J be the first subindex for which $p_J > n - \frac{r}{2}$, where r is as in (3.9). Let $I = [w_-, w_+]$ and

$$\Omega_1 := p^{-1}\left(\left(\frac{w_- + 1}{2}, \frac{np_1}{n - p_1} - \frac{\epsilon}{10}\right) \cap I\right)$$

and for $1 < k \leq J - 1$ set

$$\Omega_k := p^{-1}\left(\left(p_k, \frac{np_k}{n - p_k} - \frac{\epsilon}{10}\right) \cap I\right);$$

furthermore, define Ω_J and Ω_{J+1} as

$$\Omega_J := p^{-1}\left((n - r, n + r) \cap I\right)$$
$$\Omega_{J+1} := p^{-1}\left(\left(n + \frac{r}{2}, \infty\right) \cap I\right)$$

and let $(\chi_k)_{1 \leq k \leq J+1}$ be a partition of unity subordinated to the cover $(\Omega_k)_k$ of Ω. A straightforward argument shows that if $v \in C_0^\infty(\Omega)$ (which can be considered extended by 0 to $\mathbb{R}^n$) then, for each $k : 1 \leq k \leq J + 1$, $v\chi_k \in$

$C_0^\infty(\Omega \cap \Omega_k)$. It follows from this observation that if $v \in W_0^{1,w}(\Omega)$, then for each fixed k,

$$v\chi_k \in W_0^{1,w}(\Omega \cap \Omega_k).$$

Fix a sequence (u_j) bounded in $W_0^{1,\varphi}(\Omega)$; then inequalities (3.12) in concert with a simple calculation imply that (u_j) is bounded in $W_0^{1,w}(\Omega)$. We contend that $(u_j\chi_k)_j$ is bounded in $W_0^{1,p_k}(\Omega)$, for any subindex $k : 1 \le k \le J-1$. Denote the indicator function of any set A by I_A. Then by construction

$$w_k := wI_{\Omega\cap\Omega_k} \ge p_k I_{\Omega\cap\Omega_k}$$

so the embedding

$$W_0^{1,w_k}(\Omega) \hookrightarrow W_0^{1,p_k I_{\Omega\cap\Omega_k}}(\Omega) \tag{3.13}$$

is continuous, that is, for some positive constant C

$$\|u_j\chi_k\|_{W_0^{1,p_k I_{\Omega\cap\Omega_k}}(\Omega)} \le C\|u_j\chi_k\|_{W_0^{1,w_k}(\Omega)}.$$

On the other hand, if F_{kj} stands for any of the functions $u_j\chi_k$, $(\nabla u_j)\chi_k$ or $u_j\nabla\chi_k$

$$1 = \int_\Omega \left|\frac{|F_{kj}|}{\|F_{kj}\|_{p_k}}\right|^{p_k} = \int_\Omega \left|\frac{|F_{kj}|}{\|F_{kj}\|_{p_k}}\right|^{p_k I_{\Omega\cap\Omega_k}}$$

and with the same token

$$1 = \int_\Omega \left|\frac{|F_{kj}|}{\|F_{kj}\|_{w_k}}\right|^{w_k} = \int_\Omega \left|\frac{|F_{kj}|}{\|F_{kj}\|_{w_k}}\right|^{w} \tag{3.14}$$

the two preceding observations and (3.13) yield

$$\|F_{kj}\|_{p_k} = \|F_{kj}\|_{p_k I_{\Omega\cap\Omega_k}} \le C\|F_{kj}\|_{w_k} = \|F_{kj}\|_w,$$

whence it follows the contention.

Hence, by vitue of Theorem 3.4.2 there is no loss of generality in assuming that $(u_j\chi_k)_j$ converges in $L^{\frac{np_k}{n-p_k}-\frac{\epsilon}{20}}(\Omega)$. For simplicity, let q_k be the right-endpoint of $p(\Omega_k)$ for $1 \le k \le J$. Next, if $1 \le j \le J$, set

$$d_j := \left(q_j + \frac{\epsilon}{20}\right) I_{\Omega_j} + \left(w_+ + \frac{\epsilon}{20}\right) I_{\Omega\setminus\Omega_j}.$$

Then $d_j \ge w + \frac{\epsilon}{20}$ for all $x \in \Omega$ and one has the continuous embedding

$$L^{d_j}(\Omega) \hookrightarrow L^{w+\frac{\epsilon}{20}}(\Omega). \tag{3.15}$$

For any function $u \in W_0^{1,\varphi}(\Omega)$ and $1 \le k \le J-1$:

$$\int_\Omega |u\chi_k|^{d_k} = \int_\Omega |u\chi_k|^{d_k I_{\Omega_k}} = \int_\Omega |u\chi_k|^{q_k+\frac{\epsilon}{20}} = \int_\Omega |u\chi_k|^{\frac{np_k}{n-p_k}-\frac{\epsilon}{20}}. \tag{3.16}$$

The preceding string of inequalities yields the following observation:

If $(u_j\chi_k)_j$ is a Cauchy sequence in $L^{\frac{np_k}{n-p_k}-\frac{\epsilon}{20}}(\Omega)$, $1 \leq k \leq J-1$ then it is convergent in $L^{d_k}(\Omega)$ and by virtue of (3.15), $(u_j\chi_k)_j$ converges in $L^{w+\frac{\epsilon}{20}}(\Omega)$. We claim that the latter observation yields the convergence of $(u_j\chi_k)_j$ in $L^{\varphi}(\Omega)$ for $1 \leq k \leq J-1$. Indeed, there is no loss of generality by assuming that $(u_j\chi_k)_j$ converges pointwise $a.e$; on the other hand:

$$\int_{\Omega} \varphi(x, |u_j(x) - u_i(x)|\chi_k(x))\, dx = \tag{3.17}$$

$$\int_{\{x:|u_j(x)-u_i(x)|\chi_k(x)\leq S_0\}} \varphi(x, |u_j(x) - u_i(x)|\chi_k(x))\, dx +$$

$$\int_{\{x:|u_j(x)-u_i(x)|\chi_k(x)>S_0\}} \varphi(x, |u_j(x) - u_i(x)|\chi_k(x))\, dx.$$

Since for any fixed $x \in \Omega$ $\varphi(x, \cdot)$ is nondecreasing, the integrand in the first term above satisfies the inequality

$$\varphi(x, |u_j(x) - u_i(x)|\chi_k(x)) \leq \varphi(x, S_0).$$

The assumption of local integrability on φ and a straightforward application of Lebesgue's dominated convergence yield

$$\lim_{i,j\to\infty} \int_{\{x:|u_j(x)-u_i(x)|\chi_k(x)\leq S_0\}} \varphi(x, |u_j(x) - u_i(x)|\chi_k(x))\, dx = 0.$$

Since $S_0 > 1$, the second integral in (3.17) is dominated by

$$\int_{\{x:|u_j(x)-u_i(x)|\chi_k(x)>S_0\}} |u_j(x) - u_i(x)|^{w(x)+\frac{\epsilon}{20}} \chi_k(x)\, dx.$$

In all, $(\rho_\varphi((u_j - u_i)\chi_k)_j) \longrightarrow 0$ as $i, j \longrightarrow \infty$. Next, we observe that for C_1 as in the statement of Lemma 3.4.3 one has:

$$\rho_\varphi(C_1(u_i - u_j)\chi_k) =$$

$$\int_{\{x:|u_j(x)-u_i(x)|\chi_k(x)\leq S_0\}} \varphi(x, C_1|u_j - u_i|\chi_k))\, dx$$

$$+ \int_{\{x:|u_j(x)-u_i(x)|\chi_k(x)>S_0\}} \varphi(x, C_1|u_j - u_i|\chi_k))\, dx;$$

a straightforward application of Lebesgue's dominated convergence theorem on the first integral and the consideration of Lemma 3.4.3 in the second one easily yield

$$\rho_\varphi(C_1(u_i - u_j)\chi_k) \longrightarrow 0 \ as \ i, j \longrightarrow \infty.$$

It follows automatically by induction that for any $l \in \mathbb{N}$ one has

$$\rho_\varphi(C_1^l(u_i - u_j)\chi_k) \longrightarrow 0 \ as \ i, j \longrightarrow \infty,$$

and it is concluded from here that the sequence $(u_j\chi_k)_j$ is Cauchy in $L^\varphi(\Omega)$, as claimed.

The remaining intervals in the covering are handled similarly: define

$$w_J := wI_{\Omega\cap\Omega_J} \geq (n - r)I_{\Omega\cap\Omega_J}$$

the embedding

$$W_0^{1,w_J}(\Omega) \hookrightarrow W_0^{1,n-r}(\Omega)$$

is bounded. Retaining the notation of the above discussion, the sequence $(u_j\chi_J)_j$ is bounded in $W_0^{1,n-r}(\Omega)$; without loss of generality it can be considered convergent in $L^{\frac{n(n-r)}{r}}(\Omega)$, which by the choice of r in (3.9) is continuously embedded in $L^{n+2r}(\Omega)$. Setting

$$h = (n + 2r)I_{\Omega_J} + \left(w_+ + \frac{\epsilon}{20}\right) I_{\Omega\setminus\Omega_J}$$

it is clear that $L^h(\Omega)$ is continuously embedded in $L^{w+\frac{\epsilon}{20}}(\Omega)$. It follows immediately that $(u_j\chi_J)_j$ is Cauchy in the latter space. Theorem 2.2.1 ensures now that $(u_j\chi_J)_j$ is convergent in $L^\varphi(\Omega)$.

Finally, via the continuous embeddings

$$W_0^{1,\varphi}(\Omega) \hookrightarrow W_0^{1,w}(\Omega) \hookrightarrow W_0^{1,(n+\frac{r}{2})I_{\Omega_{J+1}}+w_-I_{\Omega\setminus\Omega_{J+1}}}(\Omega)$$

the boundedness of $(u_j\chi_{J+1})_j$ in $W_0^{1,\varphi}(\Omega)$ yields its boundedness in $W_0^{1,n+\frac{r}{2}}(\Omega)$ and by way of Theorem 3.4.2 it is readily concluded that $(u_j\chi_{J+1})_j$ can be considered convergent in $C(\overline{\Omega})$, hence convergent in $L^\varphi(\Omega)$.
In all, for $m_- < n$ any bounded sequence $(u_j)_j \subset W_0^{1,\varphi}(\Omega)$ has a subsequence that converges in $L^\varphi(\Omega)$.

Since the case $n \leq m_-$ is handled along the same lines, we only sketch the proof for this instance.

For r chosen as in (3.9) we observe that there exists $T_0 > 1$ such that uniformly on Ω and for all $t \geq T_0$:

$$t^{m(x)-\frac{r}{4}} < M(x,t) < t^{m(x)+r}.$$

As before, one can conclude that given the conditions on the index, there are positive constants $c_1 > 1$, $c_2 > 1$ and $T > 1$ for which

$$c_1 t^{n-\frac{r}{4}} \leq \varphi(x,t) < t^{m(x)+r} \tag{3.18}$$

uniformly in Ω, for all $t \geq T$. Consider a partition of unity (χ_1, χ_2) subordinated to the cover of Ω that consists of the open sets

$$\Omega_1 = p^{-1}\left((n-r, n+r) \cap I\right) \ , \ \Omega_2 = p^{-1}\left(\left(n + \frac{r}{2}, \infty\right) \cap I\right).$$

If $(u_j)_j$ is a bounded sequence in $W_0^{1,\varphi}(\Omega)$ (hence in $W_0^{1,n-\frac{r}{4}}(\Omega)$) one can set

$$q = \left(n - \frac{r}{4}\right) I_{\Omega_1} + m_- I_{\Omega \setminus \Omega_1}$$

and along the same lines as (3.13)–(3.14) conclude that $(u_j\chi_1)_j$ is bounded in $W_0^{1,n-\frac{r}{4}}(\Omega)$. Via Theorem 3.4.2 and the choice (3.9) it follows that $(u_j\chi_1)_j$ has a subsequence that converges in $L^{n+2r}(\Omega)$. If

$$t := (n+2r) I_{\Omega \cap \Omega_1} + (m_+ + 2r)\, I_{\Omega \setminus \Omega \cap \Omega_1},$$

then the obvious equality

$$\int_{\Omega} |(u_i - u_j)\chi_1|^{n+2r} = \int_{\Omega} |(u_i - u_j)\chi_1|^{t}$$

implies that the subsequence also converges in $L^t(\Omega)$ and since $m + r < t$ in Ω it converges also in $L^{m+r}(\Omega)$, hence in $L^{\varphi}(\Omega)$ via the right-hand inequality in (3.18). Still denoting this subsequence by $(u_j\chi_1)_j$, it is easy to see, that $(u_j\chi_2)_j$ is bounded in $W_0^{1,n+\frac{r}{2}}(\Omega)$; therefore from Theorem 3.4.2 it is clear that it has a subsequence (still denoted by $(u_j\chi_2)_j$) that converges in $L^{m+2r}(\Omega)$. The right-hand inequality in (3.11) yields the convergence of $(u_j\chi_2)_j$ in $L^{\varphi}(\Omega)$. A straightforward computation reveals that the above conclusion implies the compactness of the embedding (3.8) in all cases. □

It is apparent from the proof of the preceding theorem functions in $W_0^{1,\varphi}(\Omega)$ belong to a higher order integrability space than just $L^{\varphi}(\Omega)$: We state this important fact as a separate corollary:

Corollary 3.4.5. *For an MO function φ on Ω that fulfills the conditions of Theorem 3.4.4, the embedding*

$$W_0^{1,\varphi}(\Omega) \hookrightarrow L^{m(x)+\frac{\epsilon}{20}}(\Omega) \subsetneqq L^{\varphi}(\Omega)$$

is compact.

The following corollary generalizes the Poincaré Inequality to the setting of MO spaces.

Corollary 3.4.6 (Poincaré-type Inequality). *For φ satisfying the conditions of Theorem 3.4.4, there exists a positive constant C depending only on n, Ω, φ such that for any $u \in W_0^{1,\varphi}(\Omega)$*

$$\|u\|_\varphi \leq C\||\nabla u|\|_\varphi.$$

Proof. If not, it would be an elementary matter to construct a sequence

$$(v_k) \subset W_0^{1,\varphi}(\Omega)$$

with

$$\|v_k\|_{1,\varphi} = 1 \geq \|v_k\|_\varphi \geq k\||\nabla v_k|\|_\varphi \text{ for } k \in \mathbb{N}.$$

Clearly,

$$|\nabla v_k| \longrightarrow 0 \text{ in } L^\varphi(\Omega) \tag{3.19}$$

as $k \longrightarrow \infty$ and the compactness of the Sobolev embedding yields the existence of $v \in L^\varphi(\Omega)$ for which

$$v_k \longrightarrow v \text{ in } L^\varphi(\Omega).$$

Necessarily then,

$$\|v_k - v_j\|_{1,\varphi} = \|v_k - v_j\|_\varphi + \|\nabla (v_k - v_j)\|_\varphi \longrightarrow 0 \text{ as } k, j \longrightarrow \infty;$$

it follows that $(v_k)_k$ converges in $W_0^{1,\varphi}(\Omega)$ and it is obvious that the limit must be v. On the other hand, (3.19) forces $\nabla v = 0$ and hence $v = 0$, which is a contradiction.

□

Remark 3.4.7. A quick examination of the counterexamples presented before Theorem 3.4.4 reveals that in Example 3.4.1, the Matuszewska index p (being unbounded on $(0,1)$) is not the restriction to $\Omega = (0,1)$ of any continuous function on $\mathbb{R}$, whereas in Example 3.4.2 estimate (3.7) is violated.

For its connection to classical analysis and its ubiquity in current applications, we single out the following particular case, first stated in [60] (see also, [66]):

Corollary 3.4.8. *Let $\Omega \subset \mathbb{R}^n$ be a domain and $p \in C(\overline{\Omega})$ with $1 < p_- \leq p_+ < \infty$. Then the embedding*

$$W_0^{1,p}(\Omega) \hookrightarrow L^p(\Omega)$$

is compact and there exists a positive constant C depending only on p and Ω such that

$$\||\nabla u|\|_\varphi \leq C\|u\|_\varphi.$$

Proof. The proof follows immediately from the fact that under the required conditions on the exponent p, the *MO* function

$$\varphi(x,t) = t^{p(x)}$$

satisfies the necessary hypothesis of Theorem 3.4.4. □

Example 3.4.3. In addition, the embedding

$$W_0^{1,\varphi}(\Omega) \hookrightarrow L^{\varphi}(\Omega)$$

is compact for the following MO functions

$$(1 < p_- = \inf_{x\in\Omega} p(x) \le p_+ = \sup_{x\in\Omega} p(x) < \infty):$$

(*i*) $\varphi(x,t) = t^{p(x)} (\log(1+t))^{q(x)}$ for $p, q \in C(\overline{\Omega})$, $1 < p_- \le p_+ < \infty$

(*ii*) $\varphi(x,t) = t^{p(x)} (1+|x|)^{-n}$, $p \in C(\overline{\Omega})$.

(*iii*) $\varphi(x,t) = \frac{t^{p(x)}}{p(x)}$, $p \in C(\overline{\Omega})$.

The above statements can be easily proved by computing the Matuszewska-Orlicz indexes of each particular case and verifying the hypotheses of Theorem 3.4.4.

In what follows we present an example, based in Example 2.9.3, to the effect that the uniform convergence in the limits (2.51) and (2.52) can not be weakened.

Example 3.4.4. We keep the notation of 2.9.3. In particular, we consider Ω to be the Euclidean unit ball in $\mathbb{R}^6$ and for $n \in \mathbb{N}$ we set B_n to be the ball of radius 2^{-n-2} centered at $x_n = (2^{-n}, 0, 0, 0, 0, 0)$, denote the ball concentric with B_n of radius 2^{-n-3} with B_n^- and write:

$$B_n^+ = B_n \setminus B_n^-.$$

For each $h > 0$, we define the function

$$v_h : \mathbb{R}^6 \longrightarrow \mathbb{R}$$

as follows:

$$v_h(x) = \begin{cases} 1 & \text{if } |x| \le h \\ 2 - \frac{|x|}{h} & \text{if } |x| \in (h, 2h) \\ 0 & \text{if } |x| \in (2h, \infty). \end{cases}$$

Finally, we consider the sequence (u_n) is defined then as

$$u_n(x) = 2^{2(n+3)} (w_6)^{-\frac{1}{3}} v_{2^{-(n+3)}}(x - x_n),$$

where w_6 is the Lebesgue measure of the unit sphere in $\mathbb{R}^6$.

For each $n \in \mathbb{N}$ fix $z_n = 2^{3(n+3)} \approx |\nabla u_n(x)|$ and recall from Example 2.9.3 that

$$\phi_{z_n}(t) = \begin{cases} t^2 & \text{if } t \in (0, z_n) \\ 3s_n^2(t - z_n) + z_n^2 & \text{if } t \in (z_n, s_n) \\ t^3 & \text{if } t \in (s_n, \infty), \end{cases}$$

where for each $n \in \mathbb{N}$ we have chosen $z_n = 2^{3(n+3)}$ and $s_n > z_n$ is chosen in such a way that

$$s_n^3 - z_n^2 = 3s_n^2(s_n - z_n).$$

Then the MO function φ is given by:

$$\varphi(x,t) = \begin{cases} t^3 & \text{if } x \in \Omega \setminus \bigcup_n B_n^+ \\ \phi_{z_n}(t) & \text{if } x \in B_n^+. \end{cases}$$

We claim that the sequence (u_n) is bounded in $W_0^{1,\varphi}(\Omega)$. To this end observe that

$$\begin{aligned}
\rho_\varphi(u_n) &= \left(\int\limits_{B_n^-} + \int\limits_{B_n^+} \right) \varphi\left(x, u_n(x)\right) dx \\
&= \int\limits_{B_n^-} \varphi(x, 2^{2(n+3)} w_6^{-\frac{1}{3}}) dx + \int\limits_{B_n^+} \varphi\left(x, 2^{2(n+3)} w_6^{-\frac{1}{3}} \left(2 - 2^{n+3}|x - x_n|\right)\right) dx \\
&\leqq |B_n| \left(2^{2(n+3)} w_6^{-\frac{1}{3}}\right)^3 + \int\limits_{B_n^+} \varphi\left(x, 2^{2(n+3)} w_6^{-\frac{1}{3}}\right) dx \\
&\leq C + |B_n| \left(2^{2(n+3)} w_6^{-\frac{1}{3}}\right)^2 \\
&\leq C_1 + C_2
\end{aligned}$$

where C_1 is a positive constant independent of n and $C_2 \to 0$ as $n \to \infty$.

Notice that for n large enough, $2^{3(n+3)w_6^{-1/3}} > s_n$. A simple calculation shows that while $\nabla u_n(x) = 0$ on $\Omega \setminus B_n^+$, for $x \in B_n^+$ one has:

$$|\nabla u_n(x)| \approx 2^{3(n+3)} w_6^{-1/3}.$$

One derives from here that:

$$\begin{aligned}
\rho_\varphi(|\nabla u_n|) &= \left(\int\limits_{B_n^-} + \int\limits_{B_n^+} \right) \varphi(x, u_n(x))\, dx \\
&= \int\limits_{B_n^+} \varphi\left(x, 2^{2(n+3)} w_6^{-\frac{1}{3}} \left(2 - 2^{n+3}|x - x_n|\right)\right) dx \\
&\approx |B_n| \left(2^{2(n+3)} w_6^{-1/3}\right)^3 \\
&\leq C
\end{aligned}$$

for a positive constant C, independent of n. We have shown that there are positive constants k_1, k_2 such that for all $n \in \mathbb{N}$, one has:

$$\begin{aligned}
k_1 &\leq \|u_n\|_\varphi \leq k_2 \\
k_1 &\leq \||\nabla u_n|\|_\varphi \leq k_2,
\end{aligned}$$

from which it follows that the sequence (u_n) is bounded in $W_0^{1,\varphi}(\Omega)$. On the other hand, each u_n is continuous, any two different functions in the sequence (u_n) have disjoint supports and $\sup\limits_{\Omega} u_n \to \infty$ as $n \to \infty$. Hence, no subsequence of (u_n) converges in $L^\varphi(\Omega)$.

Chapter 4

Applications

4.1	Preparatory results and notation	181
4.2	Compactness of the Sobolev embedding and the modular setting	186
4.3	The variable exponent p-Laplacian	190
	4.3.1 Stability of the solutions	198
4.4	Γ-convergence	206
4.5	The eigenvalue problem for the p-Laplacian	210
4.6	Modular eigenvalues	219
4.7	Convergence properties of the eigenvalues and eigenfunctions	228
4.8	Closing remarks	243

In this Chapter we present some applications of the theoretical results developed in Chapters 1-3. In the interest of facilitating the reading, some definitions already given in previous sections will be repeated in the sequel.

4.1 Preparatory results and notation

A recollection of a few standard results from nonlinear functional analysis will facilitate the treatment of the variable exponent p-Laplacian, the operator to be studied in this chapter. We continue to use the notation from Chapter 1. Our presentation follows [96, 65, 66]. Let X, Y be real Banach spaces: all Banach spaces in this chapter will be assumed to be real. An operator

$$T : X \longrightarrow Y$$

is said to be bounded if for any bounded set $A \subset X$, the set $T(A)$ is bounded. Recall from Chapter 1 that in the particular case when $Y = X^*$, the operator T is said to be monotone if the inequality

$$\langle T(x) - T(y), x - y \rangle \geq 0$$

holds for any $x \in X$, $y \in X$.
$T : X \to Y$ is called hemicontinuous if for any fixed $x, y \in X$ the real valued

function

$$s \longmapsto \langle T(x+sy), y\rangle : \mathbb{R} \to \mathbb{R}$$

is continuous. Another class of operators is of particular importance in the sequel:

Definition 4.1.1. *An operator*

$$T : X \longrightarrow X^*$$

is said to be of type M if, for any weakly-convergent sequence $(x_n) \subset X$ with $x_n \rightharpoonup x$, the conditions

$$T(x_n) \rightharpoonup f \quad \text{and} \quad \limsup_{n\to\infty}\langle T(x_n), x_n\rangle \leq \langle f, x\rangle,$$

imply $T(x) = f$.

Theorem 4.1.1. *Let X be a reflexive Banach space and*

$$T : X \longrightarrow X^*$$

be hemicontinuous and monotone. Then T is of type M.

Proof. For fixed $y \in X$, (x_n), x and f as in Definition 4.1.1, the assumed monotonicity of T implies that

$$0 \leq \langle T(x_n) - T(y), x_n - y\rangle$$

for all n; hence

$$\langle T(y), x-y\rangle \leq \langle f, x-y\rangle .$$

In particular, for any $z \in X$,

$$\left\langle T\left(x - \frac{1}{n}z\right), z\right\rangle \leq \langle f, z\rangle ; \tag{4.1}$$

inequality (4.1) conjunction with the hemicontinuity property of T, immediately yields

$$\langle T(x), z\rangle \leq \langle f, z\rangle$$

for all $z \in X$. The last inequality implies $T(x) = f$ as claimed. □

The next lemma is of rather technical nature; its necessity will be revealed shortly:

Lemma 4.1.2. *If $F : \mathbb{R}^n \longrightarrow \mathbb{R}^n$ is continuous and there exists $\varepsilon > 0$ such that $F(u)\cdot u \geq 0$ for each u such that $\|u\| = \varepsilon$, then F has a zero.*

Proof. Denote the unit ball in $\mathbb{R}^n$ by $\mathbf{B}_n$. Assuming $F(x) \neq 0$ for each $x \in \mathbf{B}_n$, one can define the function

$$\psi : \mathbf{B}_n \longrightarrow \mathbf{B}_n$$

by

$$\psi(x) = -\frac{F(\varepsilon x)}{\|F(\varepsilon x)\|}.$$

This function is easily seen to be continuous. Let x_0 be a fixed point of ψ whose existence is guaranteed by Brouwer's theorem, and notice that $x_0 \neq 0$. Then

$$-\frac{F(\varepsilon x_0) \cdot x_0}{\|F(\varepsilon x_0)\|} = \|x_0\|^2,$$

which is a contradiction. □

From the immediately preceding results, there follows the surjectivity criterion given in the next theorem:

Theorem 4.1.3. *Consider a separable and reflexive Banach space X and let*

$$T : X \longrightarrow X^*$$

be of type M and bounded. If for some $f \in X^*$ there exists $\varepsilon > 0$ such that for every $x \in X$ with $\|x\| > \varepsilon$ one has the inequality:

$$\langle T(x), x\rangle > \langle f, x\rangle, \tag{4.2}$$

then f belongs to the range of T.

Proof. Let $\{x_k : k \in \mathbb{N}\}$ denote a basis for the linear span of a countable, dense subset $Y \subset X$. For each natural number n denote the linear span of the set $\{x_1, x_2, ..., x_n\}$ by Y_n, that is

$$Y_n = \langle\{x_1, x_2, ..., x_n\}\rangle$$

and consider the canonical isomorphism

$$i_n : \mathbb{R}^n \longrightarrow Y_n \ , \ i_n(a_1, ..., a_n) = \sum_1^n a_k x_k.$$

We first claim that for each $n \in \mathbb{N}$ there exists $u_n \in Y_n$ satisfying

$$\langle x_k, T(u_n)\rangle = f(x_k) \text{ for } k = 1, 2, ..., n. \tag{4.3}$$

This is a direct consequence of the assumption on f and of Lemma 4.1.2, for the function

$$F_n : \mathbb{R}^n \longrightarrow \mathbb{R}^n \ , \ F_n(z) = i_n^* \circ T \circ i_n(z) - i_n^* \circ f$$

satisfies the conditions of Lemma 4.1.2 and the zero u_n thereby given satisfies the equality (4.3). Consequently, for each n,

$$\langle T(u_n), u_n \rangle = f(u_n),$$

which by virtue of condition (4.2) implies that $\|u_n\| \leq \varepsilon$. Since T is bounded, the reflexivity of X yields the existence of a subsequence (v_n) of (u_n) such that

$$v_n \rightharpoonup v \in X \ , \ T(v_n) \rightharpoonup f.$$

Since T is of type M, we conclude that $T(v) = f$. □

In what follows we recall the definition of singular numbers for operators on Banach spaces. Specifically, consider a reflexive Banach space X with norm $\|\cdot\|_X$, a Banach space Y with norm $\|\cdot\|_Y$ and a compact linear operator

$$T : X \longrightarrow Y.$$

Simple functional-analytic arguments reveal that the operator norm

$$\|T\| = \|T\|_{\mathcal{B}(X,Y)} := \sup_{0 \neq x \in X} \frac{\|T(x)\|_Y}{\|x\|_X}$$

is attained in X. More precisely:

Lemma 4.1.4. *Given Banach spaces X and Y with X reflexive and a compact linear operator*

$$T : X \to Y,$$

there exists $x_0 \in X$ such that

$$\|T\| = \|T(x_0)\|_Y = \sup_{0 \neq x \in X} \frac{\|T(x)\|_Y}{\|x\|_X}.$$

Proof. By definition of $\|T\|$ there exists a sequence

$$(x_n) \subset B_X$$

with

$$\|T(x_n)\|_Y \geq \|T\| - \frac{1}{n} \text{ for each } n \in \mathbb{N};$$

from the reflexivity of X, (x_n) can be assumed to be weakly convergent to $x_0 \in B_X$; obviously, then,

$$T(x_n) \to T(x_0).$$

Invoking the compactness of T, one can extract a subsequence (y_k) of (x_n) such that $(T(y_k))$ converges strongly in Y; the sequence $(T(y_k))$, hence, must necessarily converge strongly to $T(x_0)$ in Y. As is apparent from the choice of the sequences, it holds for each natural number n that

$$\|T\| < \|T(x_n)\|_Y + \frac{1}{n},$$

whence, from the continuity of the norm one has

$$\|T\| \leq \|T(x_0)\|_Y.$$

The lemma follows at once from the latter inequality. □

The following section reveals the connection between the eigenvalues of a given compact operator and the maximum problem suggested by the previous lemma. Recall that a Banach space is called smooth if its norm is Gâteaux-differentiable at every non-zero point. We refer the reader to [12], where this connection was first observed, for further comments.

Theorem 4.1.5. *Let X, Y be smooth Banach spaces; assume that X is reflexive; let D_X and D_Y stand for the gradients of the respective norms. If*

$$T : X \longrightarrow Y$$

is a linear, compact operator (whose transpose is denoted by T^*) and $x_0 \in X\backslash\{0\}$ is maximal in the sense that

$$\|T\| = \frac{\|T(x_0)\|_Y}{\|x_0\|_X},$$

then

$$(T^* D_Y T)(x_0) = \|T\| D_X(x_0). \tag{4.4}$$

Proof. The claim follows by direct differentiation: For fixed $h \in X\backslash\{0\}$ and $t \in (-\|x_0\|_X\|h\|_X^{-1}, \|x_0\|_X\|h\|_X^{-1})$, set

$$F(t) = \frac{\|T(x_0 + th)\|_Y}{\|x_0 + th\|_X}.$$

By assumption, F has an absolute maximum at $t = 0$, and the smoothness hypothesis guarantees the differentiability of F; moreover, as shown with a simple calculation,

$$0 = F'(0) = \langle D_Y(T(x_0)), T(h)\rangle \frac{1}{\|x_0\|_X} - \frac{\|T(x_0)\|_Y}{\|x_0\|_X^2} \langle D_X(x_0), h\rangle,$$

which immediately yields (4.4). □

4.2 Compactness of the Sobolev embedding and the modular setting

This section is devoted to obtaining some further applications of the material covered in Chapters 1 and 3. As a matter of course, we retain the notation introduced in the sequel. In particular, $\Omega \subset \mathbb{R}^n$ will stand for a bounded domain and φ will denote an MO function on Ω (as in Definition 2.1.1). We also recall the notation for the modular engendered by φ:

$$\rho_\varphi(u) = \int_\Omega \varphi(x, |u(x)|)\, dx.$$

Most of the results presented in this section heavily depend on the fact that if the limits (2.51) and (2.52) are uniform, then the MO function φ satisfies the Δ_2 condition (Lemma 2.9.3). The following technical theorem will be used in the sequel:

Theorem 4.2.1. *Let φ be an MO function on Ω; assume that φ satisfies the conditions of Theorem 3.4.4; in particular, φ satisfies the Δ_2 condition 2.2.2, i.e., for some $K > 0$, $S_0 > 0$ it holds that*

$$\varphi(x, 2s) \le K\varphi(x, s) \text{ for all } s \ge S_0\,,\ x \in \Omega. \tag{4.5}$$

Then,

$$\sup\{\rho_\varphi(u) : \rho_\varphi(|\nabla u|) \le r\} < \infty. \tag{4.6}$$

Proof. It follows from (4.5) that for arbitrary $v \in L^\varphi(\Omega)$

$$\begin{aligned}\rho_\varphi(2v) &= \int_\Omega \varphi(x, 2|v(x)|)dx \\ &= \left(\int_{|v|<S_0} + \int_{|v|\ge S_0}\right) \varphi(x, 2|v(x)|)dx \\ &\le \left(\int_{|v|<S_0} \varphi(x, 2S_0)dx + K\rho_\varphi(v)\right) \\ &\le \left(\int_\Omega \varphi(x, 2S_0)dx + K\rho_\varphi(v)\right).\end{aligned}$$

If $r \ge 1$ and $u \in W_0^{1,\varphi}(\Omega)$ with

$$\rho_\varphi(|\nabla u|) \le r, \tag{4.7}$$

it is a simple matter to verify that if $|||\nabla u|||_\varphi \geq 1$

$$\begin{aligned} 1 = \rho_\varphi\left(\frac{|\nabla u|}{|||\nabla u|||_\varphi}\right) &\leq \frac{1}{|||\nabla u|||_\varphi}\rho_\varphi(|\nabla u|) \\ &\leq \frac{1}{|||\nabla u|||_\varphi} r; \end{aligned}$$

it is thus clear that if (4.7) holds, then:

$$|||\nabla u|||_\varphi \leq r.$$

Therefore,

$$|||\nabla u|||_\varphi \leq \max\{1, r\} = b.$$

On account of Poincaré inequality one has, for C as in in Corollary 3.4.6:

$$\|u\|_\varphi \leq C|||\nabla u|||_\varphi \leq Cb.$$

It follows that

$$\rho_\varphi\left(\frac{u}{Cb}\right) \leq 1. \tag{4.8}$$

If $Cb < 2$, (4.8) implies

$$\begin{aligned} \rho_\varphi(u) = \rho_\varphi\left(Cb\frac{u}{Cb}\right) &\leq \rho_\varphi\left(2\frac{u}{Cb}\right) \\ &\leq \int_\Omega \varphi(x, 2S_0)dx + K\rho_\varphi\left(\frac{u}{Cb}\right) \\ &\leq \int_\Omega \varphi(x, 2S_0)dx + 1. \end{aligned} \tag{4.9}$$

Otherwise,

$$\begin{aligned} \rho_\varphi(u) &= \rho_\varphi\left(Cb\frac{u}{Cb}\right) \\ &\leq \rho_\varphi\left(2^{[\log_2 Cb]+1}\frac{u}{Cb}\right) \\ &\leq \left(1 + K + ...K^{[\log_2 Cb]}\right)\int_\Omega \varphi(x, 2S_0)dx + K^{[\log_2 Cb]+1}. \end{aligned} \tag{4.10}$$

In all, (4.9) and (4.10) yield (4.6). □

An immediate consequence of the preceding theorem is the following functional-analytic result:

Lemma 4.2.2. *For $r > 0$, the modular ball*

$$B_r := \left\{u \in W_0^{1,\varphi}(\Omega) : \rho_\varphi(|\nabla u|) \leq r\right\}$$

is weakly closed.

Proof. A standard result in Functional Analysis ([95, Theorem 3.12]) guarantees that any convex, strongly closed subset of a locally convex topological vector space is necessarily weakly closed. The proof will thus follow by observing that B_r is both convex and strongly closed.

The convexity of B_r follows immediately from the convexity of ρ_φ. Let $u \in W_0^{1,\varphi}(\Omega)$ and consider a sequence $(u_n) \subset B_r$ with

$$\|\nabla(u_n - u)\|_\varphi \longrightarrow 0 \text{ as } n \longrightarrow \infty;$$

according to Corollary 2.1.7 one can assume that

$$\nabla u_n(x) \longrightarrow \nabla u(x) \ a.e. \text{ in } \Omega.$$

Since φ is an MO function, it is clear that, $a.e.\, x \in \Omega$:

$$\varphi(x, |\nabla u_n(x)|) \to \varphi(x, |\nabla u(x)|) \text{ as } n \to \infty.$$

Fatou's lemma now yields

$$\begin{aligned}\rho_\varphi(|\nabla u|) &= \int_\Omega \varphi(x, |\nabla u(x)|)dx \leq \liminf_n \int_\Omega \varphi(x, |\nabla u_n(x)|)dx \\ &= \liminf_n \rho_\varphi(|\nabla u_n|) \leq r.\end{aligned}$$

In all, B_r is convex and strongly closed, hence weakly closed, as claimed. □

Theorem 4.2.3. *Let $r > 0$. Then there exists a function $u_r \in W_0^{1,\varphi}(\Omega)$ that is maximal in the following sense:*

$$\rho_\varphi(u_r) = \mathbf{S}_r = \sup\{\rho_\varphi(u) : \rho_\varphi(|\nabla u|) \leq r\}$$

Proof. Pick a sequence (u_n) in B_r with

$$\mathbf{S}_r - \frac{1}{n} < \rho_\varphi(u_n).$$

The sequence (u_n) is bounded in $W_0^{1,\varphi}(\Omega)$; indeed, Theorem 4.2.1 guarantees that the numerical sequence $(\rho_\varphi(u_n))$ is bounded and it follows from here that either $\|u_n\|_\varphi \leq 1$ or

$$1 = \rho_\varphi\left(u_n/\|u_n\|_\varphi\right) \leq \|u_n\|_\varphi^{-1} \rho_\varphi(u_n), \tag{4.11}$$

so that the boundedness of the sequence $(\|u_n\|_\varphi)$ follows from that of the sequence $(\rho_\varphi(u_n))$ (4.6). Likewise, since by definition, $(\rho_\varphi\left(|\nabla u_n|\right))$ is bounded, either $\||\nabla u_n|\|_\varphi \leq 1$ or

$$1 = \rho_\varphi\left(|\nabla u_n|/\||\nabla u_n|\|_\varphi\right) \leq \||\nabla u_n|\|_\varphi^{-1} \rho_\varphi(|\nabla u_n|). \tag{4.12}$$

Inequalities (4.11) and (4.12) together with the discussions preceding them show the claimed boundedness of (u_n).

On account of the theorem of Banach-Alaoglu and the reflexivity of $W_0^{1,\varphi}(\Omega)$ (Theorem 3.3.1) there is no loss of generality by assuming that (u_n) converges weakly in $W_0^{1,\varphi}(\Omega)$; let

$$u_n \overset{W_0^{1,\varphi}(\Omega)}{\rightharpoonup} u \in W_0^{1,\varphi}(\Omega). \tag{4.13}$$

In particular, in the light of Theorem 4.2.2, statement (4.13) ensures that $u \in B_r$.

The compactness of the Sobolev embedding (Theorem 3.4.4) guarantees the strong convergence

$$u_n \overset{L^{\varphi}(\Omega)}{\longrightarrow} u.$$

In view of Sobolev's embedding theorem (Theorem 3.4.4), one must have a subsequence (still denoted by u_n) that is strongly convergent to u in the $L^{\varphi}(\Omega)$ norm. Owing to Corollary 2.1.7 one can assume that $u_n \longrightarrow u$ μ-*a.e.* in Ω. In fact, u is the sought-for maximal function. To see this, we notice that *a.e.* in Ω,

$$\varphi(x, |u_n(x)|) \longrightarrow \varphi(x, |u(x)|)$$

and that on account of convexity, for any $n \in \mathbb{N}$:

$$\varphi(x, |u_n(x)|) \leq \frac{1}{2}\varphi(x, 2|u_n(x) - u(x)|) + \frac{1}{2}\varphi(x, 2|u(x)|). \tag{4.14}$$

Select n large enough so that $2\|u - u_n\|_{\varphi} < 1$; for such n (4.14) yields:

$$\varphi(x, |u_n(x)|) \leq \|u_n - u\|_{\varphi}\varphi\left(x, \frac{|u_n(x) - u(x)|}{\|u - u_n\|_{\varphi}}\right) + \frac{1}{2}\varphi(x, 2|u(x)|). \tag{4.15}$$

Denote the left-hand side and the right-hand side of (4.15) by v_n and w_n respectively. Then the following conditions hold:

(*i*) $v_n(x) \to v(x) = \varphi(x, |u(x)|) \in L^1(\Omega)$ *a.e.* in Ω

(*ii*) $w_n(x) \to w(x) = \frac{1}{2}\varphi(x, 2|u(x)|) \in L^1(\Omega)$ *a.e.* in Ω

(*iii*) $v_n, w_n \in L^1(\Omega)$ for any $n \in \mathbb{N}$

(*iv*) $\int_\Omega w_n dx \to \int_\Omega \frac{1}{2}\varphi(x, 2|u|)dx = \frac{1}{2}\rho_{\varphi}(2u)$.

Since $w - v \geq 0$ *a.e* in Ω, Fatou's Lemma leads to:

$$\begin{aligned}\int_\Omega (w - v)dx &\leq \int_\Omega w\, dx + \liminf_n \int_\Omega (-v_n)\, dx \\ &= \int_\Omega w\, dx - \limsup_n \int_\Omega v_n\, dx\end{aligned}$$

and

$$\int_\Omega (w+v)dx \leq \int_\Omega w\,dx + \liminf_n \int_\Omega v_n\,dx.$$

The two last statements yield

$$\lim_{n\to\infty} \int_\Omega \varphi(x,|u_n(x)|)dx = \int_\Omega \varphi(x,|u(x)|)dx$$

or, equivalently

$$\rho_\varphi(u_n) \longrightarrow \rho_\varphi(u) \text{ as } n \to \infty. \tag{4.16}$$

By construction $\rho_\varphi(u_n) \longrightarrow \mathbf{S}_r$; (4.16) is therefore the desired result. □

4.3 The variable exponent p-Laplacian

At this stage we take a detour from the general theory to focus on the particular case of the Musielak-Orlicz function given by $\varphi(x,t) = t^{p(x)}$. We devote this section to present a self contained discussion on the variable exponent $p(x)$-Laplacian. We base our presentation on [38]. The main result here is the proof of Theorem 4.3.4, originally due to Fan and Zhang [38]. We start with the following lemma:

Lemma 4.3.1. *For $n \in \mathbb{N}$, $x, y \in \mathbb{R}^n$ and constant p,*

$$\begin{aligned} &\frac{1}{2}\left((|x|^{p-2} - |y|^{p-2})(|x|^2 - |y|^2) + (|x|^{p-2} + |y|^{p-2})|x-y|^2\right) \\ &= (|x|^{p-2}x - |y|^{p-2}y)\cdot(x-y). \end{aligned} \tag{4.17}$$

Proof. This follows by straightforward calculation:

$$\begin{aligned} &(|x|^2 - |y|^2)(|x|^{p-2} - |y|^{p-2}) + |x-y|^2(|x|^{p-2} + |y|^{p-2}) \\ &= (x-y)\left((x+y)(|x|^{p-2} - |y|^{p-2}) + (x-y)(|x|^{p-2} + |y|^{p-2})\right) \\ &= 2(x-y)((|x|^{p-2}x - |y|^{p-2}y)). \end{aligned}$$

□

Let $\Omega \subseteq \mathbb{R}^n$ be a domain and consider a Borel-measurable function

$$p : \Omega \longrightarrow [1,\infty).$$

For $u \in W_0^{1,p}(\Omega)$, the Dirichlet p-Laplacian $\Delta_p(u)$ is defined by

$$\Delta_p(u) = \operatorname{div}\left(|\nabla u|^{p-2}\nabla u\right).$$

In weak form, Δ_p is the (non-linear) operator

$$\Delta_p : W_0^{1,p}(\Omega) \longrightarrow \left(W_0^{1,p}(\Omega)\right)^*$$

such that for all $u, h \in W_0^{1,p}(\Omega)$,

$$\langle \Delta_p u, h\rangle = -\int_\Omega |\nabla u|^{p-2}\nabla u \cdot \nabla h \, dx.$$

A few remarks are in order. The functional

$$F : W_0^{1,p}(\Omega) \longrightarrow [0, \infty) \tag{4.18}$$

defined by

$$F(w) = \int_\Omega \frac{|\nabla w|^p}{p} dx$$

is convex, weakly lower-semicontinuous, Fréchet-differentiable and in fact (see also [19]),

$$F' = -\Delta_p. \tag{4.19}$$

For the sake of completeness, a thorough proof of (4.19) will be given in Section 4.6 in a more general setting (Lemma 4.6.1 and Remark 4.6.2).

When p is constant, it is obvious that the functional (4.18) is a multiple of the pth power of the Sobolev norm, namely

$$F(w) = \frac{1}{p}\|\nabla w\|^p.$$

For variable p the consideration of the derivative of the norm as opposed to the derivative of the modular (which is essentially the definition we have adopted for the p-Laplacian) leads to a different differential operator, as will be discussed later (Section 4.5).

Lemma 4.3.2. *Let $\Omega \subset \mathbb{R}^n$ be a bounded, Lipschitz domain and $p : \Omega \to \mathbb{R}$ a Borel-measurable function satisfying $1 < p_- \leq p(x) \leq p_+ < \infty$ a.e. in Ω. Under these conditions, the operator*

$$\Delta_p : W_0^{1,p}(\Omega) \longrightarrow \left(W_0^{1,p}(\Omega)\right)^*$$

is bounded, hemicontinuous and monotone. In addition, Δ_p is of type M.

Proof. Let $S \subset W_0^{1,p}(\Omega)$ be bounded, with

$$\sup\{\|\nabla u\|_p : u \in S\} \leq C.$$

For $u \in S$ and w in the unit ball of $W_0^{1,p}(\Omega)$, set

$$\langle \Delta_p u, w \rangle = \int_\Omega |\nabla u|^{p-2} \nabla u \cdot \nabla w \, dx. \tag{4.20}$$

Taking absolute values in equality (4.20) and invoking the variable exponent form of Hölder's inequality it is clear that

$$\sup \left\{ \|\Delta_p u\|_{(W_0^{1,p}(\Omega))^*} : u \in S \right\} \leq C \left(1 + \frac{1}{p_-} - \frac{1}{p_+} \right),$$

which shows that Δ_p is bounded.
For the proof of hemicontinuity, fix $t \in \mathbb{R}$. It is clear that for

$$|s| < |t| + \frac{1}{2} \text{ and } 1 < p \leq 2$$

it holds that

$$|\nabla(u + sv)(x)|^{p-1} \leq |\nabla u|^{p-1} + |s|^{p-1} |\nabla v(x)|^{p-1}, \tag{4.21}$$

whereas for $p > 2$ the corresponding inequality reads:

$$|\nabla(u + sv)|^{p-1} \leq 2^{p-1}(|\nabla u|^{p-1} + |s|^{p-1}|\nabla v|^{p-1}). \tag{4.22}$$

On the other hand, by definition:

$$\langle \Delta_p(u + sv), v \rangle = \int_\Omega |\nabla(u + sv)|^{p-2} \nabla(u + sv) \cdot \nabla v \, dx. \tag{4.23}$$

In view of (4.21) and (4.22) the integrand in (4.23) is bounded above by

$$|\nabla u|^{p-1}|\nabla v| + |s|^{p-1}|\nabla v|^p \, \mathbf{1}_{\{1<p\leq 2\}} \\ + 2^{p-1} \left(|\nabla u|^{p-1}|\nabla v| + |s|^{p-1}|\nabla v|^p \right) \mathbf{1}_{\{p \geq 2\}},$$

which is integrable by virtue of Hölder's inequality. A straightforward application of Lebesgue's dominated convergence theorem yields the hemicontinuity of Δ_p.
Since $W_0^{1,p}(\Omega)$ is reflexive and separable, once we have shown that Δ_p is monotone it will follow that it is of type M via Theorem 4.1.1.
The proof of monotonicity relies ultimately on the identity (4.17). For constant $p \geq 3$ and $x, y \in \mathbb{R}^n$ we have the inequality

$$|x - y|^p = |x - y|^{p-2}|x - y|^2 \leq 2^{p-3}|x - y|^2(|x|^{p-2} + |y|^{p-2}),$$

whereas if $2 \leq p < 3$ one has

$$|x - y|^p = |x - y|^2|x - y|^{p-2} \leq |x - y|^2(|x|^{p-2} + |y|^{p-2}).$$

When combined with the identity (4.17), the two last inequalities yield the estimate

$$|x-y|^p \le C(p)(|x|^{p-2}x - |y|^{p-2}y)\cdot(x-y), \tag{4.24}$$

where $C(p)=\max\{1,2^{p-2}\}>0$.

Likewise, for $1<p\le 2$ (with the obvious provision that $x\neq 0$ and $y\neq 0$) we have

$$(p-1)|x-y|^2(1+|x|+|y|)^{p-2} \le (|x|^{p-2}x-|y|^{p-2}y)(x-y), \tag{4.25}$$

which follows from

$$\begin{aligned}
&(|x|^{p-2}x-|y|^{p-2}y)(x-y)\\
&=(x-y)\int_0^1 \frac{d}{dt}\left((|y+t(x-y)|^{p-2}(y+t(x-y))\right)dt\\
&=|x-y|^2\int_0^1 |y+t(x-y)|^{p-2}\\
&\qquad+(p-2)\int_0^1 |y+t(x-y)|^{p-4}\left((y+t(x-y))(x-y)\right)^2\\
&\le (p-1)|x-y|^2\int_0^1 |y+t(x-y)|^{p-2} \ge (p-1)|x-y|^2(1+|x|+|y|)^{p-2}.
\end{aligned}$$

For fixed u and v in $W_0^{1,p}(\Omega)$, it follows straight from the definition of Δ_p that:

$$\langle \Delta_p(u)-\Delta_p(v),u-v\rangle = \int_\Omega (|\nabla u|^{p-2}\nabla u-|\nabla v|^{p-2}\nabla v)\cdot\nabla(u-v)dx. \tag{4.26}$$

The desired conclusion is obtained by splitting the integral in the right-hand side of (4.26) over $\{x: 1<p(x)\le 2\}$ and $\{x: p(x)\ge 2\}$ and applying the inequalities (4.24) and (4.25). □

Lemma 4.3.3. *[38, Theorem 3.1] Let $\Omega\subset\mathbb{R}^n$ be a bounded, Lipschitz domain and $p:\Omega\to\mathbb{R}$ a Borel-measurable function satisfying $1<p_-\le p(x)\le p_+<\infty$ a.e. in Ω. Then the operator*

$$\Delta_p : W_0^{1,p}(\Omega)\longrightarrow \left(W_0^{1,p}(\Omega)\right)^*$$

given by

$$\langle \Delta_p u, v\rangle = -\int_\Omega |\nabla u|^{p-2}\nabla u\nabla v\,dx$$

is a homeomorphism.

Proof. The surjectivity of Δ_p is derived from Theorem 4.1.3, injectivity follows from Poincaré inequality and inequalities (4.24) and (4.25). The continuity of Δ_p^{-1} ensues from a functional-analytic argument coupled with inequalities (4.24) and (4.25). We proceed to the details of the proof. Fix $f \in \left(W_0^{1,p}(\Omega)\right)^*$. For $u \in W_0^{1,p}(\Omega)$ with

$$|||\nabla u|||_p > \max\left\{1, \|f\|_{\left(W_0^{1,p}(\Omega)\right)^*}^{1/(p_- -1)}\right\},$$

one can write

$$1 = \int_\Omega \frac{|\nabla u|^p}{|||\nabla u|||_p^p} dx \leq \frac{1}{|||\nabla u|||_p^{p_-}} \int_\Omega |\nabla u|^p dx;$$

thus for such u,

$$\langle \Delta_p u, u\rangle = \int_\Omega |\nabla u|^p dx \geq |||\nabla u|||_p^{p_-} = |||\nabla u|||_p^{p_- -1} |||\nabla u|||$$
$$> \|f\|_{\left(W_0^{1,p}(\Omega)\right)^*} |||\nabla u|||_p.$$

As a direct consequence of Theorem 4.1.3 it follows that f is in the range of Δ_p, hence that the latter is surjective.

The injectivity of the map can be obtained as a consequence of inequalities (4.24) and (4.25) in conjunction with Poincaré inequality. Indeed, let $u \in W_0^{1,p}(\Omega)$ and $v \in W_0^{1,p}(\Omega)$ satisfy

$$\Delta_p(u) = \Delta_p(v).$$

In this case, we estimate $\rho_p(\nabla u - \nabla v)$ as follows:

$$\int_\Omega |\nabla u - \nabla v|^p dx =$$
$$\int_{\{p>2\}} |\nabla u - \nabla v|^p dx + \int_{\{1<p\leq 2\}} |\nabla u - \nabla v|^p dx. \tag{4.27}$$

The first integral in (4.27) is, according to (4.24), bounded by

$$\max\{2^{p_+ -2}, 1\} \int_\Omega (|\nabla u|^{p-2}\nabla u dx - |\nabla v|^{p-2}\nabla v) \cdot (\nabla u - \nabla v) dx$$
$$= \max\{2^{p_+ -2}, 1\} \langle \Delta_p(u) - \Delta_p(v), u - v\rangle = 0. \tag{4.28}$$

Denote the second integral in (4.27) by I, that is:

$$I = \int_{\{1<p<2\}} \frac{|\nabla u - \nabla v|^p (1 + |\nabla u| + |\nabla v|)^{p(2-p)/2}}{(1 + |\nabla u| + |\nabla v|)^{p(2-p)/2}} dx.$$

Utilizing the modular Hölder inequality (Theorem 2.4.2), the right-hand side in (4.28) can be seen to be bounded above by

$$2\left\| \frac{|\nabla u - \nabla v|^p}{(1+|\nabla u| + |\nabla v|)^{p(2-p)/2}} \mathbf{1}_{\{1<p<2\}} \right\|_{2/p} \times \left\| (1+|\nabla u| + |\nabla v|)^{p(2-p/2} \right\|_{2/(2-p)}.$$

Hence, from the norm-modular inequalities in Proposition 1.5.4, setting

$$C(u,v) = \left\| (1+|\nabla u| + |\nabla v|)^{p(2-p)/2} \right\|_{2/(2-p)}$$

and

$$J = \int_\Omega \frac{|\nabla u - \nabla v|^2}{(1+|\nabla u| + |\nabla v|)^{(2-p)}}\, dx,$$

we conclude that

$$I \leq C(u,v) \max\left\{ J^{\frac{p_+}{2}}, J^{\frac{p_-}{2}} \right\}. \tag{4.29}$$

Estimates (4.29) and (4.25) imply that for some positive constant α,

$$I \leq C(u,v) \left\langle \Delta_p(u) - \Delta_p(v), u - v \right\rangle^\alpha = 0. \tag{4.30}$$

It is immediate from (4.27), (4.28) and (4.30) that $u - v$ is constant. Poincaré inequality now implies that $u = v$.
It remains to verify the continuity of the inverse operator Δ_p^{-1}: to that effect set

$$\Delta_p = T$$

and suppose that for $(v_n) \subset W_0^{1,p}(\Omega)$ we have

$$\Delta_p(v_n) \to \Delta_p(u).$$

If the sequence (v_n) were unbounded, one could extract from it a subsequence (u_n) such that for all n,

$$\|u_n\|_p > 1.$$

In this case, set

$$w_n = (1/\||\nabla u_n|\|_p)\, u_n$$

and observe that for arbitrary $\phi \in W_0^{1,p}(\Omega)$ with $\||\nabla \phi|\|_p \leq 1$, the equality

$$|\langle T(w_n), \phi \rangle| = \left| \int_\Omega \frac{1}{\||\nabla u_n|\|_p^{p-1}} |\nabla u_n|^{p-2} \nabla u_n \cdot \nabla \phi\, dx \right|,$$

yields

$$|\langle T(w_n), \phi\rangle| \leq \frac{1}{\||\nabla u_n|\|_p^{p_- -1}} \left| \int_\Omega |\nabla u_n|^{p-2}\nabla u_n \cdot \nabla \phi \, dx \right|$$
$$\leq \frac{1}{\||\nabla u_n|\|_p^{p_- -1}} \|T(u_n)\|_{(W_0^{1,p}(\Omega))^*}.$$

It is easy to see that the last inequality implies that

$$\|T(w_n)\|_{(W_0^{1,p}(\Omega))^*} \to 0 \text{ as } n \to \infty. \tag{4.31}$$

On the other hand, it follows automatically by definition that

$$\|T(w_n)\|_{(W_0^{1,p}(\Omega))^*} \geq \langle T(w_n), w_n\rangle = \int_\Omega |\nabla w_n|^p dx = 1,$$

which contradicts (4.31). Consequently, the sequence (v_n) must be bounded in $W_0^{1,p}(\Omega)$.

Next we exploit inequalities (4.24) and (4.25). Observe that

$$\int_\Omega |\nabla v_n - \nabla v|^p dx = \tag{4.32}$$
$$\int_{\{p\geq 2\}} |\nabla v_n - \nabla v|^p dx + \int_{\{1<p<2\}} |\nabla v_n - \nabla v|^p dx.$$

By inequality (4.24), the first term in (4.32) is bounded by

$$\int_{\{p\geq 2\}} |\nabla v_n - \nabla v|^p dx \leq \left|\max\{2^{p_+-2}, 1\}\langle T(v_n) - T(v), v_n - v\rangle\right|$$
$$\leq \max\{2^{p_+-2}, 1\}\|T(v_n) - T(v)\|_{(W_0^{1,p}(\Omega))^*}\|v_n - v\|_{W_0^{1,p}(\Omega)}, \tag{4.33}$$

and the right hand side above tends to 0 as $n \to \infty$ since $T(v_n) \to 0$ and (v_n) is bounded.

The second term in (4.32) can be expressed as

$$\int_{\{1<p<2\}} \frac{|\nabla v_n - \nabla v|^p (1 + |\nabla v| + |\nabla v_n|)^{p(2-p)/2}}{(1 + |\nabla v| + |\nabla v_n|)^{p(2-p)/2}} dx,$$

which a straightforward application of the generalized Hölder inequality (Theorem 2.4.2) shows to be dominated by

$$\left\|(1 + |\nabla v| + |\nabla v_n|)^{p(2-p)/2}\right\|_{2/(2-p)} \left\|\frac{|\nabla v_n - \nabla v|^p}{(1 + |\nabla v| + |\nabla v_n|)^{p(2-p)/2}}\right\|_{2/p}.$$

Invoking again the boundedness of the sequence (v_n) it is immediate that the left factor above is bounded by a constant independent of n. Set

$$I_n = \int_\Omega |\nabla(v_n - v)|^2 (1 + |\nabla v| + |\nabla v_n|)^{p-2} dx$$

Choosing n so large that

$$\|T(v_n) - T(v)\|_{(W_0^{1,p}(\Omega))^*} < \frac{(p_- - 1)}{\sup_n \|v_n\|_{W_0^{1,p}(\Omega)} + \|v\|_{W_0^{1,p}(\Omega)}},$$

inequality (4.25) yields $I_n < 1$; consequently, using Proposition 1.5.4 we conclude that

$$\left\| \frac{|\nabla v_n - \nabla v|^p}{(1 + |\nabla v| + |\nabla v_n|)^{p(2-p)/2}} \right\|_{2/p} \leq \max\left\{ I_n^{p_+/2}, I_n^{p_-/2} \right\} = I_n^{p_-/2}.$$

A further application of inequality (4.25), for such n, implies

$$\begin{aligned} I_n^{\frac{p_-}{2}} &\leq \frac{1}{p_- - 1} |\langle T(v_n) - T(v), v_n - v \rangle| \\ &\leq \frac{1}{p_- - 1} \|T(v_n) - T(v)\|_{(W_0^{1,p}(\Omega))^*} \|v_n - v\|_{W_0^{1,p}(\Omega)} \\ &\leq \frac{1}{p_- - 1} \|T(v_n) - T(v)\|_{(W_0^{1,p}(\Omega))^*} (\sup_n \|v_n\|_{W_0^{1,p}(\Omega)} + \|v\|_{W_0^{1,p}(\Omega)}). \end{aligned} \tag{4.34}$$

The bounds (4.33) and (4.34) show, according to (4.32), that

$$\int_\Omega |\nabla v - \nabla v_n|^p \, dx \to 0 \text{ as } n \to \infty$$

and hence

$$\|v_n - v\|_{W_0^{1,p}(\Omega)} \to 0 \text{ as } n \to \infty.$$

Consequently,

$$\Delta_p^{-1} : \left(W_0^{1,p}(\Omega)\right)^* \longrightarrow W_0^{1,p}(\Omega)$$

is continuous. □

The preceding discussion leads directly to the main result of this section, first addressed in [60, 38].

Theorem 4.3.4. *[38, Theorem 4.2] Let $g \in \left(W_0^{1,p}(\Omega)\right)^*$. Then the Dirichlet problem*

$$\Delta_p = g$$

has a unique solution $u \in W_0^{1,p}(\Omega)$.

4.3.1 Stability of the solutions

The question of stability for the solutions of the Dirichlet problem with respect to the variable exponent p arises naturally. For constant p we refer the reader to the work by P. Lindqvist [71]. In this Section we address the question for variable p. Specifically, we fix a non-decreasing sequence $(p_k) \subset C(\overline{\Omega})$ and assume that for some $p \in C(\overline{\Omega})$, $p \leq p_k$ and

$$p_k \to q$$

uniformly in Ω.

Consider an arbitrary $f \in \left(W_0^{1,p}(\Omega)\right)^*$. For each k, let $u_k \in W_0^{1,p_k}(\Omega)$, $u_p \in W_0^{1,p}(\Omega)$ and $u_q \in W_0^{1,q}(\Omega)$ be, respectively the unique solutions of

$$\Delta_{p_k}(u_{p_k}) = f, \quad \Delta_p(u_p) = f, \quad \text{and } \Delta_q(u_q) = f \tag{4.35}$$

given by Theorem 4.3.4. The following is a variant of [34, Lemma 4.1]. As usual,

$$e = \lim_{x \to 0}(1+x)^{\frac{1}{x}}.$$

Lemma 4.3.5. *Let $0 < \varepsilon < \frac{1}{e}$ be arbitrary and N be large enough to guarantee that $\|p_i - p_j\|_\infty < \varepsilon$ if $i, j > N$. Fix k, j such that $N \leq k \leq j$. Then, for each u_{p_j} defined by (4.35),*

$$\int_\Omega \frac{|\nabla u_{p_j}|^{p_k}}{p_k}\, dx \leq \varepsilon \int_\Omega \frac{1}{p_k}\, dx + \varepsilon^{-\varepsilon} \int_\Omega \frac{|\nabla u_{p_j}|^{p_j}}{p_k}\, dx.$$

Proof.

$$\int_\Omega \frac{|\nabla u_{p_j}|^{p_k}}{p_k}\, dx \leq$$

$$\int_{|\nabla u_{p_j}|<\varepsilon} \frac{|\nabla u_{p_j}|^{p_k}}{p_k}\, dx + \int_{\varepsilon \leq |\nabla u_{p_j}|<1} \frac{|\nabla u_{p_j}|^{p_k}}{p_k}\, dx + \int_{|\nabla u_{p_j}|\geq 1} \frac{|\nabla u_{p_j}|^{p_k}}{p_k}\, dx. \tag{4.36}$$

Observe that if $\varepsilon < 1$, it holds the inequality $\varepsilon^{p_k(x)} < \varepsilon$ for any $x \in \Omega$. In other words, one has, for $k \leq j$:

$$\int_{|\nabla u_{p_j}|<\varepsilon} \frac{|\nabla u_{p_j}|^{p_k}}{p_k}\, dx < \varepsilon \int_\Omega 1/p_k\, dx.$$

Using the conditions on ε, we can deal with the second integral in (4.36) as follows:

$$\int_{\varepsilon \leq |\nabla u_{p_j}|<1} \frac{|\nabla u_{p_j}|^{p_k - p_j} |\nabla u_{p_j}|^{p_j}}{p_k}\, dx \leq \int_{|\nabla u_{p_j}|<1} \frac{\varepsilon^{-\varepsilon} |\nabla u_{p_j}|^{p_j}}{p_k}\, dx.$$

It is also immediate that

$$\int\limits_{|\nabla u_{p_j}|\geq 1} \frac{|\nabla u_{p_j}|^{p_k}}{p_k}\,dx \leq \int\limits_{|\nabla u_{p_j}|\geq 1} |\nabla u_{p_j}|^{p_j}/p_k\,dx.$$

Substituting the last three inequalities in the expression (4.36) yields the conclusion of the lemma. □

We are now ready to inspect the effect of the variation of the integrability exponent on the solution of the Dirichlet problem.

Theorem 4.3.6. *For u_{p_k}, and u_q as in the previous paragraph, there exists a subsequence (still denoted by (u_{p_k})) such that $u_{p_k} \rightharpoonup u_q$, $u_{p_k} \to u_q$ strongly in $L^p(\Omega)$ and for which*

$$\lim_{k\to\infty} \int\limits_{\Omega} |\nabla u_{p_k}|^{p_k}\,dx = -\int\limits_{\Omega} f u_q\,dx = \int\limits_{\Omega} |\nabla u_q|^q\,dx.$$

Proof. By definition of Δ_{p_k} and from inequalities (1.5.4) it immediately follows that either $|||\nabla u_{p_k}|||_{p_k} \leq 1$ or

$$|||\nabla u_{p_k}|||_{p_k}^{p_{k-}} \leq \int\limits_{\Omega} |\nabla u_{p_k}|^{p_k}\,dx = -\langle f, u_{p_k}\rangle \leq \|f\|_{(W_0^{1,p}(\Omega))^*} \cdot |||\nabla u_{p_k}|||_p.$$

Young's inequality, in concert with [34, Lemma 4.1] implies that for any $\delta > 0$,

$$\begin{aligned}
&|||\nabla u_{p_k}|||_{p_k}^{p_{k-}} \qquad\qquad (4.37)\\
&\leq \left(\left(\frac{\|f\|_{(W_0^{1,p}(\Omega))^*}}{\delta}\right)^{p_{k-}/(p_{k-}-1)} \frac{p_{k-}-1}{p_{k-}} + \frac{(|||\nabla u_{p_k}|||_p\delta)^{p_{k-}}}{p_{k-}}\right)\\
&\leq \left(\left(\frac{\|f\|_{(W_0^{1,p}(\Omega))^*}}{\delta}\right)^{p_{k-}/(p_{k-}-1)} \frac{p_{k-}-1}{p_{k-}}\right.\\
&\qquad \left. + \frac{\left(\left(\|p-q\|_\infty^{-\|p-q\|_\infty} + \|p-q\|_\infty|\Omega|\right) |||\nabla u_{p_k}|||_{p_k}\delta\right)^{p_{k-}}}{p_{k-}}\right).
\end{aligned}$$

Since δ is arbitrary, inequality (4.37) shows that $(|||\nabla u_{p_k}|||_{p_k})_{k\in\mathbb{N}}$ is bounded. As a consequence of the inequality

$$|||\nabla u_{p_k}|||_p \leq \left(\|p-q\|_\infty^{-\|p-q\|_\infty} + \|p-q\|_\infty|\Omega|\right) |||\nabla u_{p_k}|||_{p_k}$$

[34, Theorem 4.1], it is immediate that the sequence (u_{p_k}) is bounded in $W_0^{1,p(\cdot)}(\Omega)$. Thus, it has a weakly convergent subsequence, which we continue to denote by (u_{p_k}). Let u be the weak limit of (u_{p_k}) in $W_0^{1,p(\cdot)}(\Omega)$. We now show that $u = u_q$. To that goal, we fix an exponent $r \in C(\overline{\Omega})$ such that $r > p$, and choose k large enough so that $p_k > r$ and $\|q - p_k\|_\infty \leq \|q - r\|_\infty/2$ on Ω. The sequence $(u_{p_{k+j}})_j$ is bounded in $W_0^{1,r(\cdot)}(\Omega)$, since according to [34, Lemma 4.1],

$$|||\nabla u_{p_{k+j}}|||_r \leq \left(\|p - q\|_\infty^{-\|p-q\|_\infty} + \|p - q\|_\infty |\Omega|\right) |||\nabla u_{p_{k+j}}|||_{p_{k+j}};$$

it is therefore clear that $(u_{p_{k+j}})_j$ is weakly convergent in $W_0^{1,r(\cdot)}(\Omega)$. Denote the weak limit by $v \in W_0^{1,r(\cdot)}(\Omega)$. The uniqueness of the limit implies $u = v$, so $u \in \cap_{p \leq r < q} W_0^{1,r(\cdot)}(\Omega)$. Sobolev's embedding theorem allows us to assume without loss of generality that for $u_{p_k} \longrightarrow u$ strongly in $L^{p_j}(\Omega)$ for $k \geq j$.

By virtue of the weak lower semicontinuity of the functional

$$I : W_0^{1,p_k}(\Omega) \longrightarrow [0, \infty) \ , \ I(w) = \int_\Omega |\nabla w|^{p_k}\, dx,$$

for any fixed j, u is subject to the condition

$$\int_\Omega |\nabla u|^{p_k}\, dx \leq \liminf_{j \geq k} \int_\Omega |\nabla u_{p_j}|^{p_k}\, dx. \tag{4.38}$$

On the other hand, for $j \geq k$,

$$\int_\Omega |\nabla u_{p_j}|^{p_k}\, dx \leq \|p_k - p_j\|_\infty |\Omega| + \|p_k - p_j\|_\infty^{-\|p_j - p_k\|_\infty} \int_\Omega |\nabla u_{p_j}|^{p_j}\, dx \tag{4.39}$$

which is bounded independently of j by virtue of (4.37) and by the assumption on the sequence (p_k). Fatou's Lemma yields

$$\int_\Omega \liminf_k |\nabla u|^{p_k}\, dx = \int_\Omega |\nabla u|^q\, dx \leq \liminf_k \int_\Omega |\nabla u|^{p_k}\, dx. \tag{4.40}$$

In view of (4.38) and (4.39), the right-hand side of (4.40) is finite, so that $u \in W_0^{1,q}(\Omega)$. For a given $\delta > 0$, fix positive numbers θ and γ such that $(1+\theta)(1+\gamma) < 1 + \delta/2$. Since for $\varepsilon < e^{-1}$ the function $\omega(\varepsilon) = \varepsilon^{-\varepsilon}$ decreases to 1 as $\varepsilon \to 0^+$, given the assumptions on the sequence (p_k), there exists $M > 0$ such that for $M \leq k \leq j$,

$$\|p_j - p_k\|_\infty < \min\left\{\gamma, \frac{\delta}{2|\Omega|} p_-\right\} \ , \ \|p_j - p_k\|_\infty^{-\|p_k - p_j\|_\infty} < 1 + \theta. \tag{4.41}$$

Notice that for such k and j,

$$\left\|\frac{p_j}{p_k} - 1\right\|_\infty < \gamma. \tag{4.42}$$

Next, Lemma 4.3.5, (4.41), (4.42) guarantee that for $M \le k \le j$,

$$\begin{aligned}\int_\Omega \frac{|\nabla u_{p_j}|^{p_k}}{p_k}\,dx &\le \|p_k - p_j\|_\infty \int_\Omega \frac{1}{p_k}\,dx \\ &\quad + \|p_k - p_j\|_\infty^{-\|p_k - p_j\|_\infty} \int_\Omega \frac{|\nabla u_{p_j}|^{p_j}}{p_j}(1+\gamma)\,dx \\ &\le \frac{\delta}{2} + (1+\theta)(1+\gamma)\int_\Omega \frac{|\nabla u_{p_j}|^{p_j}}{p_j}\,dx \\ &\le \frac{\delta}{2} + (1+\frac{\delta}{2})\int_\Omega \frac{|\nabla u_{p_j}|^{p_j}}{p_j}\,dx. \end{aligned} \tag{4.43}$$

For fixed $k \in \mathbb{N}$ and each $M \le k$, the weak lower semicontinuity of the functional

$$\begin{aligned} &I_k : W_0^{1,p}(\Omega) \longrightarrow [0,\infty) \\ &I_k(w) = \int_\Omega \frac{|\nabla w|^{p_k}}{p_k}\,dx, \end{aligned}$$

yields

$$\int_\Omega \frac{|\nabla u|^{p_k}}{p_k}\,dx \le \liminf_{j\ge k} \int_\Omega \frac{|\nabla u_{p_j}|^{p_k}}{p_k}\,dx.$$

Hence, as one sees from (4.43):

$$\int_\Omega \frac{|\nabla u|^{p_k}}{p_k}\,dx \le \liminf_{j\ge k} \int_\Omega \frac{|\nabla u_{p_j}|^{p_k}}{p_k}\,dx \le \liminf_{j\ge k} \left(\frac{\delta}{2} + (1+\frac{\delta}{2}) \int_\Omega \frac{|\nabla u_{p_j}|^{p_j}}{p_j}\,dx \right),$$

i.e.,

$$\int_\Omega \frac{|\nabla u|^{p_k}}{p_k}\,dx \le \liminf_{j\ge k} \int_\Omega \frac{|\nabla u_{p_j}|^{p_j}}{p_j}\,dx. \tag{4.44}$$

From the fact that $u \in W_0^{1,q}(\Omega)$ it is easy to conclude, via Lebesgue's dominated convergence theorem, that

$$\lim_{k\to\infty} \int_\Omega \frac{|\nabla u|^{p_k}}{p_k}\,dx = \int_\Omega \frac{|\nabla u|^q}{q}\,dx.$$

By virtue of the minimizing property of u_{p_k} it is clear that

$$\int_\Omega \frac{|\nabla u_{p_k}|^{p_k}}{p_k}\,dx + \int_\Omega f u_{p_k}\,dx \le \int_\Omega \frac{|\nabla u|^{p_k}}{p_k}\,dx + \int_\Omega f u\,dx,$$

which automatically leads to

$$\limsup \int_\Omega \frac{|\nabla u_{p_k}|^{p_k}}{p_k}\,dx \leq \limsup \int_\Omega \frac{|\nabla u|^{p_k}}{p_k}\,dx = \int_\Omega \frac{|\nabla u|^q}{q}\,dx$$

which, in conjunction with (4.44) implies

$$\int_\Omega \frac{|\nabla u|^q}{q}\,dx = \lim \int_\Omega \frac{|\nabla u_{p_k}|^{p_k}}{p_k}\,dx. \tag{4.45}$$

Again the minimal character of u_{p_k} yields the inequality

$$\int_\Omega \frac{|\nabla u_{p_k}|^{p_k}}{p_k}\,dx + \int_\Omega f u_{p_k}\,dx \leq \int_\Omega \frac{|\nabla u_q|^{p_k}}{p_k}\,dx + \int_\Omega f u_q\,dx,$$

which is valid for each k. Passing to the limits as $k \to \infty$ taking into account (4.45) and the fact that $u_q \in W_0^{1,q}(\Omega)$, one obtains

$$\int_\Omega \frac{|\nabla u|^q}{q}\,dx + \int_\Omega f u\,dx \leq \int_\Omega \frac{|\nabla u_q|^q}{q}\,dx + \int_\Omega f u_q\,dx,$$

i.e., u minimizes the functional

$$G : W_0^{1,q}(\Omega) \to [0,\infty) \tag{4.46}$$

$$G(v) = \int_\Omega \frac{|\nabla v|^q}{q}\,dx + \int_\Omega f v\,dx. \tag{4.47}$$

In turn, (4.46) implies that $\Delta_q(u) = f$; and, since $u \in W_0^{1,q}(\Omega)$, Theorem 4.3.4 yields $u = u_q$. Hence, by letting $k \to \infty$ in

$$\int_\Omega |\nabla u_{p_k}|^{p_k}\,dx = -\int_\Omega f u_{p_k}\,dx,$$

which holds by definition, we at once obtain

$$\lim_{k\to\infty} \int_\Omega |\nabla u_{p_k}|^{p_k}\,dx = -\int_\Omega f u\,dx = \int_\Omega |\nabla u_q|^q\,dx,$$

as claimed. □

Corollary 4.3.7. *In the terminology of Theorem 4.3.4, under the assumption*

$$\inf_\Omega q \geq 2, \tag{4.48}$$

there exists a subsequence of (u_{p_j}) (still denoted by (u_{p_j})) such that

$$\int_\Omega |\nabla u_{p_j} - \nabla u_q|^{p_j}\,dx \longrightarrow 0 \ \text{ as } j \longrightarrow \infty.$$

In the particular case that $n = 1$, the restriction (4.48) is not needed.

Proof. For fixed j, weak-lower semicontinuity and convexity yield

$$\begin{aligned}\int_\Omega |\nabla u_q|^{p_j}\,dx &\le \liminf_{j\le k\to\infty}\int_\Omega \left|\frac{\nabla u_q}{2}+\frac{\nabla u_{p_k}}{2}\right|^{p_j}dx\\ &\le \frac{1}{2}\int_\Omega |\nabla u_q|^{p_j}\,dx+\frac{1}{2}\liminf_{j\le k\to\infty}\int_\Omega |\nabla u_{p_k}|^{p_j}\,dx. \qquad (4.49)\end{aligned}$$

From [34, Theorem 4.1] and due to the convergence given in Theorem 4.3.6, one obtains

$$\begin{aligned}&\liminf_{j\le k\to\infty}\int_\Omega |\nabla u_{p_k}|^{p_j}\,dx\\ &\le \liminf_{k\longrightarrow\infty}\left(\|p_j-p_k\|_\infty^{-\|p_j-p_k\|_\infty}\int_\Omega |\nabla u_{p_k}|^{p_k}\,dx+\|p_j-p_k\|_\infty|\Omega|\right)\\ &=\left(\|p_j-q\|_\infty^{-\|p_j-q\|_\infty}\int_\Omega |\nabla u_q|^{q}\,dx+\|p_j-q\|_\infty|\Omega|\right).\end{aligned}$$

Thus, letting $j\to\infty$ in (4.49) it is clear that

$$\liminf_{j\le k\to\infty}\int_\Omega \left|\frac{\nabla u_q}{2}+\frac{\nabla u_{p_k}}{2}\right|^{p_j}dx=\int_\Omega |\nabla u_q|^{q}\,dx.$$

Clarkson's inequality (see Theorem 1.4.30), valid in this case since $\inf_\Omega p_j\ge 2$, leads to the bound:

$$\begin{aligned}&\int_\Omega \left|\frac{\nabla u_q}{2}+\frac{\nabla u_{p_k}}{2}\right|^{p_j}dx+\int_\Omega \left|\frac{\nabla u_q}{2}-\frac{\nabla u_{p_k}}{2}\right|^{p_j}dx\\ &\le \frac{1}{2}\int_\Omega |\nabla u_q|^{p_j}\,dx+\frac{1}{2}\int_\Omega |\nabla u_{p_k}|^{p_j}\,dx\end{aligned}$$

from which the first part of the Corollary is obtained easily by taking the $\liminf_{j\le k\to\infty}$ of both sides.

We next set out to tackle the 1-dimensional case, observe that there exists a positive constant c such that as long as $1<s<S<\infty$ and $(\xi,r)\in(s,S)\times[-M,M]$,

$$\frac{1}{2}\left(|r+1|^\xi+|r-1|^\xi\right)-|r|^\xi\ge c. \qquad (4.50)$$

Set $A=\{x\in(a,b):|(u_{p_j}-u_q)'|>\delta|(u_{p_j}+u_q)'|\}$ and fix $\delta>0$. Then if

$x \in A$, the substitution

$$r = \frac{u_q'(x) + u_{p_j}'(x)}{u_q'(x) - u_{p_j}'(x)},$$

$\xi = p_j(x)$ in (4.128) yields, pointwise in Ω,

$$\frac{1}{2}\left(|u_q'|^{p_j} + |u_{p_j}'|^{p_j}\right) - \left|\frac{u_q' + u_{p_j}'}{2}\right|^{p_j} \geq c\left|\frac{u_q' - u_{p_j}'}{2}\right|^{p_j},$$

whence it follows by integrating on A that

$$\frac{1}{2}\left(\int_\Omega |u_q'|^{p_j}\,dx + \int_\Omega |u_{p_j}'|^{p_j}\,dx\right) - \int_\Omega \left|\frac{u_q' + u_{p_j}'}{2}\right|^{p_j} dx \tag{4.51}$$
$$\geq c\int_A \left|\frac{u_q' - u_{p_j}'}{2}\right|^{p_j} dx.$$

From Lebesgue's dominated convergence theorem and the proof of Theorem 4.3.6, it follows that

$$\lim_{j\to\infty}\int_\Omega |u_q'(x)|^{p_j}\,dx = \lim_{j\to\infty}\int_\Omega \left|\frac{u_q' + u_{p_j}'}{2}\right|^{p_j} dx = \int_\Omega |u_q'|^q\,dx.$$

Letting $j \to \infty$ in (4.130) makes it clear that

$$\int_A \left|\frac{u_q' - u_{p_j}'}{2}\right|^{p_j} dx \longrightarrow 0 \text{ as } j \to \infty$$

which through an easy calculation, yields

$$\int_A |u' - u_{p_j}'|^p dx \longrightarrow 0 \text{ as } j \longrightarrow \infty.$$

On the other hand

$$\int_{\Omega\backslash A} \left|\frac{u_q' - u_{p_j}'}{2}\right|^{p_j} dx \leq \delta^{p_-}\int_\Omega \left|\frac{u_q' + u_{p_j}'}{2}\right|^{p_j} dx \leq \delta^{p_-}.$$

Putting this all together, we have

$$\int_\Omega |u_q' - u_{p_j}'|^p dx \longrightarrow 0 \text{ as } j \longrightarrow \infty.$$

□

Consider next a non-increasing sequence (p_j).

Theorem 4.3.8. *Let $\Omega \subset \mathbb{R}^n$ be a bounded, Lipschitz domain, $(p_k) \subset C(\overline{\Omega})$ a non-increasing sequence converging to p uniformly on Ω. Let u_{p_k} for $k \in \mathbb{N}$ and u_p be as in (4.35). Assume that for some $r > p$ the higher integrability condition $\int_\Omega |\nabla u_p|^r \, dx < \infty$ holds.*[1] *Then there exists a subsequence of (u_{p_k}) (still denoted by (u_{p_k})) for which $u_{p_k} \rightharpoonup u_p$ and $u_{p_k} \to u_p$ strongly in $L^p(\Omega)$ and*

$$\int_\Omega |\nabla u_{p_k}|^{p_k} \, dx \to \int_\Omega |\nabla u_p|^p \, dx.$$

Proof. Owing to inequality (4.37) and the argument thereafter, the sequence (u_{p_j}) is bounded in $W_0^{1,p}(\Omega)$, and the numerical sequence $\left(\int_\Omega |\nabla u_{p_j}|^{p_j} \, dx \right)$ is bounded. Therefore, no generality is lost in assuming that $\left(\int_\Omega |\nabla u_{p_j}|^{p_j} \, dx \right)$ (and therefore also $\left(\int_\Omega |\nabla u_{p_j}|^{p_j} / p_j \, dx \right)$) is considered to be convergent unless it is finite, in which case the claim would follow trivially since

$$\langle f, u_{p_j} \rangle \to \langle f, u \rangle.$$

We point out that for fixed j, it holds that

$$\int_\Omega |\nabla u_{p_j}|^{p_k} \, dx \to \int_\Omega |\nabla u_{p_j}|^p \, dx \text{ as } j \leq k \to \infty.$$

Taking $k \geq j$ and invoking the estimate

$$\int_\Omega |\nabla u_{p_j}|^{p_k} \, dx \leq \|p_j - p_k\|_\infty^{-\|p_k - p_j\|_\infty} \int_\Omega |\nabla u_{p_j}|^{p_j} \, dx + \|p_j - p_k\|_\infty |\Omega|,$$

it is easily seen that

$$\int_\Omega |\nabla u_{p_j}|^p \, dx \leq \|p_j - p\|_\infty^{-\|p_j - p\|_\infty} \int_\Omega |\nabla u_{p_j}|^{p_j} \, dx + \|p_j - p\|_\infty |\Omega|.$$

Weak-lower semicontinuity yields

$$\begin{aligned} \int_\Omega |\nabla u|^p \, dx &\leq \liminf_j \int_\Omega |\nabla u_{p_j}|^p \, dx \\ &\leq \liminf_j \int_\Omega |\nabla u_{p_j}|^{p_j} \, dx = \lim_j \int_\Omega |\nabla u_{p_j}|^{p_j} \, dx. \end{aligned}$$

[1]This condition is satisfied, for example if $\mathbb{R}^n \setminus \Omega$ is p-fat (see [43, Theorem 5.1]).

In a similar fashion, via Lemma 4.3.5, one obtain

$$\int_\Omega \frac{|\nabla u|^p}{p}\,dx \le \lim_j \int_\Omega \frac{|\nabla u_{p_j}|^{p_j}}{p_j}\,dx.$$

In view of the above considerations, the minimality condition

$$\int_\Omega \frac{|\nabla u_{p_j}|^{p_j}}{p_j}\,dx + \langle f, u_{p_j}\rangle \le \int_\Omega \frac{|\nabla u_p|^{p_j}}{p_j}\,dx + \langle f, u_p\rangle, \tag{4.52}$$

implies

$$\int_\Omega \frac{|\nabla u|^p}{p}\,dx + \langle f, u\rangle \le \int_\Omega |\nabla u_p|^p\,dx + \langle f, u_p\rangle,$$

and the minimal character of u_p implies automatically that $u = u_p$. A further application of (4.52) yields

$$\lim_{j\to\infty} \int_\Omega \frac{|\nabla u_{p_j}|^{p_j}}{p_j}\,dx \le \int_\Omega \frac{|\nabla u|^p}{p}\,dx.$$

In conclusion,

$$\lim_{j\to\infty} \int_\Omega |\nabla u_{p_j}|^{p_j}\,dx \le \int_\Omega |\nabla u|^p\,dx.$$

□

Corollary 4.3.9. *In the terminology of Theorem 4.3.8, if* $\inf_\Omega p \ge 2$, *then for a subsequence of* (u_{p_j}) *(still denoted by* (u_{p_j})*):*

$$\int_\Omega |\nabla u_{p_j} - \nabla u_p|^p\,dx \longrightarrow 0 \ \text{as}\ j \longrightarrow \infty.$$

Corollary 4.3.9 can be obtained via arguments similar to those used in the proof of Corollary 4.3.7; its proof is omitted.

4.4 Γ-convergence

In this section we digress to discuss the connection of the results in the preceding paragraph with the idea of Γ-convergence. Due to the main ramifications of the topic and vastness of the literature on Γ-convergence, we content ourselves with some basic observations and examples. We refer the interested reader to the bibliographic references mentioned in the course of the exposition.

The concept of Γ-convergence was introduced by E. De Giorgi in the 1970s ([27, 28, 29]). Its flexibility and ubiquity in the direct methods of variational calculus have brought this idea to the main stage of mathematical research (see [14, 21] and the references therein). In this section, we explore the connection between the results in Section 4.3.1 and Γ-convergence.

Definition 4.4.1. *Let X be a metric space. A sequence (F_k) of functionals $F_k : X \longrightarrow [0,\infty]$ is said to* **Γ-converge** *to $F : X \longrightarrow [0,\infty]$ at $x \in X$ iff*

(*i*) *For any sequence $(x_k) \subseteq X$ with $x_k \longrightarrow x$ as $k \longrightarrow \infty$ one has:*

$$F(x) \leq \liminf_{k\to\infty} F_k(x_k).$$

(*ii*) *For any $y \in X$, there exists a sequence $(y_k) \subseteq X$, converging to y for which*

$$F(y) = \lim_{k\longrightarrow\infty} F_k(y_k).$$

A sequence (F_k) that Γ-converges to F at every $x \in X$, it is said to **Γ-converge to F on X**.
We write

$$F = \Gamma - \lim_{k\to\infty} F_k$$

to indicate that the sequence (F_k) Γ-converges to F.

A sequence (F_k) is said to be equicoercive if there exists a compact set $K \subset X$ such that $\inf_X F_k = \inf_K F_k$ for each $k \in \mathbb{N}$.

Notice that the second condition in Definition 4.4.1 is satisfied automatically if $\lim_{k\longrightarrow\infty} F_k(x) = F(x)$. For a bounded Lipschitz domain Ω and $s \in C(\overline{\Omega})$, fix $f \in \left(W_0^{1,s}(\Omega)\right)^*$; for $s \leq r \in C(\overline{\Omega})$ define

$$F_r : L^r(\Omega) \to [0,\infty]$$

by

$$F_r(u) = \begin{cases} \int_\Omega \frac{|\nabla u|^r}{r}\,dx + \langle f, u\rangle, & \text{if } u \in W^{1,r}(\Omega) \\ \infty & \text{otherwise.} \end{cases} \tag{4.53}$$

Theorem 4.4.1. *Let $\Omega \subset \mathbb{R}^n$ be a bounded, Lipschitz domain, $p, q \in C(\overline{\Omega})$ and $f \in \left(W_0^{1,p}(\Omega)\right)^*$.*

(*i*) *If* (p_j) *is a non-increasing sequence in* $C(\overline{\Omega})$ *converging uniformly in* Ω *to its infimum* p*, then the sequence* (F_{p_j}) *is equicoercive in* $L^p(\Omega)$ *and*

$$F_p(u) = \Gamma - \lim_{j\to\infty} F_{p_j}(u)$$

at every $u \in L^p(\Omega)$ *for which there exists* $\epsilon > 0$ *such that*

$$\int_\Omega |\nabla u|^{p+\epsilon}\, dx < \infty.$$

(*ii*) *If* (p_j) *is a non-decreasing sequence in* $C(\overline{\Omega})$ *converging uniformly in* Ω *to its supremum* q *with*

$$1 < \inf_{\overline{\Omega}} p \leq \sup_{\overline{\Omega}} q < \infty,$$

then the sequence (F_{p_j}) *of functionals defined on* $L^p(\Omega)$ *via (4.53) is equicoercive in* $L^p(\Omega)$ *and* Γ*-converges to* F_q *in* $L^p(\Omega)$.

Proof. For the proof of (*i*), we observe that it is clear that the infimum of each F_{p_j} is attained in $W_0^{1,p_j}(\Omega)$. It is well known that the Fréchet derivative of F_{p_j} is the functional on $W_0^{1,p_j}(\Omega)$ given by

$$h \to \int_\Omega |\nabla u|^{p_j-2}\nabla u \nabla h\, dx + \langle f, h\rangle;$$

by virtue of Theorem 4.3.4, the unique minimizer of F_{p_j} in $W_0^{1,p_j}(\Omega)$ is u_{p_j}. Owing to Sobolev's embedding theorem 3.4.4, the closure $\overline{K}$ in $L^p(\Omega)$ of the set $K = \{u_{p_j}, j \in \mathbb{N}\}$ is compact. Also it follows from the preceding discussion that

$$\inf_{L^p} F_{p_j} = F(u_{p_j}) = \inf_{\overline{K}} F_{p_j}$$

for each $j \in \mathbb{N}$. The sequence (F_{p_j}) is thus equicoercive; (see [21]).

Consider a sequence (v_j) in $L^p(\Omega)$ that converges to $v \in L^p(\Omega)$. If

$$L = \liminf_{j\to\infty} \left(\int_\Omega \frac{|\nabla v_j|^{p_j}}{p_j}\, dx + \langle f, v_j\rangle \right) < \infty$$

one can extract from (v_j) a subsequence (still denoted by (w_j)) for which,

$$\int_\Omega \frac{|\nabla w_j|^{p_j}}{p_j}\, dx + \langle f, w_j\rangle \longrightarrow L.$$

Since $p \leq p_j$ it is easy to see that

$$\int_\Omega \frac{|\nabla w_j|^{p_j}}{p_j}\,dx \leq C\left(1 + \|f\|_{\left(W_0^{1,p_j}(\Omega)\right)^*}\|\nabla w_j\|_{W_0^{1,p_j}(\Omega)}\right)$$

for some positive constant C and each $j \in \mathbb{N}$. It follows from inequalities in Proposition 1.5.4 and Young's inequality that if $\int_\Omega |\nabla w_j|^{p_j}\,dx > 1$, then

$$\||\nabla w_j|\|_{p_j}^{p_{j-}} \leq \left(C + \left(\frac{\|f\|}{\delta}\right)^{p_{j-}/(p_{j-}-1)}\frac{p_{j-}-1}{p_{j-}} + \frac{(\||\nabla w_j|\|_{p_j}\delta)^{p_{j-}}}{p_{j-}}\right)$$

for every $\delta > 0$. Thus the sequence (w_j) is bounded in $W_0^{1,p}(\Omega)$. Owing to the reflexivity of $W_0^{1,p}(\Omega)$ in conjunction with Sobolev's embedding theorem, one readily concludes that $v \in W_0^{1,p}(\Omega)$. Therefore, without loss of generality one can consider $w_j \rightharpoonup v$ in $W_0^{1,p}(\Omega)$. Moreover, $v \in W_0^{1,p_k}(\Omega)$ for sufficiently large $k \in \mathbb{N}$, which follows immediately from the higher integrability assumption on the gradient.
Weak-lower semicontinuity and the appropriate version of Lemma 4.3.5 imply

$$\begin{aligned}
\int_\Omega \frac{|\nabla v|^p}{p}\,dx \quad & \leq \liminf_{j\to\infty}\int_\Omega \frac{|\nabla w_j|^p}{p}\,dx \\
& \leq \liminf_{j\to\infty}\left(\|p-p_j\|_\infty^{-\|p-p_j\|_\infty}\int_\Omega \frac{|\nabla w_j|^{p_j}}{p_j}\,dx + \|p-p_j\|_\infty|\Omega|\right) \\
& = \liminf_{j\to\infty}\left(\|p-p_j\|_\infty^{-\|p-p_j\|_\infty}\left(\int_\Omega \frac{|\nabla w_j|^{p_j}}{p_j}\,dx + \langle f, w_j\rangle\right)\right) \\
& + \liminf_{j\to\infty}\left(\|p-p_j\|_\infty|\Omega| - \|p-p_j\|_\infty^{-\|p-p_j\|_\infty}\langle f, w_j\rangle\right) \\
& = L - \langle f, v\rangle. \qquad (4.54)
\end{aligned}$$

It is evident from the preceding argument that if $\int_\Omega |\nabla v|^p/p\,dx = \infty$, then

$$\liminf_{j\to\infty}\left(\int_\Omega \frac{|\nabla v_j|^{p_j}}{p_j}\,dx + \langle f, v_j\rangle\right) = \infty;$$

if this is the case, then obviously

$$F_p(v) = \infty = \liminf_{j\to\infty}\left(\int_\Omega \frac{|\nabla v_j|^{p_j}}{p_j}\,dx + \langle f, v_j\rangle\right).$$

It then follows from the latter and (4.54) that

$$v_j \longrightarrow v \text{ in } L^p(\Omega) \;\Rightarrow\; F_{p_j}(v_j) \longrightarrow F_p(v) \text{ in } [0,\infty].$$

The proof of the Γ-convergence is completed by observing that for each $u \in L^p(\Omega)$,

$$\lim_{j\to\infty} F_{p_j}(u) = \lim_{j\to\infty} \int\limits_{\Omega} \frac{|\nabla u|^{p_j}}{p_j}\,dx + \langle f,u\rangle = \int\limits_{\Omega} \frac{|\nabla u|^{p}}{p}\,dx + \langle f,u\rangle = F_p(u)$$

if $u \in W_0^{1,p}(\Omega)$ or else $\lim_{j\to\infty} F_{p_j}(u) = \infty = F_p(u)$. The proof of (*ii*) follows along similar lines and will be omitted. □

Theorems 4.3.6 and 4.3.8 follow from Theorem 4.4.1 via the general theory of Γ-convergence; see [21, Corollary 7.20].

For constant exponent p, Theorem 4.3.6 and 4.3.8 were treated in [71].

4.5 The eigenvalue problem for the p-Laplacian

We devote this section to the discussion of an application of the mathematical machinery hitherto developed, to the variable-exponent form of the classical eigenvalue problem for the p-Laplacian.

Fix a bounded domain $\Omega \subset \mathbb{R}^n$ and a function $p \in C(\overline{\Omega})$ with $1 < p_- \le p_+ < \infty$ on Ω. Theorem 3.3.1 guarantees the reflexivity of the Sobolev space $W_0^{1,p}(\Omega)$, whereas Theorem 3.4.4 asserts that the Sobolev embedding

$$E : W_0^{1,p}(\Omega) \hookrightarrow L^p(\Omega) \tag{4.55}$$

is in fact, compact. Theorem 2.8.8 gives an expression for the Fréchet derivative of the norm $\|\cdot\|_p$. Elementary considerations show that the action of the Fréchet derivative of the norm

$$W_0^{1,p}(\Omega) \ni u \to \|u\|_p^{(1)} = \||\nabla u|\|_p$$

on $h \in W_0^{1,p}(\Omega)$ is given by the following:

$$h \longrightarrow \int\limits_{\Omega} \frac{p\,\|\nabla u\|_p^{-p}\,|\nabla u|^{p-1}\,\mathrm{sgn}\,\nabla u \nabla h}{\int\limits_{\Omega} p\,\|\nabla u\|_p^{-p-1}\,|\nabla u|^{p}\,dx}\,dz. \tag{4.56}$$

Remarkably, when p is constant on Ω, the operator given by (4.56) takes up the form

$$-\||\nabla u|\|_p^{1-p}\mathrm{div}\left(|\nabla u|^{p-2}\nabla u\right) = -\||\nabla u|\|_p^{1-p}\Delta_p(u),$$

where $\Delta_p(u) := \text{div}\left(|\nabla u|^{p-2}\nabla u\right)$ is the much-studied p-Laplacian operator, which emerged in the treatment of some applications to fluid mechanics (see[10] for a recent survey on the history of the p-Laplacian.)

Now, in the light of Theorem 4.1.5, any extremal function u_0 for the Sobolev embedding (4.55) is a solution to the eigenvalue problem

$$(\text{grad}\,\|u\|_p) = \lambda(\text{grad}\,\|u\|_p^{(1)}), \tag{4.57}$$

with $\lambda = \|u_0\|_p/\||\nabla u_0|\|_p$.

As it is clear from Theorem 4.1.5, if $(\lambda, u) \in (0,\infty)\times W_0^{1,p}(\Omega)$ is a solution of Problem (4.57), then necessarily $\lambda \le \|E\|$.

Problem (4.57) becomes

$$-\lambda\||\nabla u|\|_p^{1-p}\text{div}\left(|\nabla u|^{p-1}\nabla u\right) = \|u\|_p^{1-p}|u|^{p-2}u,$$

which can be equivalently written as

$$-\text{div}\left(|\nabla u|^{p-1}\nabla u\right) = \lambda^{-1}\left(\||\nabla u|\|_p/\|u\|\right)_p^{p-1}|u|^{p-2}u. \tag{4.58}$$

Any $\lambda \in \mathbb{R}$ satisfying equality (4.58) for some $u \in W_0^{1,p}(\Omega)$ satisfies the inequality

$$\lambda^{-1} = \||\nabla u\|_p/\|u\|_p \ge \|E\|^{-1};$$

moreover, if $(\lambda, u) \in (0,\infty)\times W_0^{1,p}(\Omega)$ solves Problem (4.58), then $(\lambda^{-1}\left(\||\nabla u|\|_p/\|u\|\right)_p^{p-1}, u)$ is a solution for the eigenvalue problem for the p-Laplacian, namely

$$-\Delta_p u = \gamma|u|^{p-2}u \tag{4.59}$$

and conversely if (γ, v) is any solution of (4.59), then

$$(\gamma^{-1}\left(\|v\|_p/\||\nabla v|\|_p\right)^{p-1}, v)$$

is a solution of (4.58). It is apparent from the preceding argument that under the assumption of constant p, the smallest eigenvalue of the p-Laplacian is $\|E\|^p$, the p-th power of the norm of the Sobolev embedding; also its corresponding first eigenfunction (which was shown in [4] to be unique up to multiplication times constants, see also [58] for a simpler proof) is extremal for the Sobolev embedding E. The situation changes radically when p is a function on Ω. In this case, the tempting natural generalization that results from replacing p with a function in problem (4.59), while an eigenvalue problem, is not related to problem (4.57) in any useful way, and both problems, (4.57) and (4.59) must be studied separately (see [66] and the references therein).

It should be observed that this phenomenon is due to the fact that if p is constant, then any extremal function for the Rayleigh quotient $\||\nabla u|\|_p/\|u\|_p$ is also extremal for the quotient of the modulars

$$\int_\Omega |\nabla u|^p dx \Big/ \int_\Omega |u|^p dx,$$

whereas this is not the case when p is a nonconstant function on Ω.
Our next order of business is to address the question of stability of Problem (4.57) under perturbations of the variable exponent p. More specifically, let $\Omega \subset \mathbb{R}^n$ be a bounded domain with a Lipschitz boundary and $(p_j) \subset C(\overline{\Omega})$ be a nondecreasing sequence; let

$$q := \lim_{j\to\infty} p_j = \sup_j p_j$$

and suppose that p_j converges to q uniformly in Ω.
Assume further that $n < q_- = \inf_\Omega q \leq \sup_\Omega q = q_+ < \infty$. In this setting, for each $j \in \mathbb{N}$, Theorem 4.1.5 yields the existence of an extremal function u_j for the Sobolev embedding $E : W_0^{1,p_j}(\Omega) \hookrightarrow L^{p_j}(\Omega)$. More precisely, u_j minimizes the Rayleigh quotient, i.e.,

$$\||\nabla u_j|\|_{p_j} / \|u_j\|_{p_j} = \inf_{0\neq u\in W_0^{1,p_j}(\Omega)} \||\nabla u|\|_{p_j} / \|u\|_{p_j} = \lambda_j.$$

Because of the homogeneity inherent to the above quotient, it is clear that no generality is lost by assuming that $\||\nabla u_j|\|_{p_j} = 1$. Recall that $\mathcal{P}(\Omega)$ denotes the space of all real-valued, Borel-measurable functions p on Ω with $1 \leq p \leq \infty$.

Theorem 4.5.1. *In the notation of the preceding paragraph, there exist a subsequence of (p_j), still denoted by (p_j) and a function $u_q \in W_0^{1,q}(\Omega)$ with $\||\nabla u_q|\|_q = 1$ and*

$$1/\|u_q\|_q = \inf_{0\neq u\in W_0^{1,q}(\Omega)} \||\nabla u|\|_q / \|u\|_q = \lambda_q \tag{4.60}$$

such that $u_j \rightharpoonup u_q$ in $W_0^{1,p_1}(\Omega)$. Moreover,

$$\lambda_j = \|u_j\|^{-1} \longrightarrow \lambda_q \text{ as } j \to \infty.$$

Proof. By assumptions, if $j \leq i$, Lemma 2.2.6 yields the inequality

$$\||\nabla u_i|\|_{p_j} \leq (\|p_i - p_j\|_\infty^{-\|p_i - p_j\|_\infty} + |\Omega|\|p_i - p_j\|_\infty).$$

From this inequality it is clearly seen that, for any $j \in \mathbb{N}$, the sequence $(u_i)_{i\geq j}$ is bounded in $W_0^{1,p_j}(\Omega)$; by virtue of the compactness of the Sobolev embedding there is no loss of generality in assuming that $(u_i)_{i\geq j}$ is weakly convergent in $W_0^{1,p_j}(\Omega)$, say

$$u_i \overset{W_0^{1,p_j}(\Omega)}{\rightharpoonup} u_j \in W_0^{1,p_j}(\Omega) \text{ as } j \leq i \to \infty.$$

A straightforward argument shows that

$$u_i \overset{W_0^{1,p_j}(\Omega)}{\rightharpoonup} u \in \bigcap_{j=1}^{\infty} W_0^{1,p_j}(\Omega) \text{ as } j \leq i \to \infty.$$

We claim that $u \in W_0^{1,q}(\Omega)$ and that $|||\nabla u|||_q \leq 1$. To see this we recall that the functional

$$W_0^{1,q}(\Omega) \ni u \longrightarrow F(u) = \int_\Omega |\nabla u|^q dx$$

is weakly-lower semicontinuous (see [19]). This observation together with Fatou's lemma and Lemma 2.2.6 yields:

$$\int_\Omega |\nabla u|^q dx \leq \liminf_k \int_\Omega |\nabla u|^{p_k} dx \leq \liminf_k \liminf_j \int_\Omega |\nabla u_j|^{p_k} dx$$
$$\leq \liminf_k \liminf_j (\|p_j - p_k\|_\infty^{-\|p_j - p_k\|_\infty} + |\Omega| \|p_j - p_k\|_\infty).$$

This proves the above claim. From now on we write $u = u_q$. To prove (4.60) we proceed as follows: Let $v \in W_0^{1,q}(\Omega)$. By definition, there exists a sequence $(v_k) \in C_0^\infty(\Omega)$ such that

$$v_j \overset{W_0^{1,q}(\Omega)}{\longrightarrow} v,$$

i.e.,

$$v_j \overset{L^q(\Omega)}{\longrightarrow} v \text{ and } \nabla v_j \overset{L^q(\Omega)}{\longrightarrow} \nabla v$$

By virtue of Lebesgue Dominated Convergence theorem, for each $k \in \mathbb{N}$ one has

$$\int_\Omega \left(\frac{|v_k|}{\|v_k\|_q} \right)^{p_i} dx \to \int_\Omega \left(\frac{|v_k|}{\|v_k\|_q} \right)^q dx = 1 \tag{4.61}$$

and

$$\int_\Omega \left(\frac{|\nabla v_k|}{|||\nabla v_k|||_q} \right)^{p_i} dx \to \int_\Omega \left(\frac{|\nabla v_k|}{|||\nabla v_k|||_q} \right)^q dx = 1 \tag{4.62}$$

We claim that from the two preceding statements it can be inferred that

$$\|u_k\|_{p_i} \longrightarrow \|u_k\|_q \text{ as } i \to \infty \tag{4.63}$$

and that

$$|||\nabla u_k|||_{p_i} \longrightarrow |||\nabla u_k|||_q \text{ as } i \to \infty \tag{4.64}$$

To see this, observe that for any $\epsilon > 0$, (4.61) and (4.62) imply that for i sufficiently large it holds:

$$1 - \epsilon < \int_\Omega \left(\frac{|v_k|}{\|v_k\|_q} \right)^{p_i} dx < 1 + \epsilon$$

and

$$1 - \epsilon < \int_\Omega \left(\frac{|\nabla v_k|}{|||\nabla v_k|||_q} \right)^{p_i} dx < 1 + \epsilon.$$

It is apparent from the last two statements and from the fact that $p_i > 1$ and $(1+\epsilon)) > 1$, then, that for i sufficiently large one has:

$$\int_\Omega \left(\frac{|v_k|}{(1+\epsilon)\|v_k\|_q}\right)^{p_i} dx \leq (1+\epsilon)^{-1} \int_\Omega \left(\frac{|v_k|}{\|v_k\|_q}\right)^{p_i} dx \leq 1. \tag{4.65}$$

Analogously, (4.49) shows that for i sufficiently large it holds:

$$1 < (1-\epsilon)^{-1} \int_\Omega \left(\frac{|v_k|}{\|v_k\|_q}\right)^{p_i} dx \leq \int_\Omega \left(\frac{|v_k|}{(1-\epsilon)\|v_k\|_q}\right)^{p_i} dx. \tag{4.66}$$

It follows immediately by definition of the Luxemburg norm that (4.65) implies that for large enough i :

$$\|v_k\|_{p_i} \leq (1+\epsilon)\|v_k\|_q.$$

Since the function $\lambda \to \int_\Omega (\lambda^{-1}|v_k|)^{p_i}$ is non-increasing, as an immediate consequence of the definition of the Luxemburg norm and inequality (4.66) one derives, for i as above:

$$(1-\epsilon)\|v_k\|_q \leq \|v_k\|_{p_i}.$$

The last two statements show (4.63) and the exact same reasoning yields (4.64).
On account of Proposition 1.5.7, the sequence $(u_i) \subset L^{p_1}(\Omega)$ can be considered to be convergent to u *a.e.* in Ω. By assumption, $q_- > n$; it follows that for i large enough one has

$$W_0^{1,p_i}(\Omega) \hookrightarrow C(\overline{\Omega}).$$

Then there is a positive constant C, independent of i such that $\|u_i\|_\infty \leq C$, whence it follows that for any $\lambda > 0$

$$\int_\Omega |\lambda u_i|^{p_i} dx \to \int_\Omega |\lambda u|^q dx \text{ as } i \to \infty.$$

As in the proof of (4.63) and (4.64), the preceding statement leads to $\|u_i\|_{p_i} \to \|u\|_q$ as $i \to \infty$. Consequently,

$$\lambda_i = \|u_i\|_{p_i}^{-1} \to \|u\|_q^{-1} \text{ as } i \to \infty.$$

Next, (4.63), (4.64) and the extremal character of u_i yield, for any $\epsilon > 0$:

$$\|v_k\|_q / \||\nabla v_k|\|_q \leq \|v_k\|_{p_i} / \||\nabla v_k|\|_{p_i} + \epsilon \leq \|u_i\|_{p_i} + \epsilon$$

i sufficiently large and it follows from the last inequalities by letting $i \to \infty$ that

$$\|v_k\|_q / \||\nabla v_k|\|_q \leq \|u\|_q. \tag{4.67}$$

Letting $\kappa \to \infty$ in inequality (4.67) it is concluded that

$$\|v\|_q / \||\nabla v|\|_q \leq \|u\|_q. \tag{4.68}$$

In particular, for $v = u$, (4.68) shows that $\||\nabla u|\|_q \geq 1$ and since $\||\nabla u|\|_q \geq 1$ one has that, in fact, $\||\nabla u|\|_q = 1$. The claim of the theorem follows by setting $u = u_q$. □

An analogous result follows for a non-increasing sequence. Specifically, let $\Omega \subset \mathbb{R}^n$ be a regular bounded domain and $(p_i) \subset C(\overline{\Omega})$ be a non-increasing sequence of functions converging uniformly in Ω to its infimum $p \in C(\overline{\Omega})$; assume that

$$1 < p_- \leq p_+ < \infty.$$

As in the setting of Theorem 4.5.1, for each $j \in \mathbb{N}$ let $u_j \in W_0^{1,p_j}(\Omega)$ stand for an extremal function for the Sobolev embedding

$$W_0^{1,p_j}(\Omega) \hookrightarrow L^{p_j}(\Omega);$$

we normalize by requiring that $\||\nabla u_j|\|_{p_j} = 1$ and notice that according to Theorem 4.1.5, u_j is a solution of the eigenvalue problem (4.57) with $p = p_j$ with first eigenvalue $\lambda = \lambda_j$. Let λ_p be the first eigenvalue of the problem (4.57) for $p = \inf\limits_j p_j$, i.e.,

$$\lambda_p = \inf_{0 \neq u \in W_0^{1,p}(\Omega)} \||\nabla u|\|_p / \|u\|_p.$$

Theorem 4.5.2. *In the above setting, there exists an extremal function u_p for the embedding*

$$W_0^{1,p}(\Omega) \hookrightarrow L^p(\Omega)$$

and a subsequence of (p_j) (still denoted by (p_j)) such that

$$u_j \overset{W_0^{1,p}(\Omega)}{\rightharpoonup} u_p \text{ and } u_j \overset{L^p(\Omega)}{\to} u_p.$$

Moreover, $\lambda_j \to \lambda_p$ as $j \to \infty$.

Proof. The proof will be split into three mutually exclusive cases, namely, $n \leq p_-$, $n \in (p_-, p_+]$ and $p_+ < n$.
We observe first that for $p_+ < n$ the inequality

$$p(x) + \frac{1}{2}\frac{p_-^2}{n - p_-} < p(x) + \frac{p(x)^2}{n - p(x)} = \frac{np(x)}{n - p(x)}$$

holds for every $x \in \Omega$. The embedding

$$W_0^{1,p}(\Omega) \hookrightarrow L^{p+\frac{1}{2}\frac{p_-^2}{n-p_-}}(\Omega)$$

is compact (see [60, Theorem 3.9]). Hence no generality is lost by considering the sequence (u_i) strongly convergent in $L^{p+\frac{1}{2}\frac{p_-^2}{n-p_-}}(\Omega)$ to some $u \in B_{W_0^{1,p}(\Omega)}$ (the unit ball in $W_0^{1,p}(\Omega)$) and that $u_i \overset{W_0^{1,p}(\Omega)}{\rightharpoonup} u$.
By assumption, there exists i_0 such that $i \geq i_0$ implies

$$p_i - p < \frac{1}{2}\frac{p_-^2}{n-p_-}$$

uniformly in Ω. Because of the continuous inclusion (valid for $i \geq i_0$)

$$L^{p+\frac{1}{2}\frac{p_-^2}{n-p_-}}(\Omega) \hookrightarrow L^{p_i}(\Omega),$$

it is easy to see that $u \in L^{p_i}(\Omega)$ if $i \geq i_0$ and that

$$\|u_i\|_{p_i} \leq \|u_i - u\|_{p_i} + \|u\|_{p_i}. \tag{4.69}$$

We next show that in fact u is extremal for the Sobolev embedding $W_0^{1,p}(\Omega) \to L^p(\Omega)$. To this end, fix $v \in W_0^{1,p}(\Omega)$; assume $\||\nabla v|\|_p = 1$ and select a sequence $(v_k) \subset C_0^\infty(\Omega)$ such that $v_k \overset{L^p}{\to} v$ and $|\nabla v_k| \overset{L^p}{\to} |\nabla v|$ as $k \to \infty$. As in the proof of Theorem 4.5.1 it follows that for fixed k, $\|v_k\|_{p_i} \to \|v_k\|_p$ and that $\||\nabla v_k|\|_{p_i} \to \||\nabla v_k|\|_p$ as $i \to \infty$. Then, given $\epsilon > 0$, one has, for sufficiently large $k, i \in \mathbb{N}$, by virtue of the extremal character of u_i:

$$\begin{aligned}\frac{\||\nabla v|\|_p}{\|v\|_p} &\geq \frac{\||\nabla v_k|\|_p}{\|v_k\|_p} - \epsilon/2 \geq \frac{\||\nabla v_k|\|_{p_i}}{\|v_k\|_{p_i}} - \epsilon \\ &\geq \frac{\||\nabla u_i|\|_{p_i}}{\|u_i\|_{p_i}} - \epsilon = \frac{1}{\|u_i\|_{p_i}} - \epsilon.\end{aligned}$$

This inequality together with (4.69) yields for i large enough:

$$\frac{\||\nabla v|\|_p}{\|v\|_p} \geq \frac{1}{\|u_i - u\|_{p_i} + \|u\|_{p_i}} + \epsilon$$

Letting $i \to \infty$ it is concluded

$$\frac{1}{\|u\|_p} = \inf_{W_0^{1,p}(\Omega) \ni v \neq 0} \frac{\||\nabla v|\|_p}{\|v\|_p}.$$

In particular, by setting $v = u$ in the above inequality it is obtained that $\||\nabla u|\|_p \geq 1$ i.e. $\||\nabla u|\|_p = 1$. The claimed extremality of u has been proved. This settles the case $p_+ < n$.

Suppose next that $n \in (p_-, p_+]$. In the light of Theorem 3.4.4, one can infer the existence of a function $u \in W_0^{1,p}(\Omega)$ and of a subsequence (still denoted by (u_i)) of the original sequence such that $u_i \overset{W_0^{1,p}(\Omega)}{\rightharpoonup} u$ and that $u_i \overset{L^p(\Omega)}{\to} u$. The proof relies on the following lemma:

Lemma 4.5.3. *In the above setting, $u \in L^{i_0}(\Omega)$ for some $i_0 \in \mathbb{N}$.*

Based on this result (which for clarity we will prove later) we conclude that for $i \geq i_0$ one has

$$\|u_i\|_{p_i} \leq \|u_i - u\|_{p_i} + \|u\|_{p_i}$$

and

$$\|u\|_{p_i} \leq \|u_i - u\|_{p_i} + \|u_i\|_{p_i}$$

Since

$$\|u_i - u\|_{p_i} \to 0$$

and

$$\|u\|_{p_i} \to \|u\|_p$$

as $i \to \infty$ it follows that

$$\|u_i\|_{p_i} \to \|u\|_p$$

as $i \to \infty$. The same argument used in the case $p_+ < n$ yields now the extremal character of u.

We now proceed to the Proof of Lemma 4.5.3. To this effect we choose positive numbers ϵ, δ subject to the conditions:

$$\epsilon < \min\left\{\frac{1}{n-1}, \frac{p_-^2}{n-p_-}\right\}, \quad p_- - r > 1 \text{ and } n + r < \frac{n(n-r)}{r}. \tag{4.70}$$

Furthermore, define $p_1 = p_- - \delta$, and inductively, if $p_k < n$ define

$$p_k = \frac{np_{k-1}}{n - p_{k-1}} - \frac{\epsilon}{5}.$$

Clearly, if $p_{j-1} < n$, $p_j > p_1 + (j-1)\frac{4}{5}\epsilon$ and $p_{j-1} < p_j$. Let J be the first subindex for which $p_J > n - \frac{r}{2}$, where r is as in (4.70). Let $I = [w_-, p_+]$ and

$$\Omega_1 = p^{-1}\left(\left(\frac{w_- + 1}{2}, \frac{np_1}{n-p_1} - \frac{\epsilon}{10}\right)\right)$$

and for $1 < k \leq J - 1$ set

$$\Omega_k = p^{-1}\left(\left(p_k, \frac{np_k}{n-p_k} - \frac{\epsilon}{10}\right) \cap I\right);$$

$$\Omega_J = p^{-1}\left((n-r, n+r)\right)$$
$$\Omega_{J+1} = p^{-1}\left(\left(n + \frac{r}{2}, \infty\right) x\right).$$

It can be readily verified that $(\Omega_k)_k$ is an open cover of Ω; let $(\chi_k)_{1 \leq k \leq J+2}$

be a partition of unity subordinated to the cover $(\Omega_k)_k$.

For $1 \leq k \leq J-1$ the sequence $(u_i\chi_k)_i \in W_0^{1,p}(\Omega) \hookrightarrow W_0^{1,p_k}(\Omega)$ can without loss of generality be considered to be convergent in $L^{\frac{np_k}{n-p_k}-\frac{\epsilon}{20}}(\Omega)$; this is due to the compactness of the embedding

$$W_0^{1,p_k}(\Omega) \hookrightarrow\hookrightarrow L^{\frac{np_k}{n-p_k}-\frac{\epsilon}{20}}(\Omega).$$

Thus,

$$u_i\chi_k \stackrel{L^{\frac{np_k}{n-p_k}-\frac{\epsilon}{20}}(\Omega)}{\longrightarrow} v_k \in L^{\frac{np_k}{n-p_k}-\frac{\epsilon}{20}}(\Omega). \tag{4.71}$$

It is easy to see that from $u_i \stackrel{L^p(\Omega)}{\longrightarrow} u$ it follows that $u_i\chi_k \stackrel{L^p(\Omega)}{\longrightarrow} u\chi_k$. Moreover,

$$\begin{aligned}\int_\Omega |v_k|^p dx &= \left(\int_{\{|v_k|<1\}} + \int_{\{|v_k|\geq 1\}}\right)|v_k\chi_k|^p dx \\ &\leq \left(\int_{\{|v_k|<1\}} + \int_{\{|v_k|\geq 1\}}\right)|v_k|^{\frac{np_k}{n-p_k}-\frac{\epsilon}{20}} dx < \infty;\end{aligned}$$

it follows that $v_k \in L^p(\Omega)$.
Next, observe that (4.71) implies

$$\|u\chi_k - v_k\|_p \leq \|u\chi_k - u_i\chi_k\|_p + \|u_i\chi_k - v_k\|_p \to 0 \text{ as } i \to \infty. \tag{4.72}$$

Hence $v_k = u\chi_k \in L^{\frac{np_k}{n-p_k}-\frac{\epsilon}{20}}(\Omega)$. By assumption, there exists $I_k \in \mathbb{N}$ such that $p_{I_k} < \frac{np_k}{n-p_k} - \frac{\epsilon}{20}$ and for such I_k, $u\chi_k \in L^{p_{I_k}}(\Omega)$.

Consider now the sequence $(u_i\chi_J) \subset W_0^{1,p}(\Omega) \subset W_0^{1,n-r}(\Omega)$. The Sobolev embedding

$$W_0^{1,n-r}(\Omega) \hookrightarrow\hookrightarrow L^{\frac{n(n-r)}{2r}}(\Omega)$$

is compact, hence no generality is lost by assuming that

$$u_i\chi_J \stackrel{L^{\frac{n(n-r)}{2r}}(\Omega)}{\longrightarrow} v_J \in L^{\frac{n(n-r)}{2r}}(\Omega).$$

Because of the condition $n + r < \frac{n(n-r)}{2r}$ one has, as in the preceding case, $v_J \in L^p(\Omega)$ and (4.72) holds with $k = J$, which in turn implies that $u\chi_J = v_J \in L^{\frac{n(n-r)}{2r}}(\Omega)$. Choose now I_J so that

$$p_{I_J} < \frac{n(n-r)}{2r},$$

such an I' exists because of the assumptions on the sequence (p_i). It follows then that

$$u\chi_J = v_J \in L^{p_{I_J}}(\Omega).$$

Finally, $(u_i\chi_{J+1}) \subset W_0^{1,p}(\Omega) \subset W_0^{1,n+\frac{r}{2}}(\Omega)$. In this case, the embedding

$$W_0^{1,n+\frac{r}{2}}(\Omega) \hookrightarrow\hookrightarrow C(\overline{\Omega})$$

is compact. The sequence $(u_i\chi_{J+1})$ can thus be considered to be convergent in $C(\overline{\Omega})$ to $v_{J+1} \in C(\overline{\Omega})$. As before, it follows straightforwardly that $u\chi_{J+1} = v_{J+1}$, i.e., $u\chi_{J+1} \in L^{p_j}(\Omega)$ for any $j \in \mathbb{N}$.
Now writing $p^* = \min\{p_{I_1}, ..., p_{I_J}\}$ it is clear that

$$u = \sum_{1}^{J+1} u\chi_j \in L^{p^*}(\Omega).$$

The last statement concludes the proof of the lemma. The remaining case, $n < p_+$, follows along the same lines by using the compactness of the embedding

$$W_0^{1,p}(\Omega) \hookrightarrow C(\overline{\Omega}).$$

□

4.6 Modular eigenvalues

We reopen now the discussion initiated in the course of the paragraph preceding Theorem 4.5.1. To this effect, we retain the notation introduced in the preceding section. It was observed there that the minimization of the Rayleigh quotient

$$\frac{\||\nabla u|\|_p}{\|u\|_p}$$

naturally conduces to the first eigenvalue of the problem

$$(\text{grad}\, \||\nabla u|\|_p) = \lambda(\text{grad}\, \|u\|_p) \tag{4.73}$$

and that the latter is essentially the well-known eigenvalue problem for the p-Laplacian when p is considered to be constant in Ω.
Thus, (4.73) is in this sense, the generalization of the eigenvalue problem for the p-Laplacian (p constant):

$$-\text{div}\left(|\nabla u|^{p-2}\nabla u\right) = \lambda|u|^{p-2}u. \tag{4.74}$$

There is, however, another, obvious approach to the generalization of (4.74): namely, simply replace the constant p there with an admissible function $p(x)$ on Ω. The resulting eigenvalue problem is fundamentally different from (4.73). In fact, the main obstacle encountered in (4.74) is the lack of homogeneity;

by this we mean that if u is a solution of (4.74) and $\gamma > 0$ is a constant, it does not follow that γu is a solution of (4.74). As it will be shown in Lemma 4.6.1, Problem (4.74) is an equality involving the derivative of the modular $\rho_\varphi(\cdot)$. This fact should be contrasted with problem (4.73), which is an equality involving the derivative of the norm $\|\cdot\|_\varphi$. The contrast between the two problems is dramatic. In fact, whereas under certain conditions Poincaré inequality holds in the norm-form (see Theorem 3.4.4)

$$\|u\|_\varphi \leq C\||\nabla u|\|_\varphi,$$

its validity in the modular form

$$\rho_\varphi \leq C\rho_\varphi(|\nabla u|)$$

is an exception: Such a constant does not exist if p has a local extremum in Ω. We refer the reader to [39, Theorem 3.1] for the proof of this assertion. In this section we explore a specific direction in which the theoretical machinery hitherto developed can be applied to a wide class of problems. The results are inspired by the approach utilized in [67] for the study of the first eigenvalue of variable exponent p-Laplacian.

Let

$$\varphi : \Omega \times [0, \infty) \longrightarrow [0, \infty)$$

be an MO function on be a bounded domain $\Omega \subset \mathbb{R}^n$ and denote its conjugate function by φ^* (Definition 2.24). Aiming at a full description of the Fréchet derivative of the associated modular, a further assumption is imposed unto the MO function φ at this point, namely, it is from now on required that

$$\lim_{t\to 0^+} \frac{\varphi(x,t)}{t} = 0 \quad a.e. \ x \in \Omega. \tag{4.75}$$

Definition 4.6.1. *An MO function satisfying the condition (4.75) is said to be an N-function.*

Lemma 4.6.1. *Assume that φ satisfies the conditions of Theorem 3.4.4. Then at any $u \in L^\varphi(\Omega)$ for which*

$$\frac{\partial}{\partial t}\varphi(x, |u(x)|) \in L^{\varphi^*}(\Omega), \tag{4.76}$$

the functional given by the modular

$$F : L^\varphi(\Omega) \longrightarrow \mathbb{R}$$

$$F(u) = \int_\Omega \varphi(x, |u(x)|)\, dx$$

is Gâteaux-differentiable and the action of its derivative on any $h \in L^\varphi(\Omega)$ is given by:

$$\langle F'(u), h\rangle = \int_\Omega \frac{\partial\varphi}{\partial s}(x, |u(x)|)\frac{u(x)}{|u(x)|}h(x)\, dx. \tag{4.77}$$

Here we slightly abused the notation to imply that $\frac{u(x)}{|u(x)|} = 0$ *for* $u(x) = 0$.

Proof. Observe that for $u(x) \neq 0$, one has

$$\begin{aligned}\frac{|u(x)+th(x)| - |u(x)|}{t} &= \tfrac{2u(x)h(x)+t^2(h(x))^2}{|u(x)+th(x)|+|u(x)|} \\ &\longrightarrow \tfrac{u(x)h(x)}{|u(x)|} \text{ as } t \longrightarrow 0. \end{aligned} \tag{4.78}$$

Therefore, for *a.e.* $x \in \Omega$ and some $\theta \in (|u(x)|, |u(x)+th(x)|)$,

$$\begin{aligned}\frac{\varphi(x,|u(x)+th(x)|) - \varphi(x,|u(x)|)}{t} &= \tfrac{|u(x)+th(x)|-|u(x)|}{t}\tfrac{\partial\varphi}{\partial t}(x,\theta) \\ &\longrightarrow \tfrac{u(x)h(x)}{|u(x)|}\tfrac{\partial\varphi}{\partial t}(x,|u(x)|) \text{ as } t \longrightarrow 0.\end{aligned}$$

From condition (4.76) and Theorem 2.4.1 it follows then that

$$\int_\Omega \left| h(x)\frac{\partial\varphi}{\partial t}(x,|u(x)|)\right| dx < \infty,$$

which in conjunction with (4.78) yields, for any $\in L^\varphi(\Omega)$:

$$\begin{aligned}\lim_{t\longrightarrow 0} &\int_{\{x:u(x)\neq 0\}} \frac{|u(x)+th(x)|-|u(x)|}{t}\frac{\partial\varphi}{\partial t}(x,\theta(x))dx \\ &= \int_{\{x:u(x)\neq 0\}} \frac{u(x)h(x)}{|u(x)|}\frac{\partial\varphi}{\partial t}(x,|u(x)|)dx.\end{aligned}$$

On the other hand, one has by assumption, one has a.e. $x \in \Omega$:

$$\frac{\varphi(x,|th(x)|)}{t} \longrightarrow 0 \text{ as } t \longrightarrow 0.$$

Convexity yields, for $|t| \leq 1$ and *a.e.* $x \in \Omega$:

$$\frac{\varphi(x,|th(x)|)}{t} \leq \varphi(x,|h(x)|)$$

and since $h \in L^\varphi(\Omega)$, one has $\int_\Omega \varphi(x,|h(x)|)\,dx < \infty$ from which it is easily derived by way of Lebesgue's theorem that

$$\begin{aligned}&\int_{\{x:u(x)=0\}} \left|\tfrac{\varphi(x,|u(x)+th(x)|)-\varphi(x,|u(x)|)}{t}\right| dx \\ = &\int_{\{x:u(x)=0\}} \varphi(x,t|h(x)|)/t\,dx \to 0 \text{ as } t \to 0.\end{aligned}$$

In all,

$$\frac{F(u+th)-F(u)}{t} = \left(\int\limits_{\{u\neq 0\}} + \int\limits_{\{u=0\}}\right) \frac{\varphi(x,|u(x)+th(x)|)-\varphi(x,|u(x)|)}{t}dx$$
$$\to \int\limits_{\{u\neq 0\}} \frac{u(x)h(x)}{|u(x)|}\frac{\partial\varphi}{\partial t}(x,|u(x)|)dx \text{ as } t \longrightarrow 0.$$

This settles the proof of the lemma. □

Remark 4.6.2. It follows directly from Lemma 4.6.1 in concert with the chain rule that the functional

$$G : W_0^{1,\varphi}(\Omega) \longrightarrow \mathbb{R}$$
$$G(u) = \int\limits_{\Omega} \varphi(x,|\nabla u(x)|)\, dx$$

is differentiable at any $u \in W_0^{1,\varphi}(\Omega)$ for which

$$\frac{\partial\varphi}{\partial s}(x,|\nabla u(x)|) \in L^{\varphi^*}(\Omega)$$

and that its derivative at u is given by

$$\langle G'(u),v\rangle = \int\limits_{\Omega} \frac{\partial\varphi}{\partial s}(x,|\nabla u(x)|)\frac{\nabla u(x)}{|\nabla u(x)|}\nabla v(x)\, dx \tag{4.79}$$

for any $v \in W_0^{1,\varphi}(\Omega)$.

Example 4.6.1. In the particular case $\varphi(x,t) = \frac{t^{p(x)}}{p(x)}$, it emerges from Remark 4.6.2 that for $u, v \in W_0^{1,p}(\Omega)$

$$\langle G'(u),v\rangle = \int\limits_{\Omega} |\nabla u|^{p-2}\nabla u \nabla v\, dx = -\langle \Delta_p(u), v\rangle.$$

Lemma 4.6.3. *In addition to the preceding hypotheses, assume that the complementary function φ^* of φ satisfies the Δ_2 condition (see Definition 2.2.2) and is locally integrable. If for each $u \in L^{\varphi}(\Omega)$*

$$\frac{\partial}{\partial t}\varphi(\cdot,|u(\cdot)|) \in L^{\varphi^*}(\Omega) \tag{4.80}$$

and for each $u \in W_0^{1,\varphi}(\Omega)$

$$\frac{\partial}{\partial t}\varphi(\cdot,|\nabla u(\cdot)|) \in L^{\varphi^*}(\Omega), \tag{4.81}$$

then the functionals F and G are Fréchet differentiable and their corresponding Fréchet derivatives are given by (4.77) and (4.79) respectively.

Proof. First we recall the standard fact (see, for example [3, Theorem 1.9]) that the boundedness of the operators

$$\begin{aligned} \mathcal{L}_F \quad &: L^{\varphi}(\Omega) \longrightarrow (L^{\varphi}(\Omega))^* \\ &\mathcal{L}_f(u) = F'(u) \end{aligned} \tag{4.82}$$

$$\begin{aligned} \mathcal{L}_G \quad &: W_0^{1,\varphi}(\Omega) \longrightarrow \left(W_0^{1,\varphi}(\Omega)\right)^* \\ &\mathcal{L}(u) = G'(u) \end{aligned} \tag{4.83}$$

guarantees the Fréchet differentiability of F and G and the equality of the Fréchet and the Gâteaux derivatives. Hence, it suffices to prove that under the additional assumptions (4.80) and (4.81), the operators (4.82) and (4.83) are indeed bounded.

To this end, consider a convergent sequence $(u_j) \subset L^{\varphi}(\Omega)$, say $u_j \longrightarrow u$ in $L^{\varphi}(\Omega)$ as $j \longrightarrow \infty$: In the light of Corollary 2.2.3 there is no loss of generality by assuming that (u_j) converges to u almost everywhere in Ω. We start by writing

$$\begin{aligned} &\int_{\Omega} \left(\tfrac{u_j(x)}{|u_j(x)|} \tfrac{\partial \varphi}{\partial t}(x, |u_j(x)|) - \tfrac{u(x)}{|u(x)|} \tfrac{\partial \varphi}{\partial t}(x, |u(x)|)) \right) h(x)\, dx \\ &= \int_{\Omega} \tfrac{u_j(x)}{|u_j(x)|} \left(\tfrac{\partial \varphi}{\partial t}(x, |u_j(x)|) - \tfrac{\partial \varphi}{\partial t}(x, |u(x)|) \right) h(x)\, dx \\ &\quad + \int_{\Omega} \left(\tfrac{u_j(x)}{|u_j(x)|} - \tfrac{u(x)}{|u(x)|} \right) \tfrac{\partial \varphi}{\partial t}(x, |u(x)|) h(x)\, dx. \end{aligned}$$

On account of the assumption $\frac{\partial \varphi}{\partial t}(x, |u(x)|) \in L^{\varphi^*}(\Omega)$ together with Hölder's inequality, the absolute value of the first integral on the right-hand side in the preceding inequality is bounded by

$$\|h\|_{\varphi} \left\| \left(\frac{\partial \varphi}{\partial t}(\cdot, |u_j|) - \frac{\partial \varphi}{\partial t}(\cdot, |u|) \right) \right\|_{\varphi^*}. \tag{4.84}$$

Now, Theorem 2.7.2 guarantees that under assumption (4.80)

$$\left\| \left(\frac{\partial \varphi}{\partial t}(\cdot, |u_j|) - \frac{\partial \varphi}{\partial t}(\cdot, |u|) \right) \right\|_{\varphi^*} \to 0 \text{ as } j \to \infty. \tag{4.85}$$

It follows, then:

$$\int_{\Omega} \left(\frac{u_j(x)}{|u_j(x)|} \frac{\partial \varphi}{\partial t}(x, |u_j(x)|) - \frac{u(x)}{|u(x)|} \frac{\partial \varphi}{\partial t}(x, |u(x)|)) \right) h(x)\, dx \to 0 \text{ as } j \to \infty. \tag{4.86}$$

By the same token, the integrand in the second term is bounded pointwise *a.e.* in Ω by

$$2|h(x)| \left| \frac{\partial \varphi}{\partial t}(x, |u(x)|) \right|$$

The assumption $\frac{\partial\varphi}{\partial t}(x,|u(x)|) \in L^{\varphi^*}(\Omega)$ together with Hölder's inequality yields

$$\int_\Omega |h(x)| \left|\frac{\partial\varphi}{\partial t}(x,|u(x)|)\right| dx \le 2\|h\|_\varphi \left\|\frac{\partial\varphi}{\partial t}(x,|u(x)|)\right\|_{\varphi^*}; \tag{4.87}$$

a straightforward application of Lebesgue's dominated convergence theorem implies that the second term above tends to 0 as $j \to \infty$. This proves the continuity of the operator (4.82). The corresponding proof for (4.83) follows along the same lines and will be omitted. □

Corollary 4.6.4. *If $\Omega \subset \mathbb{R}^n$ is a domain and $p \in \mathcal{P}(\Omega)$ is an admissible exponent, then the functional*

$$W_0^{1,p}(\Omega) \ni u \longrightarrow G(u) = \int_\Omega |\nabla u|^p dx \in [0,\infty)$$

is Fréchet differentiable and its derivative is given by

$$G' = -\Delta_p.$$

Proof. Conditions (4.80) and (4.81) for the underlying MO function $\varphi(x,t) = t^{p(x)}/p(x)$ follow immediatly from Example 4.6.1 and Hölder's inequality (Theorem 2.4.2). □

As an immediate consequence of the preceding discussion we present an existence theorem for an eigenvalue problem involving an operator of Musielak-Orlicz type.

Theorem 4.6.5. *Let*

$$\varphi : \Omega \times [0,\infty) \longrightarrow [0,\infty)$$

be an MO function on a bounded domain $\Omega \subset \mathbb{R}^n$. Assume that φ satisfies the conditions of Theorem 3.4.4, that φ is an N-function and that

$$\frac{\partial\varphi}{\partial t}(x,t) > 0 \ \textit{for}\ t > 0 \ \textit{and a.e.}\ x \in \Omega. \tag{4.88}$$

Then there exists at least one (weak) solution (λ_0, u_0) to the problem

$$div\left(|\nabla u(\cdot)|^{-1}\nabla u \frac{\partial}{\partial t}\varphi(\cdot,|\nabla u(\cdot)|)\right) = \lambda |u(\cdot)|^{-1} u \frac{\partial}{\partial t}\varphi(\cdot,|u(\cdot)|).$$

To facilitate the proof of Theorem 4.6.5 we digress on a Lagrange-multiplier-type discussion. Let F and G be the functionals introduced in Lemma 4.6.1 and Remark 4.6.2. For any $r > 0$, we recall that Theorem 4.2.3 yields the existence of at least $u_r \in W_0^{1,\varphi}(\Omega)$ such that

$$F(u_r) = \rho_\varphi(u_r) = \mathbf{S}_r = \sup\{\rho_\varphi(u) : \rho_\varphi(|\nabla u|) = G(u) \le r\}.$$

Lemma 4.6.6. *In the notation of Theorem 4.2.3, Lemma 4.6.1 and Remark 4.6.2 it follows that*

$$\langle G'(u_r), u_r \rangle > 0. \tag{4.89}$$

Proof. Because of Remark 4.6.2, it is clear that

$$\langle G'(u_r), u_r \rangle = \int\limits_{\Omega} \frac{\partial \varphi}{\partial s}(x, |\nabla u_r(x)|)|\nabla u_r(x)| dx;$$

since $G(u_r) = \int\limits_{\Omega} \varphi(x, |\nabla u_r(x)|) dx = r > 0$ one must have $\varphi(x, |\nabla u_r(x)|) > 0$ *a.e.* $x \in \Omega$. Thus, $|\nabla u_r(x)| > 0$ *a.e.* $\in \Omega$. The claim now follows on account of assumption (4.88). □

Lemma 4.6.7. *Under the assumptions of Lemma 4.6.6, it follows that*

$$W_0^{1,\varphi}(\Omega) = \ker G'(u_r) \oplus \langle \{u_r\} \rangle. \tag{4.90}$$

Proof. The proof is elementary: Any $v \in W_0^{1,\varphi}(\Omega)$ can be written as

$$v = v - (\langle G'(u_r), v \rangle / \langle G'(u_r), u_r \rangle)\, u_r + (\langle G'(u_r), v \rangle / \langle G'(u_r), u_r \rangle)\, u_r$$

and a straightforward calculation shows this decomposition to be unique. □

Next, set

$$\begin{aligned} &\omega : \ker G'(u_r) \oplus \mathbb{R} \longrightarrow [0, \infty) \\ &\omega(h, t) = G\left((1 + t)u_r + h\right) - r. \end{aligned}$$

By definition, it is immediate that $w(0, 0) = 0$.
We claim that w is differentiable in both variables, that $\frac{\partial w}{\partial t}(0, 0) > 0$ and that $\frac{\partial w}{\partial h}(0, 0) = 0$. The differentiability of w is clear from the differentiability of G; from Remark 4.6.2 and from Lemma 4.6.6, it follows that

$$\frac{\partial w}{\partial t}(0, 0) = \langle G'(u_r), u_r \rangle > 0.$$

The last assertion follows by direct computation, namely, for any $h \in \ker G'$ (i.e., $\langle G'(u_r), h \rangle = 0$) one has:

$$\begin{aligned} \frac{w(h, 0) - w(0, 0)}{\|h\|_{1,\varphi}} &= w(h, 0)/\|h\|_{1,\varphi} = \frac{G(u_r + h) - G(u_r)}{\|h\|_{1,\varphi}} \\ &= \frac{G(u_r + h) - G(u_r) - \langle G'(u_r), h \rangle}{\|h\|_{1,\varphi}} \\ &\longrightarrow 0 \;\text{ as }\; \|h\|_{1,\varphi} \to 0. \end{aligned}$$

Lemma 4.6.8. *For w as above, the function*

$$\frac{\partial w}{\partial h} : \ker\, G'(u_r) \oplus \mathbb{R} \longrightarrow (\ker\, G'(u_r))^*$$

is continuous.

Proof. By definition, one has, for $\eta \in \ker\, G'(u_r)$:

$$\frac{\partial w}{\partial h}(h_0, t_0)(\eta) = G'(h_0 + (1+t_0)u_r)(\eta)$$
$$\int_\Omega \frac{\partial \varphi}{\partial s}(x, |\nabla u_0(x)|)\frac{\nabla u_0(x)\nabla \eta(x)}{|\nabla u_0(x)|}dx.$$

The continuity claim follows immediately from Theorem 2.7.2, and repeated applications of Hölder's inequality and Lebesgue's dominated convergence theorem, as in Lemma 4.6.3. □

In all, the implicit function theorem now applies to w. Accordingly, there exist a neighborhood of zero U in $W_0^{1,\varphi}(\Omega)$, $\epsilon > 0$ and a differentiable function

$$\psi : U \cap \ker G'(u_r) \longrightarrow (-\epsilon, \epsilon)$$

such that for any $(a, b) \in U \cap \ker G'(u_r) \times (-\epsilon, \epsilon)$ one has

$$w(a, b) = 0 \iff b = \psi(a).$$

Pick $h \in \ker G'$, let $\delta > 0$ be small enough so that $th \in U \cap \ker G'(u_r)$ for $|t| < \delta$ and set

$$\mathbf{b} : (-\delta, \delta) \longrightarrow W_0^{1,\varphi}(\Omega)$$
$$\mathbf{b}(t) = u_r + th + \psi(th).$$

Then $\mathbf{b}(0) = u_r$, $\mathbf{b}$ is differentiable at 0 (since so is ψ) and $\mathbf{b}'(0) = h$, since $\psi'(0) = 0$. Then the function

$$\xi : (-\delta, \delta) \longrightarrow [0.\infty)$$
$$\xi(t) = F(u_r + th + \psi(th))$$

is differentiable and attains a maximum at $t = 0$. It follows then that

$$\xi'(0) = F'(u_r)(h) = 0$$

and from here, by assumption on h, that

$$\ker G' \subseteq \ker F'.$$

Since both, $\ker G'$ and $\ker F'$ have codimension one and neither functional is identically zero, it is concluded that

$$\ker G'(u_r) \subseteq \ker F'(u_r),$$

and thus that there must exist a constant $\lambda \in \mathbb{R}$ such that

$$G'(u_r) = \lambda F'(u_r). \tag{4.91}$$

Proof of Theorem 4.6.5: The proof follows from Theorem 4.2.3, Lemma 4.6.1 and (4.91).

On account of Theorem 3.4.4, 4.6.1 and corollary 4.6.4, we have the following corollary:

Corollary 4.6.9. *Let $\Omega \subset \mathbb{R}^n$ be a bounded domain and $p \in C(\overline{\Omega})$. Then there exists at least a solution $(\lambda, u) \in (0, \infty) \times W_0^{1,p}(\Omega)$ of the modular eigenvalue problem*

$$-\Delta_p u = \lambda |u|^{p-2} u. \tag{4.92}$$

Proof. By virtue of Theorem 4.6.5, for each $r > 0$ there exists an eigenfunction u_r satisfying the exremality condition of Theorem 4.2.3 and a corresponding eigenvalue λ_r for which (4.92) holds. □

It is well-known that when the exponent p is constant there exists a unique positive eigenfunction $u_0 \in W_0^{1,p}(\Omega)$ corresponding to the first eigenvalue of the p-Laplacian (up to multiplication by constants). This deep result was proved first by Anane [4]. An elegant proof was given by Kawohl and Lindqvist [56]. Let $u_0 \in W_0^{1,p}(\Omega)$ be the extremal function resulting from the application of Lemma 4.1.5; that is,

$$\frac{|||\nabla u_0|||_p}{\|u_0\|_p} = \inf_{0 \neq u \in W_0^{1,p}(\Omega)} \frac{|||\nabla u|||_p}{\|u\|_p}.$$

For any $r > 0$, set $v_r = r^{\frac{1}{p}} \frac{u_0}{|||\nabla u_0|||_p}$. It follows then easily that for each $r > 0$ and any $u \in W_0^{1,p}(\Omega)$ with $|||\nabla u|||_p = r^{\frac{1}{p}}$ one has

$$\frac{r}{\|u\|^p} = \frac{|||\nabla u|||_p^p}{\|u\|^p} \geq \frac{|||\nabla u_0|||_p^p}{\|u_0\|_p^p} = \frac{r|||\nabla u_0|||_p^p}{r\|u_0\|_p^p} = \frac{r}{\|v_r\|_p^p}.$$

In other words, v_r is an extremal function as in Theorem 4.2.3. In all, the eigenvalues and eigenfunctions given by Corollary 4.6.9 coincide with the pth power of the bona-fide eigenvalues if p is constant on Ω.

4.7 Convergence properties of the eigenvalues and eigenfunctions

In this Section we undertake a stability analysis of the eigenvalues and eigenfunctions of the variable exponent p-Laplacian under perturbations of the variable exponent. Throughout this section $\Omega \subset \mathbb{R}^n$ will stand for a bounded domain and all variable exponents will be considered to be in $C(\overline{\Omega})$. As usual, it will be assumed that all variable exponents p appearing in this Section satisfy the inequalities

$$1 < p_- = \inf_{\overline{\Omega}} p \leq \sup_{\overline{\Omega}} p = p_+ < \infty.$$

Our presentation is based on our work [66].
We recall the following version of Lemma 2.2.5, which can be proved along the exact same lines [33, 34]:

Lemma 4.7.1. *If $f : \Omega \to \mathbb{R}$ is Borel-measurable, $\varepsilon > 0$ and $p, q \in C(\overline{\Omega})$ are subject to $p < q < p + \varepsilon$, then*

$$\int_\Omega |f|^p dx \leq \varepsilon|\Omega| + \varepsilon^{-\varepsilon} \int_\Omega |f|^q dx.$$

The following Lemma, of a technical nature, will be used in the sequel:

Lemma 4.7.2. *Let $(p_j) \subset C(\overline{\Omega})$ be a non-increasing, uniformly convergent sequence in Ω to its infimum p; assume $1 < p_- \leq p_+ < \infty$ and that for some constant $\theta > 0$ and $r \in C(\overline{\Omega})$*

$$p(x) + \theta < r(x) \ \ in\ \Omega.$$

If a sequence $(\psi_j) \subset L^r(\Omega)$ converges to ψ in $L^r(\Omega)$, then

$$\limsup_j \int_\Omega \frac{|\psi_j|^{p_j}}{p_j} \leq \int_\Omega \frac{|\psi|^p}{p}. \tag{4.93}$$

Proof. Fix the subindex $I \in \mathbb{N}$ so that $p_I < p + \theta < r$ uniformly in Ω; in consequence

$$\|\psi_j - \psi\|_{p_I} \longrightarrow 0 \text{ as } j \longrightarrow \infty.$$

In particular the convergence in the p_I-norm implies that the numerical sequence

$$\int_\Omega |\psi_j|^{p_I}$$

is bounded. Consequently

$$\int_\Omega |\psi_j|^{p_I} \longrightarrow \int_\Omega |\psi|^{p_I} \text{ as } j \longrightarrow \infty.$$

On account of Lemma 4.7.1, for any integer $j \geq I$ one has the estimate

$$\int_\Omega \frac{|\psi_j|^{p_j}}{p_j} \leq \|p_I - p_j\|_\infty^{-\|p_I - p_j\|_\infty} \int_\Omega (|\psi_j|^{p_I} - |\psi|^{p_I}) \frac{1}{p_j^{p_I/p_j}} + \|p_I - p_j\|_\infty^{-\|p_I - p_j\|_\infty} \int_\Omega \frac{|\psi|^{p_I}}{p_j^{p_I/p_j}} + \|p_I - p_j\|_\infty |\Omega|. \tag{4.94}$$

It is obvious that the right-hand side of (4.94) converges as $j \longrightarrow \infty$ to

$$\|p_I - p\|_\infty^{-\|p_I - p\|_\infty} \int_\Omega \frac{|\psi|^{p_I}}{p^{p_I/p}} + \|p_I - p\|_\infty |\Omega|. \tag{4.95}$$

Moreover, the inequality

$$|\psi|^{p_I} \leq |\psi|^{p_I} \chi_{\{x:|\psi(x)|<1\}} + |\psi|^r \chi_{\{x:|\psi|\geq 1\}}$$

easily implies via Lebesgue's theorem that the expression given in (4.95) tends to

$$\int_\Omega \frac{|\psi|^p}{p}$$

as $I \to \infty$. This proves the Lemma.

□

Let $\Omega \subset \mathbb{R}^n$ be a bounded domain. Consider a non-increasing sequence $(p_j) \subset C(\overline{\Omega})$ uniformly convergent in Ω to its infimum $p = \inf_j p_j$; assume $1 < p_- \leq p_+ < \infty$. Let $\lambda_{p,1}$ (resp. $\lambda_{p_j,1}$) be the first eigenvalue for the modular p-Laplacian (resp p_j-Laplacian) resulting from Corollary 4.6.9 with the particular choice of $r = 1$. Denote by u_j the maximal function associated to $\lambda_{p_j,1}$ via Theorem 4.6.5. To facilitate the flow of ideas we introduce the following convention: For each $r > 0$ let $V_{p,r}$ stands for the set of all maximal functions resulting from the application of Theorem 4.2.3.

Theorem 4.7.3. *Under the conditions of Lemma 4.7.2, one has the following conclusions:*

$$\lim_{j \longrightarrow \infty} \lambda_{p_j,1} \text{ exists} \tag{4.96}$$

and

$$(p_-/p_+)^2 \lim_{j \longrightarrow \infty} \lambda_{p_j,1} \leq \lambda_{p,1} \leq \lim_{j \longrightarrow \infty} \lambda_{p_j,1}. \tag{4.97}$$

Moreover, there exists $u \in V_{p,1}$ and a subsequence of (u_j) (still denoted by (u_j)) such that as $j \longrightarrow \infty$

$$u_j \rightharpoonup u \ \ in \ \ W_0^{1,p}(\Omega) \tag{4.98}$$

and

$$u_j \to u \ \ in \ \ L^p(\Omega). \tag{4.99}$$

Proof. By assumption, for each $j \in \mathbb{N}$, $u_j \in V_{p_j,1}$. In particular:

$$\int_\Omega \frac{|u_j|^{p_j}}{p_j} = \max_{\int_\Omega |\nabla v|^{p_j}/p_j \leq 1} \int_\Omega \frac{|v|^{p_j}}{p_j}. \tag{4.100}$$

According to Theorem 4.2.3 one must necessarily have:

$$\int_\Omega \frac{|\nabla u_j|^{p_j}}{p_j} = 1; \tag{4.101}$$

in particular, because of Lemma 4.7.1 each u_j is subject to the bound

$$\int_\Omega |\nabla u_j|^p \leq (p_{1+})\|p_j - p\|_\infty^{-\|p_j - p\|_\infty} + \|p_j - p\|_\infty |\Omega|,$$

where $p_{1+} = \sup_{\overline{\Omega}} p_1$.

Inequalities 1.5.4 imply that the sequence (u_j) is bounded in $W_0^{1,p}(\Omega)$, hence it can be assumed to be weakly convergent to $u \in W_0^{1,p}(\Omega)$; by virtue of the compactness of the Sobolev embedding (Theorem 3.4.4) there is no loss of generality if u_j is taken to be strongly convergent to u, in $L^p(\Omega)$. On account of Corollary 2.2.3, (u_j) can be considered to be pointwise almost everywhere convergent to u in Ω.

For arbitrary $\delta > 0$, pick N large enough to guarantee that $j \geq N$ imply

$$\|p_j - p\|_\infty^{-\|p_j - p\|_\infty} + \|p_j - p\|_\infty |\Omega| + < 1 + \delta$$

and that

$$\left| \frac{1}{p^{p_j/p}} - \frac{1}{p_j} \right| < \delta \ \text{ uniformly in } \Omega, \text{ for } j \geq N.$$

Then

$$\begin{aligned}
&\int_\Omega \frac{|\nabla u_j|^p}{p} \\
&\leq \|p_j - p\|_\infty^{-\|p_j - p\|_\infty} \left(\int_\Omega |\nabla u_j|^{p_j} \left(\frac{1}{p^{p_j/p}} - \frac{1}{p_j} \right) dx + \int_\Omega \frac{|\nabla u_j|^{p_j}}{p_j} dx \right) \\
&\qquad + \|p_j - p\|_\infty |\Omega| \\
&\leq (1 + \delta)(p_{1+}\delta + 1).
\end{aligned}$$

Thus, since $(u_j)_{j\geq N}$ converges weakly to u in $W_0^{1,p}(\Omega)$, the weak lower-semicontinuity of the functional

$$F : W_0^{1,p}(\Omega) \longrightarrow [0,\infty)$$
$$F(w) = \int_\Omega \frac{|\nabla w|^p}{p} dx$$

guarantees that

$$\int_\Omega \frac{|\nabla u|^p}{p} dx \leq (1+\delta)(p_{1+}\delta + 1)$$

and the arbitrariness of δ implies that

$$\int_\Omega \frac{|\nabla u|^p}{p} dx \leq 1$$

Next, it will be shown that u is maximal in the sense of Theorem 4.2.3.
Fix $\omega \in W_0^{1,p}(\Omega)$ such that

$$\int_\Omega \frac{|\nabla \omega|^p}{p} dx = 1.$$

pick a sequence of smooth functions $(\omega_k) \subset C_0^\infty(\Omega)$ converging to ω in $W_0^{1,p}(\Omega)$ and such that

$$\nabla\omega_k \to \nabla\omega \text{ and } \omega_k \to \omega \text{ a.e in } \Omega.$$

For arbitrary $\delta > 0$ let k be large enough to satisfy

$$1 - \delta/2 < \int_\Omega \frac{|\nabla\omega_k|^p}{p} dx < 1 + \delta/2.$$

Because of the particular choice of the sequence (p_j), one has:

$$\int_\Omega \frac{|\nabla\omega_k|^p}{p} dx \leq \|p_j - p\|_\infty^{-\|p_j-p\|_\infty} \int_\Omega \frac{|\nabla\omega_k|^{p_j}}{p_j} dx + \|p_j - p\|_\infty |\Omega|. \qquad (4.102)$$

Lebesgue's dominated convergence theorem yields, for each fixed k, that

$$\int_\Omega \frac{|\nabla\omega_k|^{p_j}}{p_j} dx \longrightarrow \int_\Omega \frac{|\nabla\omega_k|^p}{p} dx \text{ as } j \longrightarrow \infty.$$

Next, observe that

$$\int_\Omega \frac{|\nabla\omega_k|^p}{p} dx \longrightarrow \int_\Omega \frac{|\nabla\omega|^p}{p} dx = 1 \text{ as } k \longrightarrow \infty.$$

Thus, for k fixed, j can be chosen so large that

$$\int_\Omega \frac{|\nabla\omega_k|^{p_j}}{p_j}dx \leq 1+\delta.$$

The last inequality implies that for fixed $k \in \mathbb{N}$ there exists $J \in \mathbb{N}$ such that $j \geq J$ implies

$$\int_\Omega \frac{\left|\nabla\left(w_k/(1+\delta)^{1/p_{j-}}\right)\right|^{p_j}}{p_j}dx \leq 1,$$

whence the maximality of u_j gives for $j \geq J$:

$$\int_\Omega \frac{|w_k|^{p_j}}{(1+\delta)^{\frac{p_j}{p_{j-}}} p_j}dx \leq \int_\Omega \frac{|u_j|^{p_j}}{p_j}dx. \tag{4.103}$$

We split the remaining part of the proof the cases $p_+ < n$, $p_- \leq n \leq p_+$ and $p_- > n$.
For $p_+ < n$, $M \in \mathbb{N}$ can be chosen sufficiently large that

$$\|p_M - p\|_\infty < \frac{1}{n-1}.$$

This choice implies that, uniformly in Ω:

$$p_M \leq p + \frac{1}{n-1} < \frac{np}{n-p}.$$

Setting $r = \frac{np}{n-p} - \frac{1}{2(n-1)}$, the embedding

$$W_0^{1,p}(\Omega) \hookrightarrow\hookrightarrow L^r(\Omega)$$

is compact; no generality is lost then by assuming that

$$(u_j) \subset L^r(\Omega)$$

and that u_j converges in $L^r(\Omega)$. Clearly

$$\lim_{j\to\infty} u_j = u \text{ in } L^r(\Omega). \tag{4.104}$$

Lemma 4.7.2 coupled with (4.103) yields

$$\int_\Omega \frac{|w_k|^{p}}{(1+\delta)^{\frac{p}{p_-}} p}dx = \limsup_{j\geq J} \int_\Omega \frac{|w_k|^{p_j}}{(1+\delta)^{\frac{p_j}{p_{j-}}} p_j}dx \leq \int_\Omega \frac{|u|^p}{p}dx; \tag{4.105}$$

letting $k \to \infty$ it is apparent that

$$\int_\Omega \frac{|w|^p}{p}dx \leq (1+\delta)^{\frac{p}{p_-}} \int_\Omega \frac{|u|^p}{p}dx,$$

so that u is maximal in the desired sense.
In fact, the following stronger result holds:

$$\lim_{j\longrightarrow\infty}\int_\Omega |u_j|^{p_j}dx = \int_\Omega |u|^p dx. \tag{4.106}$$

To see this observe that, for $p_j \le r$,

$$|u|^{p_j} \le \chi_{\{x:|u(x)|>1\}}|u|^r + 1,$$

whence Lebesgue's theorem yields

$$\int_\Omega |u|^{p_j}dx \longrightarrow \int_\Omega |u|^p dx.$$

Lemma 4.7.1 yields the estimate

$$\|u-u_j\|_{p_j} \le \left(\|p_j - r\|_\infty^{-\|p_j-r\|_\infty} + \|p_j - r\|_\infty|\Omega|\right)\|u-u_j\|_r. \tag{4.107}$$

On the other hand, for $0 < t < 1$ it holds that:

$$\begin{aligned}\int_\Omega |u_j|^{p_j}dx - \int_\Omega |u|^p dx &\\ &\le \int_\Omega t^{1-p_+}|u_j - u|^{p_j}dx + (1-t)^{1-p_+}\int_\Omega |u|^{p_j}dx - \int_\Omega |u|^p dx\\ &= \int_\Omega t^{1-p_+}|u_j - u|^{p_j}dx + (1-t)^{1-p_+}\left(\int_\Omega |u|^{p_j}dx - \int_\Omega |u|^p dx\right)\\ &\quad + ((1-t)^{1-p_+} - 1)\int_\Omega |u|^p dx,\end{aligned}$$

likewise:

$$\begin{aligned}&\int_\Omega |u|^p dx - \int_\Omega |u_j|^{p_j}dx\\ &\le \int_\Omega t^{1-p_+}|u - u_j|^p dx + (1-t)^{1-p_+}\int_\Omega |u_j|^p dx - \int_\Omega |u_j|^{p_j}dx\\ &\le \int_\Omega t^{1-p_+}|u_j - u|^p dx\\ &\quad + (1-t)^{1-p_+}\left(\left(\|p-p_j\|_\infty^{-\|p-p_j\|_\infty} - 1\right)\int_\Omega |u_j|^{p_j}dx + \|p-p_j\|_\infty|\Omega|\right)\\ &\quad + ((1-t)^{1-p_+} - 1)\int_\Omega |u|^p dx.\end{aligned}$$

Letting j tend to infinity in the last two inequalities, recalling the fact that $(u_j)_{p_j \leq r}$ converges to u in $L^r(\Omega)$ and invoking (4.107), one obtains (4.106). As to the case $n \in (p_-, p_+]$, we point out that in the proof of Theorem 3.4.4 a finite collection of subdomains of Ω was obtained, namely

$$\Omega_1, \Omega_2, ...\Omega_M \text{ such that } \bigcup_1^M \Omega_k = \Omega$$

and sequences of positive real numbers $s_1, ...s_M$ and $\theta_1, ...\theta_M$ with

$$p(x) + \theta_k \leq s_k \text{ in } \Omega_k.$$

Consider a partition of unity

$$\varphi_k, \ \ 1 \leq k \leq M$$

subordinated to the open covering just described. By the assumption of uniform convergence, for sufficiently large I,

$$p_I + \theta < s_k$$

uniformly in Ω. As in the proof of Theorem 3.4.4, it follows the inclusion

$$(u_j \varphi_k) \subset L^{s_k}(\Omega)$$

and thus, for each $k = 1, 2, ...M$, the sequence

$$(u_j \varphi_k)$$

can be assumed to be convergent in $L^{s_k}(\Omega)$. Denote the limit of the sequence $(u_j \varphi_k)$ in $L^{s_k}(\Omega)$ by $v \in L^{s_k}(\Omega)$. In particular, for each k and for $j \geq I$, it holds that

$$u_j \varphi_k \in L^{p_I}(\Omega).$$

Moreover:

$$u_j \longrightarrow u \text{ in } L^p(\Omega),$$

and by reason of the inclusion

$$L^{s_k}(\Omega) \subseteq L^p(\Omega)$$

one has

$$\|u_j \varphi_k - v\|_p \to 0 \text{ as } j \longrightarrow \infty,$$

whence the inequality

$$\|v - \varphi_k u\|_p \leq \|u_j \varphi_k - v\|_p + \|(u_j - u)\varphi_k\|_p$$

implies that as $j \longrightarrow \infty$,

$$u_j \varphi_k \longrightarrow u \varphi_k \text{ in } L^{s_k}(\Omega) \hookrightarrow L^{p_I}(\Omega). \tag{4.108}$$

Consequently,

$$u = \sum_{k=1}^{M} u\varphi_k \in L^{p_I}(\Omega).$$

According to (4.108),

$$\|u - u_j\|_{p_I} \leq \sum_{k=1}^{M} \|(u - u_j)\varphi_k\|_{p_I} \to 0 \text{ as } j \to \infty; \tag{4.109}$$

from which it immediately follows that for $j \geq I$,

$$\|u - u_j\|_{p_j} \longrightarrow 0 \text{ as } j \longrightarrow \infty,$$

whence

$$\int_\Omega |u - u_j|^{p_j} dx \longrightarrow 0 \text{ as } j \longrightarrow \infty.$$

Since

$$|u(x)|^{p_i(x)} \longrightarrow |u(x)|^{p(x)} \text{ as } i \longrightarrow \infty \; a.e \text{ in } \Omega,$$

the pointwise inequality

$$|u|^{p_i} = \left|\sum_{k=1}^{M} u\varphi_k\right|^{p_i} \leq M^{p_i - 1} \sum_{k=1}^{M} |u\varphi_k|^{p_i} \leq C \sum_{k=1}^{M} |u\varphi_k|^{p_i}$$

and the pointwise estimate (which is valid in Ω for each $k = 1, 2 \ldots M$ and for any natural number $j \geq I$):

$$|u\varphi_k|^{p_j} \leq \chi_{\{x:|u\varphi_k(x)|>1\}} |u|^{p_j} + \chi_{\{x:|u\varphi_k(x)|<1\}} |u|^{p_j} \leq \chi_{\{|u(x)|>1\}} |u|^{s_k} + 1,$$

imply easily via a straightforward application of Lebesgue's theorem that

$$\int_\Omega |u|^{p_j} dx \longrightarrow \int_\Omega |u|^{p} dx \text{ as } j \longrightarrow \infty. \tag{4.110}$$

On the other hand, one has, for any $t \in (0, 1)$, and $j \geq I$,

$$\begin{aligned}
&\int_\Omega |u_j|^{p_j} dx - \int_\Omega |u|^p dx \\
&\leq \int_\Omega t^{1-p_+} |u_j - u|^{p_j} dx + (1-t)^{1-p_+} \int_\Omega |u|^{p_j} dx - \int_\Omega |u|^p dx \\
&= \int_\Omega t^{1-p_+} |u_j - u|^{p_j} dx + (1-t)^{1-p_+} \left(\int_\Omega |u|^{p_j} dx - \int_\Omega |u|^p dx \right) \\
&\quad + \left((1-t)^{1-p_+} - 1\right) \int_\Omega |u|^p dx,
\end{aligned}$$

and

$$\begin{aligned}
&\int_\Omega |u|^p dx - \int_\Omega |u_j|^{p_j} dx \\
&\quad \leq \int_\Omega t^{1-p_+}|u-u_j|^p dx + (1-t)^{1-p_+}\int_\Omega |u_j|^p dx - \int_\Omega |u_j|^{p_j} dx \\
&\quad \leq \int_\Omega t^{1-p_+}|u_j-u|^p dx \\
&\qquad + (1-t)^{1-p_+}\left(\left(\|p-p_j\|_\infty^{-\|p-p_j\|_\infty}-1\right)\int_\Omega |u_j|^{p_j} dx + \|p-p_j\|_\infty|\Omega|\right) \\
&\qquad + \left((1-t)^{1-p_+}-1\right)\int_\Omega |u|^p dx,
\end{aligned}$$

whence it transpires from the prior discussion that

$$\int_\Omega |u_j|^{p_j} dx \longrightarrow \int_\Omega |u|^p dx \text{ as } j \longrightarrow \infty. \tag{4.111}$$

Given the assumptions on p and on the sequence (p_j), (4.111) implies that

$$\int_\Omega \frac{|u_j|^{p_j}}{p_j} dx \longrightarrow \int_\Omega \frac{|u|^p}{p} dx \text{ as } j \longrightarrow \infty. \tag{4.112}$$

Letting j tend to infinity in (4.103) we obtain :

$$\limsup_j \int_\Omega \frac{|u_j|^{p_j}}{p_j} dx \leq \int_\Omega \frac{|u|^p}{p} dx.$$

In conjunction with inequality (4.103) the preceding statement yields the maximality of u in the required sense.
If $p_- \geq n$ all Sobolev spaces involved in the above proof are contained in $L^\infty(\Omega)$; the sequence (u_j), in particular, fulfills the hypothesis of Lemma 4.7.2 for any real number $r > 1$. The maximality of u follows immediately from this observation and inequality (4.103).
Next, by reason of (4.112), for sufficiently large j one has

$$\frac{p_-}{\int_\Omega |u|^p dx + 1} \leq \frac{\int_\Omega |\nabla u_j|^{p_j} dx}{\int_\Omega |u_j|^{p_j} dx} dx \leq \frac{p_{1+}}{\int_\Omega |u|^p dx + 1}$$

where $p_{j+} = \sup_{\overline{\Omega}} p_j$. There is therefore no loss of generality in assuming the sequence

$$\frac{\int_\Omega |\nabla u_j|^{p_j} dx}{\int_\Omega |u_j|^{p_j} dx}$$

to be convergent, to say L. Clearly, from (4.112):

$$\int_\Omega |\nabla u_j|^{p_j} dx \longrightarrow L \int_\Omega |u|^p dx. \tag{4.113}$$

From the weakly-lower semi-continuity of the functional

$$G : W_0^{1,p}(\Omega) \longrightarrow [0, \infty)$$
$$G(w) = \int_\Omega |\nabla w|^p dx$$

it is easily concluded that

$$\int_\Omega |\nabla u|^p dx \leq \liminf_{j \longrightarrow \infty} \int_\Omega |\nabla u_j|^p dx$$
$$\leq \liminf_{j \longrightarrow \infty} \left(\|p_j - p\|_\infty^{-\|p-p_j\|} \int_\Omega |\nabla u_j|^{p_j} dx + \|p - p_j\|_\infty |\Omega| \right);$$

from which, utilizing (4.112) it is readily concluded that

$$\frac{\int_\Omega |\nabla u|^p dx}{\int_\Omega |u|^p dx} \leq L. \tag{4.114}$$

On the other hand, for arbitrary $\delta > 0$ and large enough j,

$$\frac{\int_\Omega |\nabla u_j|^{p_j} dx}{\int_\Omega |u_j|^{p_j} dx} \leq p_{j+} \frac{\int_\Omega |\nabla u_j|^{p_j}/p_j \, dx}{\int_\Omega |u|^p dx - \delta} = p_{j+} \frac{\int_\Omega |\nabla u|^p/p \, dx}{\int_\Omega |u|^p dx - \delta} \leq p_{j+}/p_- \frac{\int_\Omega |\nabla u|^p dx}{\int_\Omega |u|^p dx - \delta}.$$

In conjuction with (4.114), the preceding inequality yields (4.97). □

In fact, the preceding convergence results can be substantially improved if $p(x) \geq 2$ in Ω. Specifically:

Theorem 4.7.4. *For $p_- \geq 2$, and u, (u_i) as in Theorem 4.7.3, we have:*

$$\lim_{j \to \infty} \int_\Omega |\nabla u - \nabla u_j|^p dx = 0.$$

Proof. Notice that since $u \in W_0^{1,p}(\Omega)$, one has

$$\lim_{k\longrightarrow\infty} \int_\Omega \frac{|\nabla u|^{p_k}}{p_k} dx = \int_\Omega \frac{|\nabla u|^p}{p} dx = 1. \tag{4.115}$$

Also, the functional

$$G : W_0^{1,p}(\Omega) \to [0, \infty)$$
$$G(v) = \int_\Omega \frac{|\nabla v|^p}{p} dx$$

is weakly lower- semicontinuous, whence

$$1 = \int_\Omega \frac{|\nabla u|^p}{p} dx \leq \liminf_{j\to\infty} \int_\Omega \frac{|\nabla u_j|^p}{p} dx \tag{4.116}$$

and

$$\int_\Omega \frac{|\nabla u|^p}{p} dx \leq \liminf_j \int_\Omega \left| \frac{\nabla(u + u_j)}{2} \right|^p \frac{1}{p} dx. \tag{4.117}$$

From the equality (valid for all natural numbers j):

$$\int_\Omega \frac{|\nabla u_j|^{p_j}}{p_j} dx = 1$$

and using the embedding estimate (4.7.1) one has, for $j \in \mathbb{N}$ and $\varepsilon = \|p_j - p\|_\infty$:

$$\begin{aligned} \int_\Omega \frac{|\nabla u_j|^p}{p} dx &\leq \varepsilon^{-\varepsilon} \int_\Omega \frac{|\nabla u_j|^{p_j}}{p^{p_j/p}} dx + \varepsilon|\Omega| \qquad (4.118) \\ &\leq \varepsilon^{-\varepsilon} \int_\Omega |\nabla u_j|^{p_j} \left(\frac{1}{p^{p_j/p}} - \frac{1}{p_j} \right) dx + \varepsilon^{-\varepsilon} \int_\Omega \frac{|\nabla u_j|^{p_j}}{p_j} dx + \varepsilon|\Omega| \\ &= \varepsilon^{-\varepsilon} \int_\Omega |\nabla u_j|^{p_j} \left(\frac{1}{p^{p_j/p}} - \frac{1}{p_j} \right) dx + \varepsilon^{-\varepsilon} + \varepsilon|\Omega|, \end{aligned}$$

which implies

$$\limsup_{j\to\infty} \int_\Omega \frac{|\nabla u_j|^p}{p} dx \leq 1. \tag{4.119}$$

Inequalities (4.116) and (4.119) yield

$$\lim_{j\to\infty} \int_\Omega \frac{|\nabla u_j|^p}{p} dx = 1. \tag{4.120}$$

Moreover,

$$\int_{\Omega} |\nabla u_j|^{p_j} dx \leq p_{j+} \leq p_+ \tag{4.121}$$

for all positive integers j and

$$\frac{1}{p^{p_j/p}} - \frac{1}{p_j} \longrightarrow 0 \text{ as } j \to \infty. \tag{4.122}$$

Consequently, for each $j \in \mathbb{N}$:

$$\begin{aligned}\int_{\Omega} \left|\frac{\nabla(u+u_j)}{2}\right|^p \frac{1}{p}\, dx &\leq \frac{1}{2}\int_{\Omega} \frac{|\nabla u|^p}{p} dx + \frac{1}{2}\int_{\Omega} \frac{|\nabla u_j|^p}{p} dx \\ &\leq \frac{1}{2}\int_{\Omega} \frac{|\nabla u|^p}{p} dx + \frac{1}{2}\left(\varepsilon^{-\varepsilon}\int_{\Omega} \frac{|\nabla u_j|^{p_j}}{p^{p_j/p}} dx + \varepsilon|\Omega|\right) \\ &= \frac{1}{2} + \frac{1}{2}\left(\varepsilon^{-\varepsilon}\int_{\Omega} \frac{|\nabla u_j|^{p_j}}{p^{p_j/p}} dx + \varepsilon|\Omega|\right).\end{aligned} \tag{4.123}$$

It follows that the expression to the right of the equal sign in (4.123) remains bounded as $j \to \infty$ and hence that

$$Q = \liminf_j \int_{\Omega} \left|\frac{\nabla(u+u_j)}{2}\right|^p \frac{1}{p}\, dx < \infty.$$

Fix $\delta > 0$ and select k so large that for $j \geq k$ one has

$$\|p - p_j\|_\infty |\Omega| < \delta$$

and

$$\|p - p_j\|_\infty^{-\|p-p_j\|_\infty} < 1 + \delta.$$

Pick I large enough so that $i \geq I$ implies

$$Q - \delta < \inf_{j \geq i} \int_{\Omega} \left|\frac{\nabla(u+u_j)}{2}\right|^p \frac{1}{p} dx.$$

Then for $j \geq I$;

$$
\begin{aligned}
1=\int_\Omega \frac{|\nabla u|^p}{p}dx &\leq \delta+\int_\Omega \left|\frac{\nabla(u+u_j)}{2}\right|^p\frac{1}{p}dx\\
&\leq \delta+\frac{1}{2}\left(1+\int_\Omega \frac{|\nabla u_j|^p}{p}dx\right)\\
&\leq \delta+\frac{1+\delta}{2}\left(1+\int_\Omega \frac{|\nabla u_j|^{p_j}}{p^{p_j/p}}dx\right)\\
&\leq \delta+\frac{1+\delta}{2}\left(1+\int_\Omega |\nabla u_j|^{p_j}dx\left(\frac{1}{p^{p_j/p}}-\frac{1}{p_j}\right)+\int_\Omega \frac{|\nabla u_j|^{p_j}}{p_j}dx\right)\\
&=\delta+\frac{1+\delta}{2}\left(1+\int_\Omega |\nabla u_j|^{p_j}\left(\frac{1}{p^{p_j/p}}-\frac{1}{p_j}\right)dx+1\right).
\end{aligned}
$$

By virtue of (4.121) and (4.122) it is apparent that the last term above tends to 1 as $j\to\infty$ and $\delta\longrightarrow 0$, whence

$$
\int_\Omega \left|\frac{\nabla(u+u_j)}{2}\right|^p\frac{1}{p}dx\longrightarrow 1 \text{ as } j\longrightarrow\infty. \tag{4.124}
$$

For any real number $t:t\geq 2$ the inequality

$$
\left|\frac{a+b}{2}\right|^t+\left|\frac{a-b}{2}\right|^t\leq\frac{1}{2}\left(|a|^t+|b|^t\right)
$$

holds for any complex numbers a and b (see [[2], Lemma 2.27]). In particular, for each $x\in\Omega$

$$
\int_\Omega \left|\frac{\nabla(u+u_j)}{2p^{1/p}}\right|^p dx+\int_\Omega \left|\frac{\nabla(u-u_j)}{2p^{1/p}}\right|^p dx\leq
$$

$$
\frac{1}{2}\left(\int_\Omega \left|\frac{\nabla u}{p^{1/p}}\right|^p dx+\int_\Omega \left|\frac{\nabla u_j}{p^{1/p}}\right| dx^p\right)
$$

whence (4.120) and (4.124) imply

$$
\int_\Omega |\nabla(u-u_j)|^p dx\leq 2^{p_+}p^{p_+/p_-}\int_\Omega \left|\frac{\nabla(u-u_j)}{2p^{1/p}}\right|^p dx\longrightarrow 0 \text{ as } j\longrightarrow\infty. \tag{4.125}
$$

This completes the proof of Theorem 4.7.4. □

Next, we single out the one dimensional case, for which the convergence statement is perhaps the best one can hope for:

Theorem 4.7.5. *For $n = 1$, $\Omega = (a, b)$, $a < b$. Let $(p_j) \subset C([a, b])$ be a non-increasing sequence of functions, uniformly convergent to $p = \inf p_j$ with $1 < p_- \leq p_+ < \infty$. For each $j \in \mathbb{N}$ let λ_j and u_j be, respectively, the first eigenvalue and a first eigenfunction of the problem*

$$\left(|u'|^{p_j-2}u'\right)' = \gamma|u|^{p_j-2}u; \tag{4.126}$$

analogously, let λ be the first modular eigenvalue of the one-dimensional p-Laplacian and u be its associated eigenfunction in the sense of Corollary 4.6.9.

Then there exists a subsequence (u_k) of (u_j) and a (first) eigenfunction u of the p-Laplacian such that

$$u_k \longrightarrow u \text{ in } W_0^{1,p}(\Omega). \tag{4.127}$$

Proof. For $p_- \geq 2$, Theorem 4.7.5 follows directly from Theorem 4.7.4. We hence focus on the case $p_- < 2$. To that end we notice the elementary fact that there exists a positive constant c such that as long as $1 < s < S < \infty$ and $(q, r) \in (s, S) \times [-M, M]$

$$\frac{1}{2}\left(|r+1|^q + |r-1|^q\right) - |r|^q \geq c. \tag{4.128}$$

Fix $\delta > 0$ and set

$$A = \{x \in (a, b) : |(u_j - u)'| > \delta|(u_j + u)'|\}.$$

Then if $x \in A$, $r = \frac{v(x)+v_j(x)}{v(x)-v_j(x)}$, $q = p_j(x)$, the substitution $v = u'/p_j^{1/p_j}$, $v_j = u_j'/p_j^{1/p_j}$ in (4.128) yields, pointwise in Ω:

$$\frac{1}{2}\left(|v|^{p_j} + |v_j|^{p_j}\right) - \left|\frac{v+v_j}{2}\right|^{p_j} \geq c\left|\frac{v-v_j}{2}\right|^{p_j}, \tag{4.129}$$

whence it follows by integrating on A that

$$\frac{1}{2}\left(\int_\Omega |v|^{p_j}\,dx + \int_\Omega |v_j|^{p_j}\,dx\right) - \int_\Omega \left|\frac{v+v_j}{2}\right|^{p_j} dx \geq c\int_A \left|\frac{v-v_j}{2}\right|^{p_j} dx. \tag{4.130}$$

As seen in Theorem 4.7.4

$$\int_\Omega \left|\frac{v+v_j}{2}\right|^{p_j} dx \longrightarrow 1 \text{ as } j \longrightarrow \infty,$$

whereas

$$\int_\Omega |v|^{p_j}\,dx \longrightarrow 1 \text{ as } j \longrightarrow \infty.$$

Letting j tend to infinity in (4.130) it is clear that

$$\int_A \left|\frac{v-v_j}{2}\right|^{p_j} dx \longrightarrow 0 \text{ as } j \to \infty \tag{4.131}$$

which through an easy calculation yields :

$$\int_A |u' - u_j'|^p dx \longrightarrow 0 \text{ as } j \longrightarrow \infty. \tag{4.132}$$

On the other hand

$$\int_{\Omega\setminus A} \left|\frac{v-v_j}{2}\right|^{p_j} dx \leq \delta^{p_-} \int_{\Omega} \left|\frac{v+v_j}{2}\right|^{p_j} dx \leq \delta^{p_-}. \tag{4.133}$$

In all,

$$\int_\Omega |u' - u_j'|^p dx \longrightarrow 0 \text{ as } j \longrightarrow \infty.$$

This last statement completes the proof. □

A result analogous to that in Theorem 4.7.3 holds for a non-decreasing sequence (p_i). Specifically:

Theorem 4.7.6. *Let (p_i) be a non-decreasing sequence of functions in $C(\overline{\Omega})$ with*

$$\lim_{i\to\infty} p_i = q \in C(\overline{\Omega}).$$

Assume furthermore that $\inf_\Omega q = q_- > n$. For each $i \in \mathbb{N}$ let $u_i \in W_0^{1,p_i}(\Omega)$ be an eigenfunction corresponding to the least eigenvalue $\lambda_{p_i,1}$ of the p_i-Laplacian. Then there exists a subsequence, still denoted by (u_i), that converges weakly to $u \in V_{q,1}(\Omega)$. Moreover

$$\int_\Omega \frac{|\nabla u|^q}{q} dx = 1.$$

and the limit

$$\lim_{i\longrightarrow\infty} \lambda_{p_i,1} = \lim_{i\to\infty} \frac{\int_\Omega |\nabla u_i|^{p_i} dx}{\int_\Omega |u_i|^{p_i} dx} = L < \infty$$

exists. The limit function u satisfies

$$\frac{q_-}{q_+} L \leq \frac{\int_\Omega |\nabla u|^q dx}{\int_\Omega |u|^q dx} \leq L, \tag{4.134}$$

and

$$\left(\frac{q_-}{q_+}\right)^2 L \leq \lambda_{q,1} \leq L.$$

Proof. The proof is similar to that of Theorem 4.7.3. We refer the reader to [33] for the details. □

4.8 Closing remarks

Various generalizations of the preceding results are possible in the context of Musielak-Orlicz spaces.
Under additional smoothness hypotheses on the variable exponent, a study of the eigenvalues of the p-Laplacian based on the abstract Ljusternik-Schnirelmann theory ([107, III, Chapter 44]) was undertaken in [39]. Specifically, log-Hölder continuity of p is assumed there, that is, that the inequality

$$|p(x) - p(y)| \leq C \log^{-1} (|x - y|^{-1} + e)$$

holds for any $x \in \Omega$, $y \in \Omega$ (see [24, Chapter 3].) Applications to other type of nonlinear problems, like Kirchhoff's equations (see [6, 18, 20], are also possible.

Bibliography

[1] Acerbi, E. and Mingione, G., *Regularity Results for a Class of Functionals with Non-standard Growth,* Arch. Rational Mech. Anal. **156**, 121–140 (2001).

[2] Adams, R., *Sobolev Spaces,* Academic Press, 1975.

[3] Ambrosetti, A. and Prodi, G., *A Primer of Nonlinear Analysis,* Cambridge studies in advanced mathematics, 1993.

[4] Anane, A., *Simplicité et isolation de la premiére valeur propre du p-Laplacien avec poids*, C.R. Acad. Sci. Paris, t. 305, Série I, 725–728 (1987).

[5] Appell, J., De Pascale, E. and Vignoli, A., *Nonlinear Spectral Theory,* Walter De Gruyter, Berlin, 2004.

[6] Autuori, G., Pucci, G. and Salvatori, M.C., *Global nonexistence for nonlinear-Kirchhoff systems,* Arch. Rat. Mech. Anal., **196**, 489-–516 (2010).

[7] Bachar, M., Bounkhel, M. and Khamsi, M., *Uniform Convexity in* $l_{p(\cdot)}$, to appear.

[8] Banach, S., *Über einige Eigenschaften der lakunären trigonometrischen Reihen,* Studia Math. **2**, 207–220 (1930).

[9] Banach, S., *Bemerkung zu der Arbeit "ber einige Eigenschaften der lakunren trigonometrischen Reihen",* Studia Math. **2**, 251 (1930).

[10] Benedikt, J., Girg, P., Kotrla, L. and Takáč. P., *Origins of the p-Laplacian and A. Missbach,* Electr. Jour. Diff. Eq. **2018** (16), 1–17 (2018).

[11] Bennet, C. and Sharpley, R., *Interpolation of Operators,* Academic Press, New York, 1988.

[12] Bennewitz, C., *Approximation numbers=singular values,* J. Comp. Appl. Math. **208**, (1), 102–110 (2007).

[13] Birnbaum, Z. and Orlicz, W., *Über die Verallgemeinerung des Begriffes der zueinander konjugierten Potenzen,* Studia Math. **3**, 1–67 (1931).

[14] Braides, A., *Local Minimization, Variational Evolution and Convergence,* Lecture Notes in Mathematics, **294**, 2013.

[15] Brezis, H., *Functional Analysis, Sobolev Spaces and Partial Differential Equations,* Universitext, Springer, 2011.

[16] Carleman, T., *Über die Fourierkoeffizienten einer stetigen Funktion,* Acta Mathematica (1918).

[17] Clarkson, J., *Uniformly convex spaces,* Trans. Am. Math. Soc. **40**(3), 396–414 (1936).

[18] Colasuonno, F. and Pucci, P., *Multiplicity of solutions for $p(x)$-polyharmonic Kirchhoff equations,* Nonlinear Anal., **74**, 5962–5974 (2011).

[19] Dacorogna, B., *Direct Methods in the Calculus of Variations,* Springer-Verlag, Berlin, 1989.

[20] Dai, G. and Hao, R., *Existence of solutions for a $p(x)$-Kirchhoff- type equation,* J. Math. Anal. Appl. **359**, 275–284 (2009).

[21] Dal Maso, G., *An introduction to Γ-convergence*, Progress in Nonlinear Differential Equations and their Applications, **8**, Basel, Birkäuser, 1993.

[22] Dankert, G., *Sobolev embedding theorems in Orlicz spaces,* Thesis, Univ. of Cologne (1966).

[23] Diening, L., *Maximal functions on generalized $L^{p(x)}(\Omega)$ spaces,* Math. Inequal. Appl. **7**, 245–253 (2004).

[24] Diening, L., Harjulehto, P., Hästö, P. and Růžička, M., *Lebesgue and Sobolev spaces with variable exponents,* Lecture notes in Mathematics, vol. 2017, Springer, Heidelberg, 2011.

[25] De La Vallée Poussin, C., *Sur L'Integrale de Lebesgue,* Trans. Amer. Math. Soc. **16** (4), 435–501 (1915).

[26] Diestel, J., *Sequences and Series in Banach Spaces,* Springer, New York, 1984.

[27] Di Giorgi, E. and Spagnolo, D., *Sulla convergenza degli integrali dell'energia per operatori ellitici del secondo ordine,* Boll. Un. Mat. It., **8**, 391–411 (1973).

[28] Di Giorgi, E. and Franzoni, T., *Su un tipo di convergenza variazionale,* Atti. Acad. Naz. Linzei. Rend. Cl. Sci. Fis. Mat. Natur., **58**, 842–850 (1975).

[29] Di Giorgi, *Sulla convergenza di alcune successioni di integrali del tipo dell'area,* Rendi Condi di Mat., **8**, 277–294 (1975).

[30] Donaldson, T. and Trudinger, N., *Orlicz–Sobolev spaces and imbedding theorems,* J. Funct. Anal., **8**, 52–75 (1971).

[31] Edmunds, D.E. and Evans, W., *Hardy operators, functions spaces and embeddings,* Springer, Berlin-Heidelberg, New York, 2004.

[32] Edmunds, D.E. and Evans, W., *Representation of Linear Operators Between Banach Spaces,* Operator Theory Advances and Applications, Birkhäuser, Basel, Heidelberg, New York, Dordrecht, London, 2013.

[33] Edmunds, D.E., Lang, J., Méndez, O. *Differential operators on spaces of variable integrability,* World Scientific, 2014.

[34] Edmunds, D.E., Lang, J. and Nekvinda, A., *Some $s-$numbers of an integral operator of Hardy type on $L^{p(\cdot)}$ spaces*, J. Functional Anal. **257**, 219–242 (2009).

[35] Edmunds, D.E., Lang, J. and Nekvinda, A., *On $L^{p(x)}$ norms*, Proc. Roy. Soc. London A, **455**, 219–225 (1999).

[36] Evans, L. C. and Gariepy, R. F., *Measure theory and fine properties of functions,* CRC Press, Boca Raton, Florida, 1992.

[37] Fabian, M., Habala, P., Santalucía, V.M., Pelant, J. and Zizler, V., *Functional analysis and infinite-dimensional geometry,* Springer-Verlag, Berlin-Heidelberg-New York, 2001.

[38] Fan, X. and Zhang, Q., *Existence of solutions for $p(x)$-Laplacian Dirichlet Problem,* Nonlinear Anal. 522, 1843–1852 (2003).

[39] Fan, X., Zhang, Q. and Zhao, D. , *Eigenvalues for $p(x)$-Laplacian Dirichlet Problem,* J. Math. Anal. Appl., **302**, 306–317 (2005).

[40] García Márquez, G. *Cien años de soledad,* Buenos Aires: Editorial Sudamericana, 1967.

[41] Garkavi, A. L., *The best possible net and the best possible cross-section of a set in a normed space*, AMS Transl. II, Ser. **39**, 111–132 (1964) (Translation from Izv. Akad. Nauk SSSR, Ser. Mat. **26**, 87–106 (1962)).

[42] Gilbarg, D. and Trudinger, N., *Elliptic Partial Differential Equations of Second Order,* Springer, Berlin-Heidelberg, New York, Tokyo, 1983.

[43] Harjulehto, P., Hästo, P. and Latvala, V., *Minimizer of the variable exponent non-uniformly convex Dirichlet energy*, J. Math. Pures Appl. (9) **89**, 174–197 (2008).

[44] Hudzik, H., *A generalization of Sobolev spaces I,* Funct. Approx. Comment. Math. **II**, 67–73 (1976).

[45] Hudzik, H., *A generalization of Sobolev spaces II,* Funct. Approx. Comment. Math., **III**, 77–85 (1976).

[46] Hudzik, H., *On generalized Orlicz-Sobolev spaces,* Funct. Approx. Comment. Math., **IV**, 37–51 (1976).

[47] Hudzik, H., *On continuity of the embedding operation from* $W^k_{M_1}(\Omega)$ *into* $W^k_{M_2}(\Omega)$, Funct. Approx. Comment. Math., **VI**, 112–118 (1978).

[48] Hudzik, H., *The problems of separability, duality, reflexivity and of comparison for generalized Orlicz-Sobolev spaces* $W^k_M(\Omega)$, Comment. Math. Parce Mat. **21**, 315–324 (1979).

[49] Hudzik, H., *Strict convexity of Musielak-Orlicz spaces with Luxemburg norm,* Bull. Acad. Sci. Pol. Math. XXXIX **5,6**, 235–247 (1981).

[50] Hudzik, H., *Convexity in Musielak-Orlicz spaces,* Hokkaido Math. J. **14**, no. 1, 85–96 (1985).

[51] Hudzik, H., *Characteristic of convexity of Musielak-Orlicz function spaces equipped with the Luxemburg norm,* Comment. Math. Univ. Carolinae **33**, 4, 615–621 (1992).

[52] James, R., *Uniformly non-square Banach spaces,* Ann. Math. **80**, 542–550 (1964).

[53] Kamińska, A., *Convexity, concavity and indices in Musielak-Orlicz spaces,* Functiones Math., special volume on the 70th birthday of J.Musielak **26**, 67–84 (1998).

[54] Khamsi, M., Kozlowski, W. and Reich, S., *Fixed point theory in modular function spaces,* Nonlinear Analysis, **14**, 935–953 (1990).

[55] Khamsi, M. and Kozlowski, W., *Fixed Point Theory in Modular Function Spaces,* Birkäuser, 2015.

[56] Kawohl, B. and Lindqvist, P. *Positive eigenfunctions for the p−Laplace operator revisited,* Analysis (Munich), **26**(4), 545–550 (2006).

[57] Meyer, P., *Probabilities and Potentials,* Blaisdell, Waltham, Mass., (1966).

[58] Kawohl, B. and Lindqvist, P., *Positive eigenfunctions for the p-Laplace operator revisited,* Analysis **26** No 4, 545–550 (2006).

[59] Kolmogorov, A., *Une contribution á l'étude de la convergence de séries de Fourier* , Fund. Math. **5**, 126–127 (1924).

[60] Kováčik, O. and Rákosník, J., *On spaces* $L^{p(x)}, W^{k,p(x)}$, Czechoslovak Mathematical Journal **41**(4), 592–618 (1991).

[61] Krasnoselskii, M., *Topological methods in the theory of nonlinear integral equations,* Pergamon, Elsford, NY, (1964).

[62] Kufner, A., *Function Spaces,* Noordhof International Publishing, Leyden, 1977.

[63] Kufner, A., John, O. and Fučik, S., *Function spaces*, Academia, Prague, 1977.

[64] Lang, J. and Méndez, O., *Sharp conditions for the compactness of the Sobolev embedding on Musielak-Orlicz spaces*, Mathematische Nachrichten, 1–12 (2018).

[65] Lang, J. and Méndez, O., *Γ-convergence of the energy functionals for the variable exponent $p(\cdot)$-Laplacian and stability of the minimizers with respect to integrability*, J. Anal. Math. **134**, no. 2, 575–596 (2018).

[66] Lang, J. and Méndez, O., *Extension of a result by Lindquist to Lebesgue spaces with variable exponents,* J. Differential Equations 259, no. 2, 562–595 (2015).

[67] Lang, J. and Méndez, O., *Modular eigenvalues of the Dirichlet $p(\cdot)$-Laplacian and their stability,* Spectral theory, function spaces and inequalities, 125137, Oper. Theory Adv. Appl., **219**, Birkhäuser/Springer Basel AG.

[68] Lang, J. and Nekvinda, A., *A difference between the continuous and the absolutely continuous norms in Banach function spaces,* Czech. Math. J. **47**, 221–232 (1993).

[69] Lavrentiev, M., *Sur quelques problemes du calcul des variations,* Ann. Math. Pura et. Appl. **4**, 7–28 (1926).

[70] Lindenstrauss, J. and Tzafriri, L., *Classical Banach spaces I and II*, Springer-Verlag, Berlin-Heidelberg-New York, 1977–1979.

[71] Lindqvist, P., *Stability for the solutions of $div(|\nabla u|^{p-2}\nabla u) = f$ with varying p,* Journal Math. Anal. Appl. **127**, 1, 93–102 (1987).

[72] Lindqvist, P., *On non-Linear Rayleigh quotients,* Pot. Anal. **2** 199–218 (1993).

[73] Lindqvist, P., *On the equation $div(|\nabla u|^{p-2}\nabla u) + \lambda|u|^{p-2}u = 0$,* Proc. Amer. Math. Soc. **109** 1, 157–164 (1990).

[74] Lindqvist, P., *Stability for the solutions of $div(|\nabla u|^{p-2}\nabla u) = f$ with varying p,* Journal Math. Anal. Appl. **127** 1, 93–102 (1987).

[75] Lions, J.-L., *Quelques méthodes de résolution des problèmes aux limites non linéaires,* Dunod, Paris, 1969.

[76] Lukeš, J., Pick, L. and Pokorný, D., *On geometric properties of the spaces* $L^{p(x)}(\Omega)$, Rev. Mat. Complut. **24**, 115–130 (2011).

[77] Luxemburg, W., *Banach function spaces,* Ph.D. Thesis, Delft Institute of Technology, Assen (Netherlands) (1955).

[78] Maligranda, L., *Orlicz Spaces and Interpolation,* Seminars in Math. University of Campinas, Campinas, SP, Brazil, 1989.

[79] Matuszewska, W., and Orlicz, W., *On certain properties of* φ*-functions,* Bull. Acad. Polon. Sci., Ser. Sci. Math. Astronom. Phys. **8**, 430–443 (1960).

[80] Mazur, S. and Orlicz, W., *On some classes of linear spaces,* Studia Math. **17**, 97–119 (1958).

[81] Musielak, J. and Orlicz, W., *On Modular Spaces,* Studia Math. **18** (1), 49–65 (1959).

[82] Musielak, J., *Orlicz Spaces and Modular Spaces,* Lecture Notes in Math., **1034**, 1983.

[83] Nakano, H., *Modulared semi-ordered linear spaces,* Maruzen, Tokyo, 1950.

[84] Orlicz, W., *Beiträge zur Theorie der Orthogonalentwicklungen,* Studia Math. **1** (1), 1–39 (1929).

[85] Orlicz, W., *Über konjugierte Exponentenfolgen,* Studia Math. **3**, 200–211 (1931).

[86] Orlicz, W., *Über eine gewisse Klasse von Räumen vom Typus B,* Bull. Int. Acad. Polon. Sci. A 207–222 (1932).

[87] Orlicz, W., *Beiträge zur Theorie der Orthogonalentwicklungen III,* Bull. Int. Acad. Polon. Scl. A 229–238 (1934).

[88] Orlicz, W., *Beiträge zur Theorie der Orthogonalentwicklungen IV,* Studia Math. **5** (1), 1–14 (1934).

[89] Orlicz, W., *Beiträge zur Theorie der Orthogonalentwicklungen V,* Studia Math. **6** (1), 20–38 (1936).

[90] Orlicz, W., *Beiträge zur Theorie der Orthogonalentwicklungen VI,* Studia Math. **8** (1), 141–147 (1939).

[91] Rao, M. and Ren, Z., *Theory of Orlicz spaces,* Marcel Dekker, New York, 1991.

[92] Rao, M. and Ren, Z., *Applications of Orlicz spaces,* Marcel Dekker, New York, 2002.

[93] Riesz, F., *Über Systeme integrierbarer Funktionen,* Math. Ann. **69**, 449–497 (1910).

[94] Riesz, F., *Les systémes d'équations linéaires á une infinité d'inconnues,* Collection Borel, Paris (1913).

[95] Rudin, W., *Functional Analysis,* Mc Graw-Hill Book Co., New York, Second Edition, 1991.

[96] Showalter, R.E., *Monotone Operators in Banach Space and Nonlinear Partial Differential Equations,* AMS Mathematical Surveys and Monographs, vol. 49, 1996.

[97] Sidon, S., *Ein Satz über die absolute Konvergenz von Fourierreihen in denen sehr viele Glieder fehlen,* Math Ann. **96**, 418–419 (1927).

[98] Sidon, S., *Verallgemeinerung eines Satzes über die absolute Konvergenz von Fourierreihen mit Lücken,* Math Ann. **97**, 675–676 (1927).

[99] Sundaresan, K., *Uniform convexity of Banach spaces $l(\{p_i\})$,* Studia Math. **39**, 401–407 (1971).

[100] Taylor, A.E., *Introduction to functional analysis,* Wiley, New York, 1958.

[101] Waterman, D., Ito, T., Barber, F. and Rattil, J., *Reflexivity and Summability: The Nakano $l(p_i)$ spaces,* Studia Math. **331** (87), 141–146 (1969).

[102] Yang, Y., *Relationship between Matuszewska-Orlicz, Semenov and Simonenko indices of φ-functions,* Publications de l'Institut Mathmatique **73** (87), 139–147 (2003).

[103] Yosida, K. and Hewitt, E., *Finitely additive measures,* Trans. Am. Math. Soc. **72** (1), 46–66 (1952).

[104] Young, W., *On a class of parametric integrals and their application to the theory of Fourier Series,* Proc. R. Soc. Lond. Ser. A Math. Phys. Eng. Sci. **85** (580), 401–414 (1911).

[105] Young, W., *On Classes of Summable Functions and their Fourier Series,* Proc. R. Soc. Lond. Ser. A Math. Phys. Eng. Sci. **87**, (594), 225–229 (1912).

[106] Young, W.H., *On Successions with Subsequences Converging to an Integral,* Proc. London Math. Soc. (2), **24**, (1925), no. 1, 120.

[107] Zeidler, H., *Nonlinear Functional Analysis and its Applications III-Variational Methods and Optimization,* Springer-Verlag, New York, 1985.

[108] Ziemer, W., *Weakly Differentiable Functions,* Springer-Verlag, New York, 1989.

[109] Zhikov, V.V., *On some variational problems,* Russian J. Math. Phys. 5 (1997), no. 1, 105–116 (1998).

[110] Zygmund, A., *Sur les séries trigonométriques lacunaires,* Proc. London Math. Soc., **5**, 138–145 (1930).

[111] Zygmund, A., *Trigonometric Series,* Cambridge University Press, 1959.

Author index

Acerbi, E., 5
Adams, R., 85, 134, 167
Adams, T., 160
Ambrosetti, A., 223
Anane, A., 211
Appell, J., 5

Bachar, M., 38, 63, 71–73, 75, 77, 80
Banach, S., 82, 83, 158
Barber, F. , 58
Benedikt, J., 211
Bennet, C., 4, 8, 85, 138, 151
Bennewitz, C., 185
Birnbaum, Z., 95, 157, 158
Bounkhel, M., 38, 63, 71–73, 75, 77, 80
Braides, A., 207
Brezis, H., 167

Carleman, T., 82, 83
Čebyšev, P., 77
Clarkson, J., 21, 63

Dacorogna, B., 191
Dal Maso, G., 207, 208, 210
Dankert, G., 160, 167
De Giorgi, E., 207
de la Vallée Poussin, C.J., 156
De Pascale, E., 5
Diening, L., 38, 100, 129, 134, 145, 147
Donaldson, T., 160, 167

Edmunds, D.E., 108, 140, 198–200, 203
Evans, L.C., 89

Fabian, M., 22, 23
Fan, X., 190, 193, 197, 220
Fatou, P., 155
Fučik, S., 108

Gariepy, R.F., 89
Garkavi, A.L., 40
Gilbarg, D., 85, 167
Girg, P., 211

Habala, P., 22, 23
Harjulehto, P., 38, 100, 129, 134, 145, 147, 205
Hästö, P., 38, 100, 129, 134, 145, 147, 205
Hudzik, H., 160, 165

Ito, T., 58

James, R., 21
John, O., 108

Kamińska, A., 153
Kawohl, B., 211
Khamsi, M., 30, 38, 39, 63, 70–73, 75, 77, 80
Kolmogorov, A., 83
Kotrla, L., 211
Kováčik, O., 106, 144, 167, 176, 216
Kozlowski, W., 30, 38, 39, 70
Krasnoselskii, M., 132
Kufner, A., 85, 108

Lang, J., 108, 140, 167, 176, 181, 198–200, 203, 211, 220

Latvala, V., 205
Lindenstrauss, J., 24
Lindqvist, P., 210, 211
Lions, J.-L., 25
Lukeš, J., 58, 137
Luxemburg, W., 8, 28

Méndez, O., 167, 176, 181, 211, 220
Maligranda, L., 151, 153
Matuszewska, W., 153
Mazur, S., 159, 160
Meyer, P., 157
Mingione, G., 5
Musielak, J., 38, 100, 101, 129, 159, 160

Nakano, H., 38, 40, 160
Nekvinda, A., 108, 140, 198–200, 203

Orlicz, W., 40, 49, 82, 83, 85, 95, 153, 157–160

Pelant, J., 22, 23
Pick, L., 58, 137
Pokorný, D., 58, 137
Prodi, G., 223

Rákosník, J., 106, 144, 167, 176, 216
Rao, M., 85, 95, 153
Rattil, J., 58
Reich, S., 30
Ren, Z., 85, 95, 153
Riesz, F., 95, 159
Rudin, W., 188
Růžička, M., 38, 100, 129, 134, 145, 147

Santalucía, V.M., 22, 23
Sharpley, R., 4, 8, 85, 138, 151
Showalter, R.E., 181
Sidon, S., 83
Sundaresan, K., 63

Takáč, P., 211
Taylor, A.E., 26
Trudinger, N., 85, 160, 167
Tzafriri, L., 24

Vignoli, A., 5

Waterman, D., 58
Weierstrass, K., 82

Yang, Y., 153
Young, W., 95, 155–157

Zhang, Q., 190, 193, 197, 220
Zhao, D., 220
Ziemer, W., 167
Zizler, V., 22, 23
Zygmund, A., 82, 83

Subject index

absolute continuity of the norm, 10
absolutely continuous norm, 12, 20, 113
AO property, 66
associate space, 6, 10, 120

Banach function space, 2
Banach-Alaoglu Theorem, 189
Borel class, 2
Borel measurable, 2
bounded map, 181
bounded variation, 55

Čebyšev center, 78
Čebyšev subset, 77
canonical isomorphism, 183
Carathéodory function, 132
characteristic function, 2
coercive map, 26
compactly embedded, 2
conjugate function, 119, 123
continuous modular, 27
continuous norm, 140
continuously embedded, 2
convex modular, 27, 30

Δ_2 condition, 32, 110
de la Vallée Poussin-Lemma, 156
decreasing sequence, 11
demicontinuous map, 26
derivative in direction, 22
Dirichlet problem, 197
distribution function, 3
distributional derivative, 162
dual space, 10
duality map, 25, 150

equicoercive sequence, 207
equimeasurable functions, 3

F-space, 160
finitely additive measure, 55
Fourier coefficients, 84
Fourier series, 82, 84
Fréchet derivative, 22, 222

Γ-convergence, 207
Gâteaux derivative, 22, 223
gauge function, 150
gauge map, 25

Hölder conjugates, 52
Hölder's inequality, 6
Hahn Banach Theorem, 166
Hahn-Banach Theorem, 9
Hardy-Littlewood maximal operator, 146
hemicontinuous map, 181

isometry, 10

Kirk's fixed point theorem, 80

Lebesgue n-measure, 2
Lebesgue measure, 2
Lebesgue spaces, 5
left-continuous modular, 27, 32
locally integrable MO function, 92
locally integrable Musielak-Orlicz function, 169
lower semicontinuous modular, 32

Luxemburg norm, 28, 33, 38, 85, 94, 97
Luxemburg norm convergence, 31

Matuszewska index, 152, 176
Matuszewska-Orlicz index, 152, 168, 169
MO function, 91
MO space, 94
modular, 27
modular convergence, 30, 31
modular space, 29
modulus of smoothness, 23
mollified function, 127
monotone convergence theorem, 9
monotone map, 25
Musielak-Orlicz function, 38, 91, 97
Musielak-Orlicz space, 40, 85, 94
Musielak-Orlicz spaces, 5
Musielak-Orlicz-Sobolev space, 162

(norm)-convergent sequence of functions, 3
N-function, 134, 220
Nakano class, 160
Nemytskii-type operator, 132
norm convergence, 31
norm-convergent sequence, 32

open ball, 1
operator norm, 184
order ideal of a Banach function space, 16
Orlicz class, 85
Orlicz function, 84
Orlicz function spaces, 38
Orlicz modular, 95
Orlicz space, 5, 85

φ-Orlicz space, 95
Parseval's Theorem, 155
Poincare-type Inequality for MO spaces, 176
proper Musielak-Orlicz function, 96, 98

ρ-Cauchy sequence, 30
ρ-bounded set, 30
ρ-closed, 30
ρ-convergent sequence, 30
ρ-limit, 30
ρ_p-normal structure property, 75
ρ_{p_n}-Lipschitzian map, 80
ρ_{p_n}-bounded sequence, 70
ρ_{p_n}-closed set, 70
ρ_{p_n}-nonexpansive map, 80
(R) property, 77
Rayleigh quotient, 219
rearrangement-invariant space, 3
reflexive space, 20
Riesz-Fischer theorem, 82
right-continuous modular, 27

s-convex modular, 27
SC modular, 40, 66
second associate space, 8
singular functional, 58
Sobolev spaces, 5
space of all real-valued continuous functions, 31
space of distributions, 162
strictly convex Banach space, 21
strictly convex dual, 26
strictly convex modular, 40
strictly convex space, 21, 39
strictly monotone, 26
strictly monotone map, 26
subadditive function, 151
supporting functional, 22

total variation of μ , 55
transpose map, 185
type M map, 182

UCC2 condition, 63
UCED modular, 40
uniform convexity, 139
uniform convexity type-property, 39

uniformly convex Luxemburg norm, 129
uniformly convex MO function, 128
uniformly convex modular, 128
uniformly convex modular in every direction, 40
uniformly convex space, 21
uniformly Fréchet-differentiable space, 23
uniformly integrable family, 156
uniformly smooth space, 23
UUC2, 39
UUC2 modular, 70
UUCED modular, 40, 66

weak convergence, 2
weak solution, 224
weak*-convergence, 2
weakly-convergent sequence, 182

Young's inequality, 52, 145

Zorn's Lemma, 76

Notation index

$(L^{\varphi}(\Omega))'$, 120
$(Mu)(x)$, 146
$(W^{k,\varphi}(\Omega))^*$, 166
$(X')'$, 8
$(X_b)^*$, 17
$(\Omega, \mathcal{A}, \mu)$, 2
$(\mathbb{N}, 2^{\mathbb{N}}, \mu)$, 13, 40
$(\mathbb{N}, 2^{\mathbb{N}})$, 57
$B(X, Y)$, 1
$B(x, r)$, 1
B_r, 187
B_X, 1, 2
$C([0, 1])$, 31
$C(\Omega)$, 126
$C_0^{\infty}(\Omega)$, 126
$D(r, \epsilon)$, 38
$D^{\alpha}T$, 162
$Fix(T)$, 80
$J(x)$, 25
L^M, 159
$L^p(\Omega)$, 12
$L^M(\Omega)$, 5
$L^{\infty}(\Omega)$, 12
$L^{\phi}(\Omega)$, 5
$L^{\varphi^*}(\Omega)$, 120
L^{φ}, 94
$L^{\varphi}(\Omega)$, 85, 163
$L_a^{\varphi}(\Omega)$, 113
$L_b^{\varphi}(\Omega)$, 114
$L^p(I)$, 1
M, 123
$M^{(\varphi)}$, 152
M_{φ}, 152, 168
S, 92
S^{α}, 159
S_X, 1
T_{φ_t}, 134
T_w, 132
$V_{\rho_{\alpha}^k}$, 163
$V_{\rho_{j,p_n}}$, 41
V_{ρ}, 28
V_{ρ}^*, 33
$W^{0,\varphi}(\Omega)$, 163
$W^{k,\varphi_p}(\Omega)$, 164
$W^{k,\varphi}(\Omega)$, 162
$W_0^{k,\varphi}(\Omega)$, 163
X, 1
X^*, 10, 17
X^*, 2, 10
X'', 8
X', 6, 7, 17
X_a, 12, 16, 17, 113, 140
X_b, 15–17, 113, 140
X_c, 140
X_{ρ}, 31
Δ_2, 110
Δ_p, 193, 211
Δ_{p_k}, 198
$\Gamma - \lim$, 208
Ω_0, 86
Ω_1, 86
Ω_{∞}, 86
$\triangle$, 115
$\delta(r, \epsilon)$, 39
δ_X, 21
δ_{ρ}, 30
$\dim X$, 1
ℓ^{∞}, 13
ℓ_a^{∞}, 13
ℓ^{p_n}, 42

$\ell_a^{p_n}$, 58
$\eta(s,\epsilon)$, 39
$\hookrightarrow$, 2
$\hookrightarrow\hookrightarrow$, 2
$\|\cdot\|_\rho$, 33
$\|\cdot\|_\rho$, 28
$\|\cdot\|$, 1
$\|\cdot\|_{p,I}$, 1
$\|\cdot\|_p$, 1
$\|\cdot\|_X$, 1
$\langle x^*, x\rangle$, 2
$\langle x^*, x\rangle_X$, 2
$|\Omega|$, 2
$|\Omega|_n$, 2
$\mathbf{c}_0$, 13
$\mathcal{P}$, 146
$\otimes$, 2
μ, 2
μ_u, 3
$\mu_u(\lambda)$, 3
grad, 22
grad $\|u\|_p$, 150
$\overset{*}{\rightharpoonup}$, 2
$\overset{\rho}{\to}$, 30
ρ_p, 86
$\rho(.)$, 27
ρ_φ^k, 163
$\rho_{1,p_n}((a_n))$, 40
$\rho_{2,p_n}((a_n))$, 41
ρ_H, 24
ρ_{X^*}, 23
ρ_X, 23
$\rho_\infty((a_n))$, 41
$\rightharpoonup$, 2
$\sim$, 153
$\simeq$, 52
$\tilde{L}^1$, 8
φ^*, 119, 123
$d_{\rho_{p_n}}$, 71
m_φ, 152, 168
p_+, 86
p_-, 86
$x^*(x)$, 33
$\mathbf{e}_j$, 50
sgn, 16
supp, 18
$\mathbf{1}_S$, 2
$\mathbf{S}_r$, 188
$\mathcal{A}$, 2
$\mathcal{B}$, 95, 167
$\mathcal{D}'(\Omega)$, 162
$\mathcal{F}$, 46
$\mathcal{I}$, 10
$\mathcal{M}(\Omega)$, 2
$\mathcal{M}(\Omega)_{\rho_p}$, 86
$\mathcal{M}(\mathbb{N})$, 40
$\mathcal{S}$, 2, 16, 124
$\mathcal{S}_\infty$, 2, 56